Kalman Filtering and Information Fusion

Hongbin Ma · Liping Yan · Yuanqing Xia ·
Mengyin Fu

Kalman Filtering and Information Fusion

Science Press
Beijing

Springer

Hongbin Ma
School of Automation
Beijing Institute of Technology
Beijing, China

Liping Yan
School of Automation
Beijing Institute of Technology
Beijing, China

Yuanqing Xia
School of Automation
Beijing Institute of Technology
Beijing, China

Mengyin Fu
School of Automation
Beijing Institute of Technology
Beijing, China

ISBN 978-981-15-0805-9 ISBN 978-981-15-0806-6 (eBook)
https://doi.org/10.1007/978-981-15-0806-6

Jointly published with Science Press
The print edition is not for sale in Mainland of China. Customers from Mainland of China please order the print book from: Science Press.

This Springer imprint is published by the registered company Springer Nature Singapore Pte Ltd.
The registered company address is: 152 Beach Road, #21-01/04 Gateway East, Singapore 189721, Singapore

The ultimate value of life
And the awakening of the ability to think,
Rather than just to survive
—Aristotle

This book is dedicated to our families,
for their endless love and support.

This book is also dedicated to our teachers,
who taught us love and wisdom.

Preface

This book will address one key technology, named as *Kalman filtering*, for digital information processing, which was ever regarded as one of the greatest discoveries in the twentieth century. Kalman filtering, also known as linear-quadratic estimation (LQE), is one of the most successful and widely used estimation methods. Standard Kalman filter can be derived from least squares algorithm, which bridges theoretical science and practical world in a large. The application of Kalman filter is the combination of mathematics and the physical world. It has been used as the key element of the algorithm for many applications in aircraft/ship/ground vehicle navigation, spacecraft attitude determination, orbit determination, missile guidance and control, RF antenna/laser terminal target acquisition/tracking, RF/optical signal acquisition and tracking, seismic data processing, medical signal processing, and other fields in the industry.

Standard Kalman filter is optimal in the sense of minimum mean squared errors and maximum likelihood estimation, provided that the system model is linear and precisely known a priori and the process and measurement noises are zero mean, completely unrelated and jointly Gaussian with known covariance matrices. However, in practice, these requirements can seldom be completely satisfied due to the following reasons:

(a) The practical systems are usually nonlinear although many of them may be approximated by linear systems. Absolutely linear systems in state-space seldom exist in practical engineering applications.
(b) Even if the practical system in consideration is linear, the system model may not be exactly known with accurate parameters. In practice, model parameters may be approximately identified by applying some system identification methods offline through the data obtained via extensive experiments. However, this approach is usually expensive and does not guarantee accurate system identification, which makes that standard Kalman filter may not work very well for the identified model. Furthermore, if the practical system is in fact time-varying, then the approach of system identification will usually fail.

(c) Standard Kalman filter requires that the process and measurement noises are zero-mean random noises. However, in some applications, the noise may be biased and its mean or mathematical expectation may be unknown. In such cases, further noise modeling is often needed and it is possible to use Kalman filter by augmenting the mean of the noise as an extra state.
(d) In most cases, we cannot have the covariance matrix of the unknown process and measurement noises a priori. Therefore, to apply the standard Kalman filter, we must first try to obtain the statistical properties of the process and measurement noises, which are usually calculated from extensive practical experiments. To deal with this problem, an alternative approach is to simply use larger covariance matrix to represent the a priori knowledge on the process and measurement noises.
(e) In practice, the probability distribution of the process noise or measurement noise may not be normal distribution, and this case is often termed as non-Gaussian system, which often results that the performance of standard Kalman filter may degrade much.

Besides the issues of various uncertainties in a single plant, which will be addressed in detail and some corresponding solutions based on adaptive estimation will be introduced in this book, more issues could be naturally raised when Kalman filtering technology is applied in multisensor systems and/or multi-agent systems. We notice that in modern engineering systems, there is a trend that systems tend to become bigger or more complex with more components. Especially, various sensors are used in systems like intelligent robots, autonomous cars, smart homes, smart buildings, etc., which call for the techniques of multisensor information fusion, which will be extensively discussed in this book. Furthermore, when multiple agents (subsystems) interact with one another, coupling uncertainties would also be one challenging issue to handle with, which is also addressed in this book by some novel decentralized adaptive filtering techniques.

To address possible uncertainty in system model or the statistical properties of the process and observation noises, some existing extensions and generalizations have been proposed to obtain good denoising and estimating effects when the model uncertainty is relatively small in sense that only small model perturbation or unknown covariance of the noise is considered. Meanwhile, the filtering and prediction problems when system models have large uncertainty or when the covariance matrices of both the process and measurement noises were completely unknown are seldom addressed. To address these problems, some novel ideas are introduced in this book. Since the number of the systems may increase, how to estimate the state with multiple sets of data is another important topic in this book. In this book, a novel real-time filter, named as fast minimum norm filtering algorithm, has been proposed to deal with the case when the covariance matrices of the process and measurement noises were unknown in the linear time-invariant systems with state-space model. A general framework of finite-model Kalman filter is introduced. A new discretized method is introduced to discretize continuous systems under nonuniformly sampling intervals is discussed. In summary, the objective of this book is to present an advanced technology used for various applications by addressing various challenging issues such as nonlinearity, uncertainties, and complexity.

This book contains four main parts. Part I provides preliminaries of Kalman filtering including brief introduction to Kalman filtering, challenges of Kalman filtering, and introduction to information fusion. Part II focuses on Kalman filtering for uncertain systems, where the uncertainties addressed include noise covariance, model uncertainty, parameter uncertainty, etc. In Part III, Kalman filtering for multisensor systems is extensively addressed, and optimal information fusion techniques are proposed for various multisensor systems. In Part IV, Kalman filtering for multi-agent systems is discussed by addressing several challenging issues such as coupling uncertainties and communication limits.

This book is featured with a number of attractive and original research studies, including a general framework of finite-model adaptive Kalman filtering, wireless sensor network localization, a recursive covariance estimation (RCE) algorithm, decentralized adaptive filtering, optimal multisensor information fusion, and so on. The book is primarily intended for researches and engineers in estimation and control community. This book will bring fresh new ideas into education and will benefit students by exposing them to the very forefront of filtering research.

The authors would also like to thank the help from our students and collaborators.

Beijing, China
July 2017

Hongbin Ma
Liping Yan
Yuanqing Xia
Mengyin Fu

Acknowledgements The authors would like to acknowledge support from National Natural Science Foundation of China (NSFC) grants 61004059, 61473120, 61473038, and 91648117, Beijing Outstanding Talents Programme (2012D009011000003). The funders had no role in study design, data collection and analysis, decision to publish, or preparation of the manuscript. The authors would also like to thank the help from Mei Wu, Nannan Li, Xiaofei Zhang, and Shan Li during the preparation of this book.

The authors would also like to thank the generous and egoless loves from our families. Without their continuous support and forgiveness, this book would not appear in its current form. During the days working with uncertainties and challenges, their encouragement, enquiries, and enduring to us always accompanied with us, which make ourselves not feel lonely in those hard times.

Contents

Part I
Kalman Filtering: Preliminaries

Chapter 1
Introduction to Kalman Filtering

Kalman filter is regarded as a great discovery in last century. In fact, filtering has a long history and has been widely used in many areas. In this chapter, we briefly introduce the background of filtering, especially Wiener filter and Kalman filter with some historical remarks. Basic formation of Kalman filter, one approach of optimal linear-quadratic estimation, is highlighted for the ease of later in-depth discussions.

1.1 What Is Filtering?

Filtering plays an important role in signal processing and people's life. All filters share a common goal: to let something pass through while something else does not. Originally, a filter or a sieve is a device for separating or isolating wanted elements from unwanted material. One common example is a device for separating impurities (solids) from a fluid or other substance (Fig. 1.1).

Fig. 1.1 Sieves

H. Ma et al., *Kalman Filtering and Information Fusion*,
https://doi.org/10.1007/978-981-15-0806-6_1

Fig. 1.2 An elementary filter topology introduces a capacitor into the feedback path of an op-amp to achieve an unbalanced active implementation of a low-pass transfer function

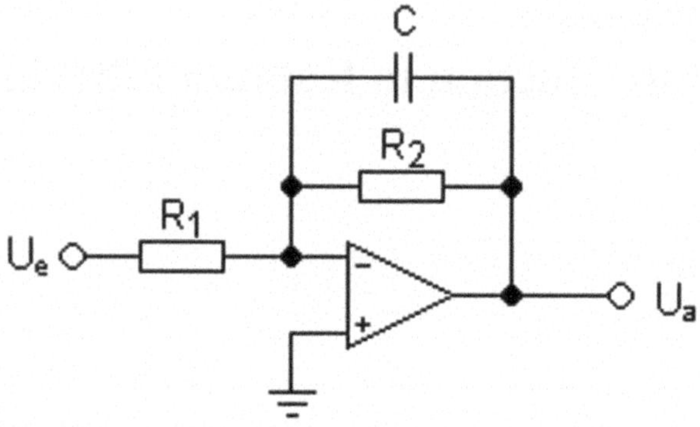

This concept was later extended to electronic filter, which refers to an electronic circuit which processes signals, for example, to remove unwanted frequency components. For instance, you can think over a low-pass filter, which lets low frequencies pass through while attenuating high frequencies. An electronic filter can be either an analog filter, a basic building block of signal processing much used in electronics, or a digital filter, a system that performs mathematical operations to reduce or enhance certain aspects of a signal through digital electronic circuit (Fig. 1.2).

With the rapid development of information techniques especially computers, the concept of filter in the form of hardware was soon generalized to form of software, i.e, a computer program to process a data stream. Mathematically speaking, a filter can refer to any procedure or algorithm which can process noisy input data and generate relatively smooth output data. One popular example is taking moving average over a sliding window for a noisy signal. For a digital filter, if the output signal is linear function of the input signal, the filter is said to be a linear filter; otherwise, it can be called a nonlinear filter. A linear digital filter may be classified as a finite impulse response (FIR) filter or an infinite impulse response (IIR) filter. There are various types of digital filters, including low-pass filters, high-pass filters, band-pass filters, and so on. Usually, a filter taking a noisy signal as its input can provide a smoothed version of the original signal as its output, and thus the filtered signal is relatively easier to use since uncertainties have been removed or reduced to some extent. For example, moving average indicators can be used to smooth stock prices (Fig. 1.3).

Among the vast types of digital filters, the Wiener filter and the Kalman filter are two milestones of signal processing, among which the former was introduced by Norbert Wiener during the 1940s and published in 1949, while the latter was named after Hungarian Rudolf E. Kálmán. The Wiener filter is an optimal linear filter used to produce an estimate of a desired or target random process by linear time-invariant (LTI) filtering of an observed noisy process, assuming known stationary signal and noise spectra, and additive noise. The Wiener filter minimizes the mean square error between the estimated random process and the desired process. Before the introduction of the Kalman filter in 1960, the Wiener filter plays important role in applications of optimal signal processing. And Kalman filter, as an approximating

Fig. 1.3 Two moving average indicators (simple and exponential) added to a typical stock chart

algorithm in optimal estimation applications and problems, has been considered to be one great technology that changes the world. In fact, successful applications of Kalman filter can be found in many areas, such as navigation, image processing, robotics, aircraft, and so on.

1.2 Historical Remarks

Norbert Wiener (November 26, 1894—March 18, 1964) was an American mathematician and philosopher. He was a professor of mathematics at the Massachusetts Institute of Technology (MIT). A child prodigy, Wiener, later became an early researcher in stochastic and mathematical noise processes, contributing work relevant to electronic engineering, electronic communication, and control systems. Wiener is considered the originator of cybernetics, a formalization of the notion of feedback, with implications for engineering, systems control, computer science, biology, neuroscience, philosophy, and the organization of society.

Wiener filter, as a filter to reduce the amount of noise present in a signal by comparison with an estimate of the desired noiseless signal, was proposed by Wiener during the 1940s and published in 1942 as a classified document. Wiener developed the filter at the Radiation Laboratory at MIT to predict the position of German bombers from radar reflections. It is necessary to predict the future, because by the time the shell reaches the vicinity of the target, the target has moved and may have changed direction slightly. They even modeled the muscle response of the pilot, which led eventually to cybernetics. The unmanned V1's were particularly easy to model, and on a good day, American guns fitted with Wiener filters would shoot down 99 out of 100 V1's as they entered Britain from the English channel, on their way to London. What emerged was a mathematical theory of great generality—a theory for predicting the future as best one can on the basis of incomplete information about the past. It was a statistical theory that included applications that did not, strictly speaking, predict the future, but only tried to remove noise. It made use of Wiener's earlier work on integral equations and Fourier transforms.

The discrete-time equivalent of Wiener's work was derived independently by Andrey Kolmogorov and published in 1941. Hence, the theory is often called the Wiener–Kolmogorov filtering theory (cf. Kriging). The Wiener filter was the first statistically designed filter to be proposed and subsequently gave rise to many others including the Kalman filter.

The Kalman filter is named after Rudolf E. Kálmán, partially because successful application of Kalman filter in the Apollo program. In fact, before Kalman's work, Thorvald Nicolai Thiele [1], and Peter Swerling developed a similar algorithm earlier. Richard S. Bucy of the University of Southern California contributed to the theory, leading to it sometimes being called the Kalman–Bucy filter. Stanley F. Schmidt is generally credited with developing the first implementation of a Kalman filter. He realized that the filter could be divided into two distinct parts, with one part for time periods between sensor outputs and another part for incorporating measurements. It was during a visit by Kálmán to the NASA Ames Research Center that Schmidt saw the applicability of Kálmán's ideas to the nonlinear problem of trajectory estimation for the Apollo program leading to its incorporation in the Apollo navigation computer.

The Apollo computer used 2k of magnetic core RAM and 36k wire rope. The CPU was built from ICs. Clock speed was under 100 kHz. The fact that the MIT engineers were able to pack such good software (one of the very first applications of the Kalman filter) into such a tiny computer is truly remarkable. Kalman filters have been vital in the implementation of the navigation systems of U.S. Navy nuclear ballistic missile submarines, and in the guidance and navigation systems of cruise missiles such as the U.S. Navy's Tomahawk missile and the U.S. Air Force's Air-Launched Cruise Missile. They are also used in the guidance and navigation systems of reusable launch vehicles and the attitude control and navigation systems of spacecraft which dock at the International Space Station [2].

This digital filter is sometimes called the Stratonovich–Kalman–Bucy filter because it is a special case of a more general, nonlinear filter developed somewhat earlier by the Soviet mathematician Ruslan Stratonovich [3]. In fact, some of the special case linear filter's equations appeared in these papers by Stratonovich that were published before summer 1960, when Kalman met with Stratonovich during a conference in Moscow [4].

1.3 Wiener Filter

The goal of the Wiener filter is to compute a statistical estimate of an unknown signal using a related signal as an input and filtering that known signal to produce the estimate as an output. For example, the known signal might consist of an unknown signal of interest that has been corrupted by additive noise. The Wiener filter can be used to filter out the noise from the corrupted signal to provide an estimate of the underlying signal of interest. The Wiener filter is based on a statistical approach, and a more statistical account of the theory is given in the minimum mean square error (MMSE) estimator article.

Typical deterministic filters are designed for a desired frequency response. However, the design of the Wiener filter takes a different approach. One is assumed to have knowledge of the spectral properties of the original signal and the noise, and one seeks the linear time-invariant filter whose output would come as close to the original signal as possible. Wiener filter has the following characteristics:

1. Assumption: Signal and (additive) noise are stationary linear stochastic processes with known spectral characteristics or known autocorrelation and cross-correlation.
2. Requirement: The filter must be physically realizable/causal (this requirement can be dropped, resulting in a noncausal solution).
3. Performance criterion: Minimum mean square error (MMSE).

The Wiener filter problem has solutions for three possible cases: one where a noncausal filter is acceptable (requiring an infinite amount of both past and future data), the case where a causal filter is desired (using an infinite amount of past data), and the finite impulse response (FIR) case where only input data is used (i.e., the result or output is not fed back into the filter as in the IIR case). The first case is simple to solve but is not suited for real-time applications. Wiener's main accomplishment was solving the case where the causality requirement is in effect, and in an appendix of Wiener's book Levinson gave the FIR solution.

1.4 Kalman Filter

Kalman filters are ideal for systems which are continuously changing. They have the advantage that they are light on memory (they don't need to keep any history other than the previous state), and they are very fast, making them well suited for real-time problems and embedded systems.

The Kalman filter keeps track of the estimated state of the system and the variance or uncertainty of the estimate. The estimate is updated using a state transition model and measurements. In statistics and control theory, Kalman filtering, also known as linear-quadratic estimation (LQE), is an algorithm that uses a series of measurements observed over time, containing statistical noise and other inaccuracies, and produces estimates of unknown variables that tend to be more accurate than those based on a single measurement alone, by estimating a joint probability distribution over the variables for each time frame.

The Kalman filter has numerous applications in technology. A common application is for guidance, navigation, and control of vehicles, particularly aircraft and spacecraft. Furthermore, the Kalman filter is a widely applied concept in time series analysis used in fields such as signal processing and econometrics. Kalman filters also are one of the main topics in the field of robotic motion planning and control, and they are sometimes included in trajectory optimization. The Kalman filter also works for modeling the central nervous system's control of movement. Due to the time delay between issuing motor commands and receiving sensory feedback, use of the Kalman filter supports a realistic model for making estimates of the current state of the motor system and issuing updated commands.

The algorithm works in a two-step process. In the prediction step, the Kalman filter produces estimates of the current state variables, along with their uncertainties. Once the outcome of the next measurement (necessarily corrupted with some amount of error, including random noise) is observed, these estimates are updated using a weighted average, with more weight being given to estimates with higher certainty. The algorithm is recursive. It can run in real time, using only the present input measurements and the previously calculated state and its uncertainty matrix; no additional past information is required.

As an example application, consider the problem of determining the precise location of a truck. The truck can be equipped with a GPS unit that provides an estimate of the position within a few meters. The GPS estimate is likely to be noisy; readings "jump around" rapidly, though remaining within a few meters of the real position. In addition, since the truck is expected to follow the laws of physics, its position can also be estimated by integrating its velocity over time, determined by keeping track of wheel revolutions and the angle of the steering wheel. This is a technique known as dead reckoning. Typically, the dead reckoning will provide a very smooth estimate of the truck's position, but it will drift over time as small errors accumulate.

In this example, the Kalman filter can be thought of as operating in two distinct phases: predict and update. In the prediction phase, the truck's old position will be modified according to the physical laws of motion (the dynamic or "state transition" model). Not only will a new position estimate be calculated, but a new covariance will be calculated as well. Perhaps, the covariance is proportional to the speed of the truck because we are more uncertain about the accuracy of the dead reckoning position estimate at high speeds but very certain about the position estimate when moving slowly. Next, in the update phase, a measurement of the truck's position is taken from the GPS unit. Along with this measurement comes some amount of uncertainty, and its covariance relative to that of the prediction from the previous phase determines how much the new measurement will affect the updated prediction. Ideally, as the dead reckoning estimates tend to drift away from the real position, the GPS measurement should pull the position estimate back toward the real position but not disturb it to the point of becoming rapidly jumping and noisy.

1.5 Conclusion

This chapter discusses the basic problem of "What is filtering?", which sheds a light to understand classic Wiener filter and modern Kalman filter. Some historical remarks are given to illustrate the evolution of filtering techniques. The Wiener filter is an optimal linear filter for stationary signal with known signal and noise spectra, in terms of minimizing the mean square error between the estimated random process and the desired process. And the Kalman filter is more powerful since it can even cope with nonstationary signals with dynamics, and Wiener filter is in fact special case of Kalman filter.

References

1. S.L. Lauritzen, *Thiele Pioneer in Statistics* (Oxford University Press, New York, 2002)
2. D. Gaylor, E.G. Lightsey, GPS/INS Kalman filter design for spacecraft operating in the proximity of international space station. J. Neurochem. (2006)
3. R.L. Stratonovich, Conditional markov processes. Theory Probab. Its Appl. **5**(2), 172–195 (2006)
4. O.A. Stepanov, Kalman filtering: past and present. An outlook from Russia. (on the occasion of the 80th birthday of Rudolf Emil Kalman). Gyroscopy Navig. **2**(2), 99–110 (2011)

Chapter 2
Challenges in Kalman Filtering

Despite the wide applications of Kalman filtering, various challenges still exist in real applications. In this chapter, we introduce some known typical challenges found in the theoretical studies and practical applications of Kalman filtering. Standard Kalman filter is optimal when exact linear Gaussian dynamical model with known statistics of process noise and measurement noise is available. However, model mismatch or model uncertainty can inevitably downgrade the performance of Kalman filter or even make the filter fail. This chapter briefly addresses these challenges, which will be coped with in later chapters.

2.1 Standard Kalman Filter

The standard Kalman filter model assumes the true state at time k is evolved from the state at $(k-1)$ according to

$$x_k = F_k x_{k-1} + B_k u_k + w_k \tag{2.1}$$

where

- F_k is the state transition model which is applied to the previous state x_{k-1};
- B_k is the control input model which is applied to the control vector u_k;
- w_k is the process noise which is assumed to be drawn from a zero-mean multivariate normal distribution, $\mathcal{N}$, with covariance, Q_k: $w_k \sim \mathcal{N}(0, Q_k)$.
- At time k, an observation (or measurement) z_k of the true state x_k is made according to

$$z_k = H_k x_k + v_k \tag{2.2}$$

where

H. Ma et al., *Kalman Filtering and Information Fusion*,
https://doi.org/10.1007/978-981-15-0806-6_2

- H_k is the observation model which maps the true state space into the observed space and
- v_k is the observation noise which is assumed to be zero-mean Gaussian white noise with covariance R_k:

$$v_k \sim \mathcal{N}(0, R_k) \tag{2.3}$$

- The initial state and the noise vectors at each step $\{x_0, w_1, \ldots, w_k, v_1, \ldots, v_k\}$ are all assumed to be mutually independent.

Many real dynamical systems do not exactly fit this model. In fact, unmodeled dynamics can seriously degrade the filter performance, even when it was supposed to work with unknown stochastic signals as inputs. The reason for this is that the effect of unmodeled dynamics depends on the input, and, therefore, can bring the estimation algorithm to instability (it diverges). On the other hand, independent white noise signals will not make the algorithm diverge. The problem of distinguishing between measurement noise and unmodeled dynamics is a difficult one and is treated in control theory under the framework of robust control.

The Kalman filter is a recursive estimator. This means that only the estimated state from the previous time step and the current measurement are needed to compute the estimate for the current state. In contrast to batch estimation techniques, no history of observations and/or estimates is required. In what follows, the notation $\hat{x}_{k'|k}$ represents the estimate of x at time k' given observations up to and including at time $k \leq k'$.

The state of the filter is represented by two variables:

- $\hat{x}_{k|k}$, the a posteriori state estimate at time k given observations up to and including at time k;
- $P_{k|k}$, the a posteriori error covariance matrix (a measure of the estimated accuracy of the state estimate).

The Kalman filter can be written as a single equation; however, it is most often conceptualized as two distinct phases: "Predict" and "Update". The predict phase uses the state estimate from the previous time step to produce an estimate of the state at the current time step. This predicted state estimate is also known as the a priori state estimate because, although it is an estimate of the state at the current time step, it does not include observation information from the current time step. In the update phase, the current a priori prediction is combined with current observation information to refine the state estimate. This improved estimate is termed the a posteriori state estimate.

Typically, the two phases alternate, with the prediction advancing the state until the next scheduled observation and the update incorporating the observation. However, this is not necessary; if an observation is unavailable for some reason, the update may be skipped and multiple prediction steps performed. Likewise, if multiple independent observations are available at the same time, multiple update steps may be performed (typically with different observation matrices H_k).

Predict

Predicted (a priori) state estimate

$$\hat{x}_{k|k-1} = F_k \hat{x}_{k-1|k-1} + B_k u_k \tag{2.4}$$

Predicted (a priori) error covariance

$$P_{k|k-1} = F_k P_{k-1|k-1} F_k^{\mathsf{T}} + Q_k \tag{2.5}$$

Update

Innovation or measurement pre-fit residual

$$\tilde{y}_k = z_k - H_k \hat{x}_{k|k-1} \tag{2.6}$$

Innovation (or pre-fit residual) covariance

$$S_k = R_k + H_k P_{k|k-1} H_k^{\mathsf{T}} \tag{2.7}$$

Optimal Kalman gain

$$K_k = P_{k|k-1} H_k^{\mathsf{T}} S_k^{-1} \tag{2.8}$$

Updated (a posteriori) state estimate

$$\hat{x}_{k|k} = \hat{x}_{k|k-1} + K_k \tilde{y}_k \tag{2.9}$$

Updated (a posteriori) estimate covariance

$$P_{k|k} = (I - K_k H_k)\, P_{k|k-1}\, (I - K_k H_k)^{\mathsf{T}} + K_k R_k K_k^{\mathsf{T}} \tag{2.10}$$

Measurement post-fit residual

$$\tilde{y}_{k|k} = z_k - H_k \hat{x}_{k|k} \tag{2.11}$$

The formula for the updated (a posteriori) estimate covariance above is valid for any gain K_k and is sometimes called the Joseph form. For the optimal Kalman gain, the formula further simplifies to

$$P_{k|k} = (I - K_k H_k)\, P_{k|k-1} \tag{2.12}$$

in which form it is most widely used in applications. However, one must keep in mind that it is valid only for the optimal gain that minimizes the residual error. Proof of the formulae is found in the derivations section.

2.2 Requirements of Standard Kalman Filtering

Standard Kalman filter is optimal in the sense of minimum mean squared errors and maximum likelihood estimation, provided that the system model is linear and precisely known a priori and the process and measurement noises are zero mean, completely unrelated, and jointly Gaussian with known covariance matrices. However, in practice, these requirements can seldom be completely satisfied due to the following reasons:

(a) The practical systems are usually nonlinear although many of them may be approximated by linear systems. Absolutely linear systems in state-space seldom exist in practical engineering applications.
(b) Even if the practical system in consideration is linear, the system model may not be exactly known with accurate parameters. In practice, model parameters may be approximately identified by applying some system identification methods offline through the data obtained via extensive experiments. However, this approach is usually expensive and does not guarantee accurate system identification, which makes that standard Kalman filter may not work very well for the identified model. Furthermore, if the practical system is in fact time-varying, then the approach of system identification will usually fail.
(c) Standard Kalman filter requires that the process and measurement noises are zero-mean random noises. However, in some applications, the noise may be biased and its mean or mathematical expectation may be unknown. In such cases, further noise modeling is often needed and it is possible to use Kalman filter by augmenting the mean of the noise as an extra state.
(d) In most cases, we cannot have the covariance matrix of the unknown process and measurement noises a priori. Therefore, to apply the standard Kalman filter, we must first try to obtain the statistical properties of the process and measurement noises, which are usually calculated from extensive practical experiments. To deal with this problem, an alternative approach is to simply use larger covariance matrix to represent the a priori knowledge on the process and measurement noises.
(e) In practice, the probability distribution of the process noise or measurement noise may not be normal distribution, and this case is often termed as *non-Gaussian* system, which often results that the performance of standard Kalman filter may degrade much.

Due to the above issues, to resolve these practical problems to some extent, some extensions or variants of standard Kalman filter have been proposed in the literature. For example, for issue (a), EKF is one approach to apply technique of Kalman filter to nonlinear systems based on the idea of Taylor expansion, while UKF [1] and particle filters (PF) [2, 3] are two typical approaches to handle filtering problem for nonlinear/non-Gaussian systems addressed by issue (a) or issue (e). Comparing with other issues, issue (c) may be relatively easy to resolve, and one typical approach has been mentioned above if we have some physical or a priori knowledge on the sources of possible errors in the sensors. As to issue (d), there exists one approach called AKF [4–7], whose idea is to adaptively estimate the uncertain statistical properties of the noises and combine the Kalman filter with the modified covariance matrices. Kalman filter based on support vector machine (SVMKF) [8] may be regarded as another example to deal with issue (d). As to challenging issue (b), a general framework of finite-model Kalman filtering (FMKF) was introduced based on the idea that the large model uncertainty may be restricted by a finite set of known different models, and under this framework, a simple yet effective finite-model Kalman filter, termed as MVDP-FMKF, was discussed extensively according to the idea of adaptive switching via the principle of minimum output prediction error [9].

2.3 Effects of System Uncertainties

For the linear systems with state-space model, Kalman filter provides optimal state estimates in the sense of minimum mean squared errors and maximum likelihood estimation. The Kalman filtering has demonstrated its power in state estimation, system identification, adaptive control, signal processing, and has been widely used in many engineering problems since it was originally developed in the 1960s. But only with accurate system parameters and noise statistical properties, the estimation obtained by standard Kalman filter is the optimal state estimate.

- The covariance matrix of process noise is harder to be determined than that of the measurement noise by routine experiments since the statistical property of process noise cannot be obtained directly by collecting a large number of sensor data due to the intrinsic coupling of process noise and system dynamics.
- In Chaps. 3 and 4, one algorithm, named as recursive covariance estimation, is introduced to estimate the covariance matrix when there is no information of process noise covariance matrix in the linear time-invariant systems.
- In Chap. 5, a modified Kalman filter with recursive covariance estimation is presented to cope with the general case when the covariance matrices of process and measurement noises in the physical model cannot be obtained a priori and the rigorous stability analysis is given to show that this algorithm is optimal in sense that the covariance matrix and state estimations are asymptotically consistent with the ideal Kalman filter using the exact covariance matrices.
- In actual systems, most of time, the exact noise statistical properties could not be obtained as apriori information or even wrong statistical properties may be captured by the offline method. In Chap. 7, a real-time state estimator without noise covariance matrices knowledge is introduced.

- In actual systems, there are model uncertainty where it is expensive to have large number of high-cost experiments or even impossible to obtain the exact system model, and large model uncertainty make standard Kalman filters does not work well or even diverge. In Chap. 7, we introduced a framework of finite-model Kalman filtering.
- In actual systems, there are many various degrees of parametric uncertainty, and those parametric uncertainties make it difficult to using standard Kalman filtering. In Chap. 8, some Kalman filters for continuous parametric uncertain systems are investigated.

2.4 Effects of Multiple Sensors

In this part, we introduced some Kalman filter algorithm for multisensor systems. These algorithms are concerned with issues of Kalman filter of multisensor systems, including the distributed fusion, the sequential fusion, and the centralized fusion of systems with cross-correlated noises, and the distributed fusion of multirate multisensor systems, and the fusion estimation of nonlinear systems with unreliable observations. Part III is organized as follows:

In Chap. 9, we discussed how to find the optimal fusion of sensors with cross-correlated sensor noises. The centralized fusion, the distributed fusion, and the sequential fusion algorithms are introduced. The performances of different fusion algorithms are analyzed and compared through a target tracking example.

In Chap. 10, an optimal state estimation for distributed fusion of multirate multisensor linear dynamic systems is introduced, where the noises of different sensors are cross-correlated and also coupled with the system noise of the previous step.

In Chap. 11, we studied the state estimation of a kind of nonlinear dynamic systems, where multiple sensors observing the same target with different sampling rates and the observations are randomly unreliable. Based on online checking of the reliability of the measurements, the cubature Kalman filter (CKF) is improved, which has better robustness and stability. Based on the improved CKF, an effective state estimation algorithm is presented, and the multirate multisensor information are effectively fused. The numerical simulation shows the feasibility and the effectiveness of the presented algorithms.

2.5 Effects of System Couplings

In practice, many situations of complex systems usually contain uncertain coupling parameters since some uncertain interactions are inevitable, for example, nobody can successfully grow up without learning to adapt to the environment and other people. This problem is important and fundamental for further studies on controlling multi-agent systems. For every agent, only noisy measurement information is available, but there are uncertain interactions among agents, which make the evolution of agents become complicated and difficult to predict.

The situations where there are unknown or uncertain relations between agents are basically ignored for decades. However, these situations have important effects on the studies of complex systems which often exist in the practical applications. In such situations, the agents must try to cope with the uncertainties occurred in local interactions with only limited available history information and local information.

The key point is to simultaneously estimate the states and unknown coupling parameters. Because of the information constraints, without knowing the coupling gains of the local interactions, it is not easy for each agent to estimate its states by traditional Kalman filter or other state observers. Based on the key idea of state augmentation and the certainty equivalence principle borrowed from the area of adaptive control, we introduce several filtering methods to resolve the fundamental problem.

2.6 Conclusion

In this chapter, several challenges in Kalman filtering have been briefly addressed, which are to be detailedly investigated in later parts. Uncertainty is everywhere. Kalman filter is one powerful tool to cope with uncertainty. However, the performance of Kalman filter relies on exact dynamical model, which covers issues of linearity, additive noise, noise mean, noise covariance, system matrix, model parameters, and so on. Any uncertainty in the model description may lead to model mismatch and performance downgrade of standard Kalman filter. This is why introducing various ideas of adaptation is crucial so as to cope with model uncertainty when applying filtering techniques.

References

1. S.J. Julier, J.K. Uhlmann, Unscented filtering and nonlinear estimation. Proc. IEEE **92**(3), 644 401–422 (2004)
2. A. Doucet, N. De Freitas, N. Gordon, *Sequential Monte Carlo Methods in Practice* (Springer, Berlin, 2001)
3. J. Georgy, A. Noureldin, M. Korenberg, M. Bayoumi, Low-cost three-dimensional navigation solution for RISS/GPS integration using mixture particle filter. IEEE Trans. Veh. Technol. **59**(2), 599–615 (2010)
4. R.K. Mehra, On the identification of variances and adaptive Kalman filtering. IEEE Trans. Autom. Control **15**(2), 175–184 (1970)
5. O. Yeste Ojeda, J. Grajal, Adaptive-fresh filters for compensation of cycle-frequency errors. IEEE Trans. Signal Process. **58**(1), 1–10 (2010)
6. A.H. Mohamed, K.P. Schwarz, Adaptive Kalman filtering for INS/GPS. J. Geod. **73**(4), 193–203 (1999)
7. M.R. Azimi-Sadjadi, R.R. Xiao, X. Yu, Neural network decision directed edge-adaptive Kalman filter for image estimation. IEEE Trans. Image Process. **8**(4), 589–592 (1999)

8. X. Xiao, B. Feng, B. Wang, On-line realization of SVM Kalman filter for MEMS gyro, in *Proceedings of the 3rd International Conference on Measuring Technology and Mechatronics Automation*, pp. 768–770
9. B. Feng, H.B. Ma, M.Y. Fu, S.T. Wang, A framework of finite-model kalman filter with case study: MVDP-FMKF algorithm. Acta Automatica Sinica **39**(8), 1246–1256 (2013)

Part II
Kalman Filtering for Uncertain Systems

Chapter 3
Kalman Filter with Recursive Process Noise Covariance Estimation

Kalman filter has been found to be useful in vast areas. However, it is well known that successful use of standard Kalman filter is greatly restricted by the strict requirements on a priori information of the model structure and statistics information of the process and measurement noises. Generally speaking, the covariance matrix of process noise is harder to be determined than that of the measurement noise by routine experiments since the statistical property of process noise cannot be obtained directly by collecting a large number of sensor data due to the intrinsic coupling of process noise and system dynamics. Considering such background of wide applications, this chapter introduces one algorithm, named as *recursive covariance estimation (RCE)* algorithm, to estimate the unknown covariance matrix of noise from a sample of signals corrupted with the noise. Based on this idea, for a class of discrete-time linear time-invariant (LTI) systems where the covariance matrix of process noise is completely unknown, a new Kalman filtering algorithm named as *Kalman filter with recursive covariance estimation (KF-RCE)* is presented to resolve this challenging problem of state estimation without the statistical information of process noise, and the rigorous stability analysis is given to show that this algorithm is optimal in sense that the covariance matrix and state estimations are asymptotically consistent with the ideal Kalman filter when the exact covariance matrix of process noise is completely known a priori. Extensive simulation studies have also verified the theoretical results and the effectiveness of the proposed algorithm.

3.1 Introduction

Kalman filtering techniques have been widely used in many applications, such as navigation [1–4], communication [5–7], and fault detecting [8]. In his pioneering work [9, 10], Kalman showed state estimating and prediction problems; he also

H. Ma et al., *Kalman Filtering and Information Fusion*,
https://doi.org/10.1007/978-981-15-0806-6_3

proposed a method for the solution of this common problems when system model is linear, precisely known and noise statistical properties are obtained.

Many extensions and generalizations followed Kalman's basic work. Stanley F. Schmidt introduces extend Kalman filter (EKF) [11] algorithm to solve filtering and prediction problems of nonlinear systems encountered in Apollo program in the 1960s. Based on Taylor expansion, EKF is trying to approximate nonlinear systems by linear model and then use standard Kalman filter to estimate system states. Such an idea was proved to be very useful in many different applications [12, 13]; hence, EKF was ever regarded as one standard filtering method for nonlinear systems for a long time and it attracted amount of research interests [14, 15]. Then Carlson [16] presented federal Kalman filter (FKF) [17] algorithm to deal with multisystem optimal data fusion.

When system model is nonlinear, unscented Kalman filter (UKF) [18] was introduced by S. J. Julier and J. K. Uhlmann in 1997, and later applied in various areas such as induction motor drives [19], UAV attitude estimation [20], and so on. In recent years, some variants of UKF were proposed in the literature [21]. Robust Kalman filter (RKF) [22–24] was presented to solve the problems that the system model cannot be obtained accurately. Then adaptive Kalman filter (AKF) [25, 26] was used to deal with the problem that process and measurement noise statistical properties are uncertain.

In this chapter, we will present a new algorithm to estimate process noise covariance matrix in real time from measurement sequence for the discrete-time linear time-invariant (LTI) systems. A new Kalman filter with recursive covariance estimation (RCE), abbreviated as KF-RCE, is presented to deal with the problem that the covariance matrix of process noise in Kalman filter is completely unknown. Then stability analysis is presented to show that state estimation obtained by the new algorithm is convergent to the optimal estimation by the Kalman filter with exact system parameters under certain system conditions. Finally, three simulation examples were presented to illustrate that the new algorithm can effectively deal with the problem of the completely unknown process noise covariance matrix.

The rest of this chapter is organized as follows: in Sect. 3.2, the problem will be formulated which is dealt with by the KF-RCE algorithm and standard Kalman filter is reviewed with mathematical equations for convenience of later discussions; then, to solve how to obtain the covariance matrix of a vector from the sample sequence corrupted with a noise in real time; in Sect. 3.3, a new algorithm named `RecursiveCovarianceEstimation` is presented, and a simulation example will be presented to illustrate that the new algorithm is practical and effective; in Sect. 3.4, the new algorithm named KF-RCE will be introduced to show how to cope with the case, while the covariance matrix of process noise is completely unknown for the discrete-time LTI systems; in Sect. 3.5, the stability analysis of the KF-RCE will be presented. In Sect. 3.6, three simulation examples will be presented to illustrate the effectiveness of the KF-RCE algorithm; finally, in Sect. 3.7 concluding remarks are presented to summarize KF-RCE algorithm and our further work about the improvement of the algorithm.

3.2 Problem Formulation

3.2.1 *Standard Kalman Filter*

The Kalman filter is a set of mathematical equations that provides an efficient computational (recursive) means to estimate the state of a process, in a way that minimizes the mean of the squared error. This filter is very powerful since it supports estimations of past, present, and even future states [9, 10]. Due in large part to advantages in digital computing, the Kalman filter has been the subject of extensive research and application, particularly in the area of autonomous or assisted navigation [27].

It is the core of the program to design Kalman filter's observation update and state update. Suppose that estimated state x_k is controlled by system noise sequence w_k on time t_k, the system state equation is as follows:

$$\begin{cases} x_k = Ax_{k-1} + Bw_{k-1} \\ y_k = Cx_k + v_k \end{cases} \tag{3.1}$$

where A is state transfer matrix, B is the input matrix, C is observation matrix, v_k is measurement noise sequence, and y_k represents observable measurements at time t_k.

If the process and measurement noises are zero mean, Gaussian, and uncorrelated, then $\hat{x}_k$, the estimate of x_k, can be recursively obtained by the following equations [9, 10]:

$$\begin{cases} \hat{x}_{k,k-1} = A\hat{x}_{k-1} \\ \hat{x}_k = \hat{x}_{k,k-1} + K_k(y_k - C\hat{x}_{k,k-1}) \\ P_{k,k-1} = AP_{k-1}A^T + Q_{k-1} \\ K_k = P_{k,k-1}C^T(CP_{k,k-1}C^T + R_k)^{-1} \\ P_k = (I - K_kC)P_{k,k-1} \end{cases} \tag{3.2}$$

where $\hat{x}_{k-1}$ is the optimal estimation of state x_{k-1} based on the measurements up to and including time $k-1$; P_{k-1} denotes the covariance of estimation error of $\hat{x}_{k-1}$; Q_k denotes the known covariance of the process noise Bw_k at the kth time; and R_k is the known covariance matrix of the measurement noise v_k at time k.

3.2.2 *Problem To Be Resolved*

It is well known that standard Kalman filter works well when exact a priori knowledge on the system structure (i.e., matrices A, B, C) as well as the statistical properties, mainly Q_k and R_k, of the process and measurement noises are available. However, in practice, such ideal assumptions usually cannot be satisfied due to various reasons such as the modeling error, difficulties in experiments, and so on. Especially, we usually cannot obtain the exact statistical properties of the noise because there is no way to directly isolate the noise from the measurement signals and in most cases we may not be able to completely understand the mechanism of the noise. Previous studies [28, 29] have shown that using wrong values of Q_k and R_k in Kalman filter may degrade the performance of state estimation or even make the filter not work at all. Thus, to successfully apply standard Kalman filter in practice, it is often required to do much work to obtain the statistical properties of the noise or alternatively, regarding the upper bound Q of Q_k in the filter loop to describe the uncertainty of Q_k, which reduces the accuracy of state estimation.

Remark 3.1 As in Eq. (3.1), the state transfer matrix A can be obtained by system state kinetics relationship or sample data processing. The measurement matrix C is determined by the linear mathematical relationship between measurement signal and the chosen system states. The measurement noise statistical property can be obtained by the sensor capacity which is used to obtain measurement signals or by the process of receiving measurement data. In general, the process noise statistical property is determined by the system state uncertainty. But the state uncertainty always cannot be obtained by a priori means in the engineering systems. The process noise statistical property plays an important role in Kalman filter. The uncertainty of process noise covariance matrix Q may make the filter not work well or even diverge [28, 29]. In this chapter, we will present a new method to estimate the process noise covariance matrix in real time for the discrete-time LTI systems.

Noticing the above issues, it can be seen that correct value of Q_k plays crucial role in the applications of Kalman filter. To overcome the side effects of uncertainty in Q_k, the original idea of adaptive Kalman filter (AKF) was first introduced by R. K. Mehra in 1970 [25], where the author constructed an innovation sequence according to the property of an optimal filter so as to test the optimality of covariance estimation based on the state estimation, which makes the closed-loop stability analysis very difficult due to the intrinsic couplings between the state estimation and covariance estimation. With the unknown noise statistics, a kind of adaptive sequential estimation algorithm, named adaptive limited memory filter (ALMF), was presented by Kenneth Myers and Byron Tapley in 1976 [30]. It is a good attempt to deal with the parameter uncertainties in the noise statistics under the framework of Kalman filter algorithm, and ALMF algorithm was useful for the application of an orbit determination problem for a near-earth satellite with significant modeling errors [30]. However, the estimation of covariance matrix of process noise was related with the state estimation, which may make the mathematical stability analysis very difficult. Besides, one may notice that most of AKF algorithms were computationally complex and they usually cannot satisfy the real-time requirement, which was almost one of the basic demands in all engineering systems.

Summarizing existing approaches reported in the literature, to the best knowledge of the authors, we think it is necessary to develop new algorithms to revisit the problem of dealing with the uncertainty in Q_k in the filter design, especially when we have no any a priori knowledge on the process noise except that the process noise is zero mean and uncorrelated with measurement noise. That is to say, if we do not know the range of Q_k or nominal values of Q_k, is it possible to design an algorithm for filter design which can not only deal with large uncertainty of the process noise but also guarantee the filtering accuracy and stability?

Mathematically speaking, in this chapter we are going to investigate this fundamental problem under the following technical assumptions:

Assumption 3.1 For the discrete-time linear time-invariant (LTI) systems as in Eq. (3.1), supposing that A, B, and C are constant and known, and the LTI system is controllable and observable.

Remark 3.2 Controllability and observability are the basic requirements to satisfy the stability of the standard Kalman filter. For simplicity and ease of problem formulation, the system matrices of A, B, C are supposed to be constant and known in this chapter.

Assumption 3.2 The system measurement y_k is bounded.

Remark 3.3 This assumption is reasonable and could be easily satisfied by most practical systems.

Assumption 3.3 The measurement matrix C has left inverse matrix, $[C^T C]^{-1} C^T$.

Remark 3.4 This technical assumption of the measurement matrix C will be used later to ensure the stability of algorithm. This condition may be weaken further, which will be addressed in our future work. From this assumption, it can be obtained that the algorithm could cope with the case that the number of measurement dimension is more than or equal to the number of the state dimension. With the wide use of multiple sensors, it can be seen that such a technical condition can hold in a number of practical applications.

Assumption 3.4 The autonomous system modeled as follows:

$$\eta_{k+1} = (I - \hat{K}_k C) A \eta_k \tag{3.3}$$

is exponentially stable, where $\hat{K}_k$ is the Kalman gain obtained by the KF-RCE algorithm.

Remark 3.5 This technical condition is not a strong condition and it is reasonable to guarantee the closed-loop stability. To illustrate what does it mean, we just consider the ideal case of standard Kalman filter when the statistics information of the process noise and the measurement noise is exactly known. In such a case, if the system is controllable and observable in the standard Kalman filter, it is easy to see that we have

$$\begin{aligned}
& x_{k+1} - \hat{x}_{k+1} \\
&= [Ax_k + Bw_{k+1}] - [\hat{x}_{k+1,k} + K_{k+1}(y_{k+1} - C\hat{x}_{k+1,k})] \\
&= [Ax_k + Bw_{k+1}] \\
&\quad -[A\hat{x}_k + K_{k+1}(Cx_{k+1} + v_{k+1} - CA\hat{x}_k)] \\
&= [Ax_k + Bw_{k+1}] \\
&\quad -[A\hat{x}_k + K_{k+1}(CAx_k + CBw_{k+1} + v_{k+1} - CA\hat{x}_k)] \\
&= [A - K_{k+1}CA](x_k - \hat{x}_k) \\
&\quad +[Bw_{k+1} - K_{k+1}CBw_{k+1} - K_{k+1}v_{k+1}]
\end{aligned} \tag{3.4}$$

whose autonomous system is exactly

$$\eta_{k+1} = (I - K_k C)A\eta_k \tag{3.5}$$

Hence, Assumption 3.4 means that $\tilde{x}_k = x_k - \hat{x}_k$ can be controlled by its autonomous system, which converges to zero exponentially, which is natural for the stability of Kalman filter and the convergence of state estimation.

Assumption 3.5 Suppose that $Q_k = Q$ is constant yet completely unknown. The process and measurement noises are uncorrelated, and measurement noise covariance matrix R has been obtained.

Remark 3.6 In practice, most engineering systems are formulated by LTI system models. In Kalman filter algorithm, designers always suppose that the process and measurement noise covariance matrices are constant since time-varying covariance matrices are hardly to be determined. Besides, the Kalman filter with upper value of Q_k always could obtain acceptable estimating precision.

Problem 3.1 *Under Assumptions 3.1–3.5, how to design a filter which is able to cope with completely unknown covariance matrix Q of the process noise?*

Generally speaking, such statistical properties should be obtained via many samples of the process noise. However, we may encounter two difficulties: (a) It is almost impossible to isolate the process noise from the output data generated by the system dynamics. (b) Even if the process noise can be isolated from the output data, according to statistical knowledge, many samples of output data are needed so as to generate enough samples of process noise for the purpose of estimating the statistical properties of the process noise. Facing these challenges, we hope to use only the data in one sample to estimate the covariance matrix Q online rather than doing many experiments offline for the purpose of obtaining statistical properties of process noise.

3.3 Basic Idea: Estimating Covariance Matrix

For the convenience of later discussions, we need to introduce first the basic idea of estimating covariance matrix for a random variable, which is the basis of constructing the filter algorithm proposed in this chapter.

We consider the following simple yet nontrivial problem:

Problem 3.2 *For a random variable w, it is supposed that*

$$\xi = w + v \tag{3.6}$$

and $E(w) = 0$, $E(v) = 0$, $cov(v) = R$, where $E(\cdot)$ is the mathematical expectation and $cov(\cdot)$ is the covariance matrix. Then, provided that we have an i.i.d sample

$\{\xi_i, i = 1, 2, 3, \ldots\}$ of ξ, how to obtain variance or covariance of random variable w from the sample sequence of random variable ξ in real time?

Now we try to answer this basic question.

Case I: Scalar case. When ξ is a scalar, obviously, since w and v are uncorrelated, according to Eq. (3.6), we have

$$\mathrm{D}(w) = \mathrm{D}(\xi) - R \tag{3.7}$$

The left problem is how to obtain $\mathrm{D}(\xi)$.

According to the basic statistical knowledge, we have that $\mathrm{D}(\xi)$ is calculated as follows:

$$\mathrm{D}(\xi) = E(\xi^2) - E^2(\xi) \tag{3.8}$$

Since $E(w) = 0$ and $E(v) = 0$, it can be obtained that $E(\xi) = E(w) + E(v) = 0$. Then we have

$$\mathrm{D}(\xi) = E(\xi^2) \tag{3.9}$$

According to the law of large numbers, when $E(\xi^2) < \infty$, it is obtained

$$E(\xi^2) = \lim_{n\to\infty} \frac{1}{n} \sum_{i=1}^{n} \xi_i^2 \quad a.s. \tag{3.10}$$

Therefore, we can eventually obtain precise value of $\mathrm{D}(\xi)$ almost surely as $n \to \infty$ or we have enough samples of the random variable ξ. Then using Eq. (3.7), $\mathrm{D}(w)$ can be calculated by generating an i.i.d sample of ξ which in turn asymptotically provides the correct value of $\mathrm{D}(w)$ almost surely as $n \to \infty$.

Case II: Vector case. When ξ is a vector, obviously the covariance matrix cannot be obtained as simple as a scalar variable. Obviously, the diagonal elements of covariance matrix can be obtained just as the scalar variable. The left problem is how to calculate nondiagonal terms.

As is well known, nondiagonal terms denote the covariances of the different pairs of elements in a vector. From the basic knowledge of statistics, it is obtained

$$\mathrm{cov}(x, y) = E(xy) - E(x)E(y) \tag{3.11}$$

Consequently, if $E(x) = 0$ and $E(y) = 0$,

$$\mathrm{cov}(x, y) = E(xy) \tag{3.12}$$

If we can obtain enough samples of zero-mean random variables x and y, the covariance between x and y can be calculated as follows:

$$\mathrm{cov}(x, y) = E(xy) = \lim_{n\to\infty} \frac{1}{n} \sum_{i=1}^{n} x_i y_i \tag{3.13}$$

Summarizing the above discussions, the following result is obtained:

Theorem 3.1 *Suppose that $\xi \in R^{m\times 1}$ be a random variable and $E(\xi) = 0$, the covariance matrix of the random variable ξ can be obtained from its samples as follows:*

$$cov(\xi) = E(\xi\xi^T) = \lim_{n\to\infty} cov_n(\xi) = \lim_{n\to\infty} \frac{1}{n} \sum_{i=1}^{n} \xi_i \xi_i^T \tag{3.14}$$

and consequently covariance matrix of random variable w can be obtained by

$$cov(w) = \lim_{n\to\infty} \frac{1}{n} \sum_{i=1}^{n} \xi_i \xi_i^T - R \tag{3.15}$$

where R is the covariance matrix of random variable v and the relationship among ξ, w and v is as in Eq. (3.6).

However, the calculation method indicated above is not suitable for online parameter estimating since we cannot obtain all the vector samples at a certain time. If we want to use similar method on the real-time parameter fusion, a recursive form algorithm must be restructured to implement this estimation method.

Supposing that $\mathrm{cov}_n(\xi)$ is the estimated covariance matrix from the samples up to and including the time n, with the arrival of a new sample, the $(n+1)$th sample should be used to modify $\mathrm{cov}_n(\xi)$ in order to obtain new estimation $\mathrm{cov}_{n+1}(\xi)$.

As in Eq. (3.13), it can be obtained that

$$\begin{aligned} \mathrm{cov}_{n+1}(\xi) &= \tfrac{1}{n+1} \sum_{i=1}^{n+1} \xi_i \xi_i^T \\ &= \tfrac{1}{n+1} \left(\sum_{i=1}^{n} \xi_i \xi_i^T + \xi_{n+1} \xi_{n+1}^T \right) \\ &= \tfrac{n}{n+1} \mathrm{cov}_n(\xi) + \tfrac{1}{n+1} \xi_{n+1} \xi_{n+1}^T \end{aligned} \tag{3.16}$$

Therefore, the covariance matrix of random variable ξ can be calculated in real time with increasing sample sequence. According to the law of large numbers, it can be obtained that $\mathrm{cov}_n(\xi) - \mathrm{cov}(\xi) \to 0$ almost surely as $n \to \infty$, where $\mathrm{cov}(\xi)$ is the real covariance matrix of the random variable ξ. Then it can be obtained that

$$\mathrm{cov}_k(w) = \mathrm{cov}_k(\xi) - R \tag{3.17}$$

Therefore, Algorithm 1 can be used to resolve Problem 3.2.

Algorithm 1 RCE($\{\xi_i\}$, R)

Initialization: $\mathrm{cov}_0(\xi) = 0$, $\mathrm{cov}_0(w) = 0$, R
Input: the sample sequence $\{\xi_k\}$ of random variable ξ
Output: estimated covariance matrix $\{\mathrm{cov}_k(w)\}$ of the random variable w
1: **for** $k = 1$ to n **do**
2: Using the sample data ξ_k to update the estimated covariance matrix $\mathrm{cov}_k(\xi)$ just as in Eq. (3.16)
3: Using the method as in Eq. (3.17) to calculate the estimated covariance matrix $\mathrm{cov}_k(w)$
4: **end for**
5: Return $\{\mathrm{cov}_k(w)\}$

Now an example is presented to illustrate the simulation results of Algorithm 1, which verify the effectiveness of this recursive covariance estimation algorithm. In this simulation, it is supposed that the vector $\xi \in R^{4\times 1}$ and

$$\mathrm{cov}(\xi) = Q = \begin{bmatrix} 0.25 & 0.04 & 0.04 & 0.04 \\ 0.04 & 0.25 & 0.04 & 0.04 \\ 0.04 & 0.04 & 0.25 & 0.04 \\ 0.04 & 0.04 & 0.04 & 0.25 \end{bmatrix} \tag{3.18}$$

$$\mathrm{cov}(v) = R = \mathrm{diag}\{0.25 \quad 0.25 \quad 0.25 \quad 0.25\} \tag{3.19}$$

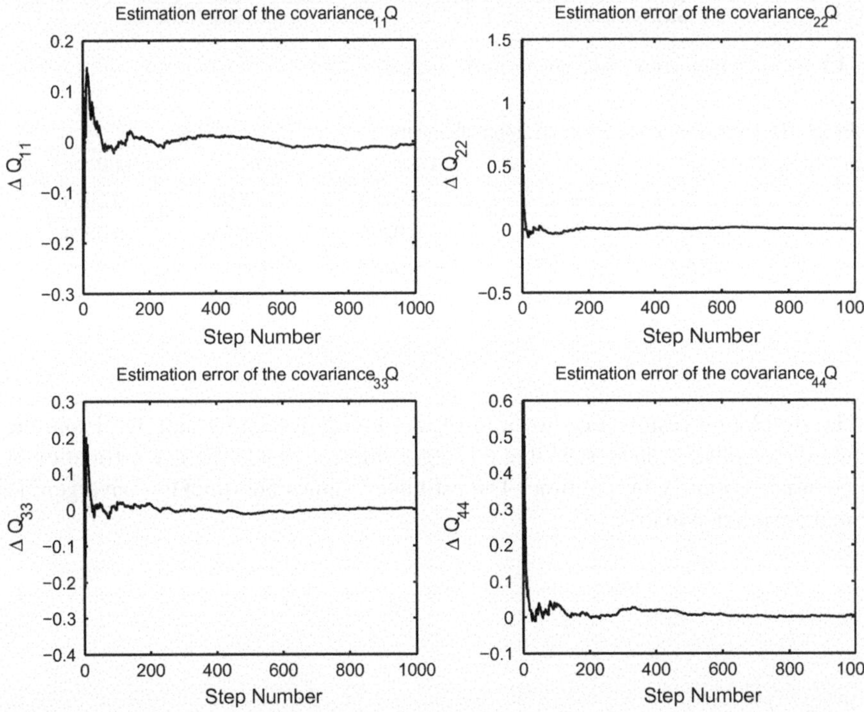

Fig. 3.1 Q estimation error of the diagonal elements

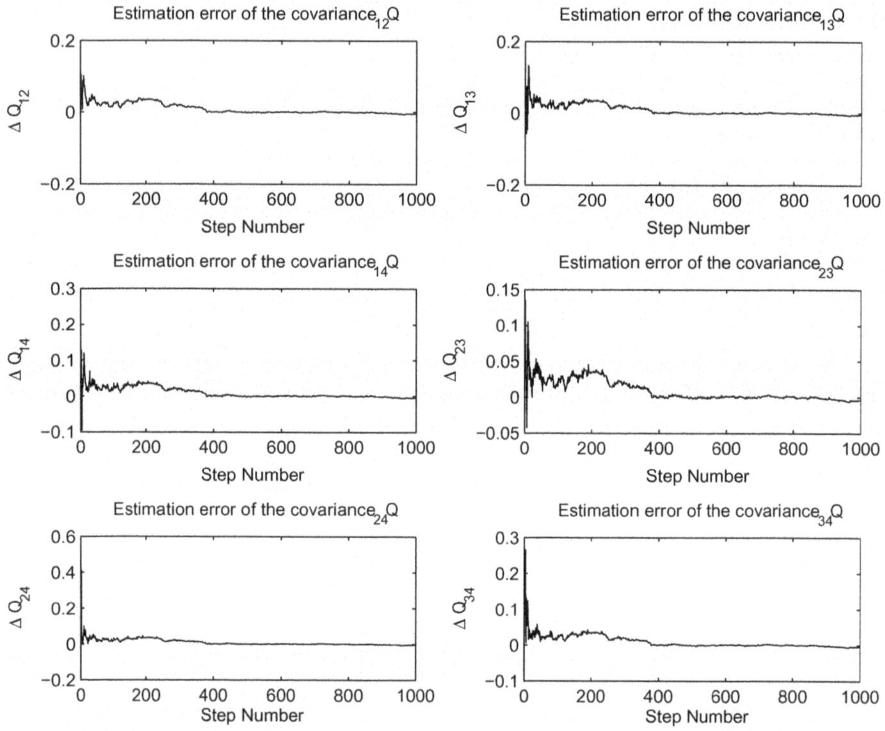

Fig. 3.2 Q estimation error of the nondiagonal elements

Table 3.1 Q estimation error of the diagonal elements

Step No.	1	10	100	500	1000
$\hat{Q}_{11} - Q_{11}$	−0.2016	0.1138	−0.0155	0.0093	−0.0073
$\hat{Q}_{22} - Q_{22}$	−0.0773	0.0555	−0.0295	0.0055	0.0004
$\hat{Q}_{33} - Q_{33}$	0.2583	0.1454	0.0225	−0.0119	0.0007
$\hat{Q}_{44} - Q_{44}$	0.1129	0.0784	0.0276	0.0065	0.0042

The simulation results are shown in Figs. 3.1, 3.2, Tables 3.1 and 3.2. From the simulation results, it can be seen that this algorithm can yield effective estimation of the covariance matrix in real time. The estimated values are quickly convergent to the real covariance matrix.

Table 3.2 Q estimation error of the nondiagonal elements

Step No.	1	10	100	500	1000
$\hat{Q}_{12} - Q_{12}$	0.0057	0.1013	0.0196	−0.0010	−0.0041
$\hat{Q}_{13} - Q_{13}$	0.0613	0.1328	0.0246	−0.0011	−0.0041
$\hat{Q}_{14} - Q_{14}$	0.0491	0.1197	0.0187	−0.0011	−0.0042
$\hat{Q}_{23} - Q_{23}$	0.1310	0.1053	0.0252	−0.0014	−0.0042
$\hat{Q}_{24} - Q_{24}$	0.1080	0.1004	0.0187	−0.0000	−0.0041
$\hat{Q}_{34} - Q_{34}$	0.1976	0.1011	0.0180	−0.0014	−0.0043

3.4 Kalman Filter Based on Algorithm `RecursiveCovarianceEstimation`

In the sequel, we are going to apply the basic idea shown in Algorithm 1 to the discrete-time linear time-invariant (LTI) systems, which can be described as follows:

$$\begin{cases} x_k = Ax_{k-1} + Bw_{k-1} \\ y_k = Cx_k + v_k \end{cases} \tag{3.20}$$

where x_k is the state at time k, y_k is the output at time k, w_k is the process noise at time k, and v_k is the measurement noise at time k. Here $A \in R^{n\times n}$ is state transfer matrix, a constant matrix; B is the control input matrix and time-invariant; $C \in R^{m\times n}$ is the measurement matrix; and $E(w_k) = 0$, $E(v_k) = 0$, $\text{cov}(Bw_k) = Q$, $\text{cov}(v_k) = R$, w_k and v_k are uncorrelated.

From Eq. (3.20), it can be obtained that

$$\begin{aligned} y_k &= Cx_k + v_k \\ &= C[Ax_{k-1} + Bw_{k-1}] + v_k \\ &= CAx_{k-1} + CBw_{k-1} + v_k \end{aligned} \tag{3.21}$$

and

$$y_{k-1} = Cx_{k-1} + v_{k-1} \tag{3.22}$$

Then we have

$$Cx_{k-1} = y_{k-1} - v_{k-1} \tag{3.23}$$

From Assumption 3.3, we have

$$[C^T C]^{-1} C^T C = I \tag{3.24}$$

By multiplying $[C^T C]^{-1} C^T$ to both sides of Eq. (3.23), we have

$$x_{k-1} = [C^T C]^{-1} C^T y_{k-1} - [C^T C]^{-1} C^T v_{k-1} \tag{3.25}$$

Then substituting Eq. (3.25) into Eq. (3.21), it can be obtained that

$$y_k = CAMy_{k-1} - CAMv_{k-1} + CBw_{k-1} + v_k \tag{3.26}$$

where $M = [C^T C]^{-1} C^T$.

Multiplying M to Eq. (3.26), we obtain

$$My_k - AMy_{k-1} = Bw_{k-1} + Mv_k - AMv_{k-1} \tag{3.27}$$

Introducing the notation

$$\xi_k = My_k - AMy_{k-1} \tag{3.28}$$

$$V_k = Mv_k - AMv_{k-1} \tag{3.29}$$

Since $\mathrm{cov}(Bw)$ is a constant covariance matrix, from Eq. (3.27), it can be obtained that

$$\xi_k = Bw_{k-1} + V_k \tag{3.30}$$

Since A, C and $\mathrm{cov}(v_k)$ are known a priori, $\mathrm{cov}(V)$ can be calculated in a priori. Using the definition in Eq. (3.28), one more sample ξ_k can be obtained with the arrival of measurement y_k. Then using the new sample, the estimated covariance matrix $\mathrm{cov}_k(Bw)$ can be obtained as in Algorithm 1.

$$\mathrm{cov}_k(Bw) = \mathrm{cov}_k(\xi) - \mathrm{cov}(V) \tag{3.31}$$

Using this method, the estimated covariance matrix $\mathrm{cov}_k(Bw)$ can be obtained in real time, which can serve as a basis for applying standard Kalman filter to generate system state estimates

Algorithm 2 KF-RCE($\{y_k\}, R, A, B, C$)

Initialization: A, B, C, R, $\hat{x}_0$ the initial state estimate,
P_0 the initial state error covariance matrix,
$\mathrm{cov}_0(\xi) = 0$, $\mathrm{cov}_0(Bw) = 0$
Using R, A and C to calculate the covariance matrix $\mathrm{cov}(V)$ of the notion defined by Eq. (3.29)
Input: measurement sequence $\{y_k\}$
Output: real-time system state estimate $\{\hat{x}_k\}$, state estimation error covariance matrix $\{P_k\}$
for $k = 1$ to n **do**
  Using the measurement y_k to calculate $\xi(k)$ as in Eq. (3.28)
  Using the sample $\xi(k)$ to update the estimated covariance matrix $\mathrm{cov}_k(\xi)$ as in Eq. (3.16)
  Using $\mathrm{cov}_k(\xi)$ and $\mathrm{cov}(V)$ to calculate the estimated process noise matrix $\mathrm{cov}_k(Bw)$ as in Eq. (3.31)
  $\hat{x}_{k,k-1} = A\hat{x}_{k-1}$
  $\hat{P}_{k,k-1} = AP_{k-1}A^T + \mathrm{cov}_k(Bw)$
  $\hat{K}_k = \hat{P}_{k,k-1}C^T(C\hat{P}_{k,k-1}C^T + R)^{-1}$
  $\hat{x}_k = \hat{x}_{k,k-1} + \hat{K}_k(y_k - C\hat{x}_{k,k-1})$
  $P_k = (I - \hat{K}_k C)\hat{P}_{k,k-1}$
end for
Return $\{\mathrm{cov}_k(Bw)\}$, $\{\hat{x}_k\}$ and $\{P_k\}$

3.5 Stability Analysis

In this section, the stability analysis of the KF-RCE algorithm is presented to show that KF-RCE algorithm is stable and the state estimate obtained by KF-RCE is convergent to the ideal Kalman filter with exact system parameters.

The following lemmas are critical to establish the stability of the dynamic error system.

Lemma 3.1 (The bounded input-to-bounded output stability) *Consider a time-varying dynamic error system*

$$\delta_k = F_k \delta_{k-1} + u_k \tag{3.32}$$

where $k = 1, 2, \ldots,$ the output (dynamic error) $\delta_k \in R^m$, the input $u(k) \in R^m$. Assume that $F_k \to F$ as $k \to \infty$, where F is a stable matrix (i.e., all eigenvalues of F lie inside the unit circle), and the input $u(k)$ is bounded. Then the output δ_k is bounded [31–33].

Proof Let $\lambda_1, \ldots, \lambda_m$ be the eigenvalues of F. Since F is a stable matrix, then $|\lambda_i| < 1$, $i = 1, \ldots, m$, so that its spectral radius $\rho = max(|\lambda_1|, \ldots, |\lambda_m|) < 1$. Applying the matrix theory (Roger & Johnson 1985), there exists a matrix norm $\|\cdot\|$ such that $\|F\| = \rho + \mu = \rho_0 < 1, \mu > 0$. Setting $F_k = F + \Delta F_k$, from $F_k \to F$ we have that $\Delta F \to 0$, as $k \to \infty$. Hence taking $\alpha > 0$ such that $\alpha < 1 - \rho_0$, there exists $k_0 > 0$ such that $\|\Delta F_k\| < \alpha$, as $k > k_0$. Defining $\rho_m = \rho_0 + \alpha$, we have that $0 < \rho_m < 1$, and $\|F_k\| \le \|F\| + \|\Delta F_k\| < \rho_0 + \alpha = \rho_m < 1$, as $k > k_0$. By the iteration for Eq. (3.32), we obtain that

$$\delta_k = F_{k,k_0} \delta_{k_0} + \sum_{i=k_0+1}^{k} F_{k,i} u_i \tag{3.33}$$

with the definitions that $F_{k,k} = I_n$, $F_{k,i} = F_k F_{k-1} \cdots F_{i+1}$, $k > i$. Thus $\left\| F_{k,i} \right\| \le \|F_k\| \cdots \|F_{i+1}\| < \rho_m^{k-i}$. The boundedness of u_t yields that $\|u_t\| \le c_1, t \ge 0$. When $k > i > k_0$, from Eq. (3.33), we have

$$\begin{aligned} \|\delta_k\| &\le \left\| F_{k,k_0} \right\| \left\| \delta_{k_0} \right\| + \sum_{i=k_0+1}^{k} \left\| F_{k,i} \right\| \|u_i\| \\ &\le \rho_m^{k-k_0} \left\| \delta_{k_0} \right\| + \sum_{i=k_0}^{k} \rho_m^{k-i} c_1 \end{aligned} \tag{3.34}$$

Notice that $0 < \rho < 1$, we have that $0 < \rho_m^{k-k_0} < 1$, and

$$\sum_{i=k_0+1}^{k} \rho_m^{k-i} = \sum_{j=0}^{k-k_0-1} \rho_m^j = \frac{1 - \rho_m^{k-k_0}}{1 - \rho_m} < \frac{1}{1 - \rho_m} \tag{3.35}$$

which yields the boundedness of δ_k. □

Lemma 3.2 (The bounded input-to-bounded output stability and the infinitesimal input-to-infinitesimal output stability) *Consider a dynamic error system*

$$\delta_k = F_k \delta_{k-1} + u_k \tag{3.36}$$

where $k = 1, 2, \ldots,$ $\delta_k \in \boldsymbol{R}^m$ *is the output, and* $u_k \in \boldsymbol{R}^m$ *is the input. Assume that* $F_k \to F$ *as* $k \to \infty$, *where* F *is a stable matrix. If* u_k *is bounded, then* δ_k *is bounded. If* $u_k \to 0$, *as* $k \to \infty$, *then* $\delta_k \to 0$, *as* $k \to \infty$ *[31–33].*

Proof Iterating Eq. (3.36) yields

$$\delta_k = F_{k,0}\delta_0 + \sum_{j=0}^{k-1} F_{k,j} u_{k-j} \tag{3.37}$$

with the definitions that $F_{k,k} = I_n$, $F_{k,i} = F_k F_{k-1} \cdots F_{i+1}$, $k > i$. Because F_k is always a stable matrix, it has the spectral radius $\rho, 0 \le \rho < 1$, and there exists a matrix norm $\|\cdot\|$ such that $\|F_k\| = \rho + \mu = \rho_0 < 1, \mu > 0$. From Eq. (3.37), it follows that

$$\|\delta_k\| \le \rho_0^k \|\delta_0\| + \sum_{j-0}^{k-1} \rho_0^j \|u_{k-j}\| \tag{3.38}$$

If u_k is bounded, i.e., $\|u_k\| \le c_1, k \ge 0$, then

$$\|\delta_k\| \le \|\delta_0\| + c_1 \sum_{j=0}^{\infty} \rho_0^j = \|\delta_0\| + \frac{c_1}{1-\rho_0}, \quad k \ge 0 \tag{3.39}$$

which yields the boundedness of δ_k. If $u_k \to 0$ as $k \to \infty$, then $\|u_k\| \to 0$ as $k \to \infty$, which yields $\|u_k\| \le c_2$, $k \ge 0$. Noting $0 < \rho_0 < 1$, it follows that $\rho_0^k \to 0$, as $k \to \infty$. Therefore, for arbitrarily small $\varepsilon > 0$, there exist k_ρ and k_u such that $\rho_0^k < \varepsilon$, as $k > k_\rho$, $\|u_k\| < \varepsilon$, as $k > k_u$. Here ρ_0 is a fixed number, so that k_ρ only depends on ε. Consider the decomposition

$$\sum_{j=0}^{k-1} \rho_0^j \|u_{k-j}\| = \sum_{j=0}^{k_\rho} \rho_0^j \|u_{k-j}\| + \sum_{j=k_\rho+1}^{k-1} \rho_0^j \|u_{k-j}\| \tag{3.40}$$

When $k > k_\delta = k_u + k_\rho$, we have

$$\begin{aligned}
\sum_{j=0}^{k_\rho} \rho_0^j \left\| u_{k-j} \right\| &< \varepsilon \sum_{j=0}^{k_\rho} \rho_0^j < \tfrac{\varepsilon}{1-\rho_0} \\
\sum_{j=k_\rho+1}^{k-1} \rho_0^j \left\| u_{k-j} \right\| &< c_2 \sum_{j=k_\rho+1}^{k-1} \rho_0^j \\
&= \tfrac{c_2 \rho_0^{k_\rho+1}(1-\rho_0^{k-k_\rho-1})}{1-\rho_0} < \tfrac{c_2 \varepsilon}{1-\rho_0}
\end{aligned} \tag{3.41}$$

Taking $k > k_\delta$, from Eqs. (3.38), (3.40), and (3.41), we have $\|\delta_k\| \leq c_3\varepsilon$, with constant $c_3 = \|\delta(0)\| + [(1 + c_2)/(1 - \rho_0)]$ which is independent of ε. Since $\varepsilon > 0$ can be taken as an arbitrarily small number, then $\|\delta_k\|$ can be arbitrarily small for sufficiently large k, such that $k > k_\delta$, i.e., $\delta(k) \to 0$, as $t \to \infty$ □

Theorem 3.2 *For the discrete-time LTI systems as in Eq. (3.20), using the method in Algorithm 2, the estimated process noise covariance matrix $cov_k(Bw)$ is convergent to the real covariance matrix $cov(Bw)$ as $k \to \infty$, i.e.,*

$$cov_k(Bw) - cov(Bw) \to 0 \quad as\ k \to \infty \quad a.s. \tag{3.42}$$

if there exists a matrix norm satisfying $\|cov(Bw)\| < \infty$, where $cov_k(Bw)$ is the estimated covariance matrix obtained by Eq. (3.31) and $cov(Bw)$ is the real covariance matrix of process noise covariance matrix, i.e., Q.

Proof From the method stated in Algorithm 1, it can be obtained that

$$\text{cov}_k(\xi) - \text{cov}(\xi) \to 0 \quad as \quad k \to \infty \quad a.s. \tag{3.43}$$

From the method stated in Algorithm 1, we obtained that

$$\begin{aligned}
\text{cov}_k(Bw) &= \text{cov}_k(\xi) - \text{cov}(V) \\
&= \text{cov}(\xi) - \text{cov}(V) + (\text{cov}_k(\xi) - \text{cov}(\xi)) \\
&= \text{cov}(Bw) + (\text{cov}_k(\xi) - \text{cov}(\xi))
\end{aligned} \tag{3.44}$$

Then

$$\text{cov}_k(Bw) - \text{cov}(Bw) = \text{cov}_k(\xi) - \text{cov}(\xi) \tag{3.45}$$

Since

$$\text{cov}_k(\xi) - \text{cov}(\xi) \to 0 \quad as \quad k \to \infty \quad a.s. \tag{3.46}$$

consequently

$$\text{cov}_k(Bw) - \text{cov}(Bw) \to 0 \quad as \quad k \to \infty \quad a.s. \tag{3.47}$$

□

Theorem 3.3 *For the discrete-time LTI systems given in Eq. (3.20), using the method in Algorithm 2, it can be obtained that* $\limsup_{k\to\infty} \left\| \hat{P}_{k,k-1} \right\| < \infty$, *and* $\hat{P}_{k,k-1} -$

$P_{k,k-1} \to 0$ as $k \to \infty$, where $\hat{P}_{k,k-1}$ is the priori *state estimating covariance matrix in the KF-RCE algorithm and $P_{k,k-1}$ is the* priori *state estimating covariance matrix in the ideal Kalman filter with exact system parameters.*

Proof Introducing a notion

$$\begin{aligned} R_c(P) &= APA^T - APC^T(CPC^T + R)^{-1}CPA^T \\ &= AP(I - C^T(CPC^T + R)^{-1}CP)A^T \\ &= AP(I - C^T((CPC^TR^{-1} + I)R)^{-1}CP)A^T \\ &= AP(I - C^TR^{-1}(CPC^TR^{-1} + I)^{-1}CP)A^T \\ &= AP(I + C^TR^{-1}CP)^{-1}A^T \\ &= A(I + PC^TR^{-1}C)^{-1}PA^T \end{aligned} \tag{3.48}$$

In the following analysis, we say that $S_1 > S_2$ (or $S_1 < S_2$) if $S_1 - S_2$ is positive definite (or negative definite). For arbitrary positive matrices P_1 and P_2 with $R_c(P_1) > R_c(P_2)$, it can be obtained that

$$A(I + P_1C^TR^{-1}C)^{-1}P_1A^T > A(I + P_2C^TR^{-1}C)^{-1}P_2A^T \tag{3.49}$$

Obviously,

$$(I + P_1C^TR^{-1}C)^{-1}P_1 > (I + P_2C^TR^{-1}C)^{-1}P_2 \tag{3.50}$$

Calculating the inverse of Eq. (3.50), it can be obtained that

$$P_1^{-1}(I + P_1C^TR^{-1}C) < P_2^{-1}(I + P_2C^TR^{-1}C) \tag{3.51}$$

Then

$$P_1^{-1} < P_2^{-1} \tag{3.52}$$

Since P_1 and P_2 were covariance matrices obtained by Kalman-like filtering algorithm and positive definite, from Eq. (3.52) it can be obtained that

$$P_1 > P_2 \tag{3.53}$$

For arbitrary positive matrices P_1 and P_2, $R_c(P_1) > R_c(P_2)$ can be obtained if and only if $P_1 > P_2$.

From Theorem 3.2, it can be obtained that

$$\mathrm{cov}_k(Bw) - \mathrm{cov}(Bw) \to 0 \quad as \quad k \to \infty \tag{3.54}$$

Since $\mathrm{cov}(Bw) > 0$ and $\mathrm{cov}_k(Bw) > 0$, there exists a positive matrix Q_M, which satisfies that

$$\mathrm{cov}_k(Bw) \le Q_M \tag{3.55}$$

Supposing that there exists a sequence $\{\hat{P}_{k,k-1}, k = 1, 2, \ldots\}$, which is generated from the initial value P_0 with the following equation:

$$\hat{P}_{k+1,k} = R_c(\hat{P}_{k,k-1}) + \mathrm{cov}_k(Bw) \tag{3.56}$$

Supposing that there is a sequence $\{\hat{P}^M_{k,k-1}, k = 1, 2, \ldots\}$, which is generated from the initial value P_0 with the following equation:

$$\hat{P}^M_{k+1,k} = R_c(\hat{P}^M_{k,k-1}) + Q_M \tag{3.57}$$

Since the sequence $\{\hat{P}_{k,k-1}\}$ and $\{\hat{P}^M_{k,k-1}\}$ have the same initial value P_0 and the function $R_c(\cdot)$ is monotonically increasing, it can be obtained that

$$\hat{P}_{k,k-1} \leq \hat{P}^M_{k,k-1} \quad k = 1, 2, \ldots \tag{3.58}$$

Since the systems as in Eq. (3.20) are controllable and observable, the sequence $\{\hat{P}^M_{k,k-1}, k = 1, 2, \ldots\}$ is a convergent sequence. Since $\hat{P}_{k,k-1} > 0$ and $\hat{P}_{k,k-1} \leq \hat{P}^M_{k,k-1}$, it can be obtained that the sequence $\{\hat{P}_{k,k-1}, k = 1, 2, \ldots\}$ is a convergent sequence.

That is to say, $\lim\limits_{k\to\infty} \hat{P}_{k+1,k}$ exists. Then it can be obtained that

$$\begin{aligned}
\lim_{k\to\infty} \hat{P}_{k+1,k} &= \lim_{k\to\infty} (R_c(\hat{P}_{k,k-1}) + \mathrm{cov}_k(Bw)) \\
&= \lim_{k\to\infty} R_c(\hat{P}_{k,k-1}) + \lim_{k\to\infty} \mathrm{cov}_k(Bw) \\
&= \lim_{k\to\infty} (R_c(\hat{P}_{k,k-1})) + \mathrm{cov}(Bw) \\
&= \lim_{k\to\infty} (R_c(\hat{P}_{k,k-1}) + \mathrm{cov}(Bw))
\end{aligned} \tag{3.59}$$

Since the LTI system as in Eq. (3.20) is controllable and observable, there exists P_{ass} which satisfies the Riccati equation

$$P_{ass} = R_c(P_{ass}) + \mathrm{cov}(Bw) \tag{3.60}$$

Supposing that

$$\lim_{k\to\infty} \hat{P}_{k+1,k} = P'_{ass} \tag{3.61}$$

from Eq. (3.59), it can be obtained that

$$P'_{ass} = R_c(P'_{ass}) + \mathrm{cov}(Bw) \tag{3.62}$$

Since the solution of Riccati equation as in Eq. (3.60) is unique, we know that

$$P'_{ass} = P_{ass} \tag{3.63}$$

Obviously, it can be obtained that

$$\lim_{k\to\infty}(\hat{P}_{k+1,k}-P_{ass})=0 \tag{3.64}$$

From Eq. (3.64), it can be obtained that

$$\limsup_{k\to\infty}\left\|\hat{P}_{k,k-1}\right\|<\infty \tag{3.65}$$

and

$$\hat{P}_{k,k-1}-P_{ass}\to 0 \quad as \quad k\to\infty \tag{3.66}$$

Therefore, there exists a positive integer M for arbitrary small positive δ, which satisfies

$$\left\|\hat{P}_{k,k-1}-P_{ass}\right\|\leq\delta \text{ for arbitrary } k>M \tag{3.67}$$

Since the LTI systems as in Eq. (3.20) are controllable and observable, it can be obtained that

$$P_{k,k-1}-P_{ass}\to 0 \quad as \quad k\to\infty \tag{3.68}$$

That is to say, for arbitrary small positive ζ, there exists a positive integer H, which satisfies

$$\left\|P_{k,k-1}-P_{ass}\right\|\leq\zeta \text{ for arbitrary } k>H \tag{3.69}$$

Introducing the definitions $M'=\max\{M,H\}$ and $\tau=\delta+\zeta$, it can be obtained that

$$\left\|\hat{P}_{k,k-1}-P_{k,k-1}\right\|\leq\left\|\hat{P}_{k,k-1}-P_{ass}\right\|+\left\|P_{k,k-1}-P_{ass}\right\|\leq\tau \tag{3.70}$$

for arbitrary $k\geq M'$. That is to say,

$$\hat{P}_{k,k-1}-P_{k,k-1}\to 0 \quad as \quad k\to\infty \tag{3.71}$$

□

Theorem 3.4 *For the discrete-time LTI systems given in Eq. (3.20), using the method in Algorithm 2, the Kalman gain $\hat{K}_k-K_k\to 0$ as $k\to\infty$, where $\hat{K}_k$ is the Kalman gain given by the KF-RCE algorithm and K_k by the ideal Kalman filter with exact system parameters .*

Proof Since the LTI systems are controllable and observable, the Kalman filter is stable and the Riccati equation as in Eq. (3.60) has stable solution. Introducing the notion

$$K_{ass}=P_{ass}C^T(CP_{ass}C^T+R)^{-1} \tag{3.72}$$

where P_{ass} is the solution of the Riccati equation (3.60). From the KF-RCE algorithm, it can be obtained that

$$\hat{K}_k = \hat{P}_{k,k-1}C^T(C\hat{P}_{k,k-1}C^T + R)^{-1} \tag{3.73}$$

From the ideal Kalman filter algorithm, it can be obtained that

$$K_k = P_{k,k-1}C^T(CP_{k,k-1}C^T + R)^{-1} \tag{3.74}$$

Since $\hat{P}_{k,k-1} - P_{ass} \to 0$ and $P_{k,k-1} - P_{ass} \to 0$ as $k \to \infty$, it can be obtained that

$$\hat{K}_k - K_{ass} \to 0 \tag{3.75}$$

$$K_k - K_{ass} \to 0 \tag{3.76}$$

as $k \to \infty$.

Therefore,

$$\hat{K}_k - K_k \to 0 \quad as \quad k \to \infty \tag{3.77}$$

□

Theorem 3.5 *For the discrete-time LTI systems given in Eq. (3.20), using the method in Algorithm 2, as* $k \to \infty$, $\hat{x}_k$ *obtained from KF-RCE algorithm is convergent to the* $\hat{x}_k^{op}$ *obtained from the ideal Kalman filter with exact system parameters. That is to say,* $\hat{x}_k - \hat{x}_k^{op} \to 0$ *as* $k \to \infty$.

Proof For the LTI systems as in Eq. (3.20), real-time state estimating can be obtained by Kalman filter algorithm. State estimating of Kalman filter with $\mathrm{cov}_k(Bw)$ is as follows:

$$\hat{x}_k = A\hat{x}_{k-1} + \hat{K}_k[y_k - CA\hat{x}_{k-1}] \tag{3.78}$$

State estimating of Kalman filter with $\mathrm{cov}(Bw)$ is as follows:

$$\hat{x}_k^{op} = A\hat{x}_{k-1}^{op} + K_k[y_k - CA\hat{x}_{k-1}^{op}] \tag{3.79}$$

Then from Eqs. (3.78) to (3.79), we have

$$\begin{aligned} \hat{x}_k - \hat{x}_k^{op} &= (I - K_kC)A(\hat{x}_{k-1} - \hat{x}_{k-1}^{op}) \\ &+ (\hat{K}_k - K_k)(y_k + CA\hat{x}_{k-1}) \end{aligned} \tag{3.80}$$

Introducing the notion

$$\delta_k = \hat{x}_k - \hat{x}_k^{op} \tag{3.81}$$

$$\Phi_k = (I - K_kC)A \tag{3.82}$$

$$\Phi = (I - K_{ass}C)A \tag{3.83}$$

$$u_k = (\hat{K}_k - K_k)(y_k + CA\hat{x}_{k-1}) \tag{3.84}$$

Then we can obtain

$$\delta_k = \Phi_k \delta_{k-1} + u_k \tag{3.85}$$

Since the LTI system as in Eq. (3.20) is controllable and observable, Φ_k is stable and $\Phi_k \to \Phi$, as $k \to \infty$.

From Eq. (3.20) and Algorithm 2, it can be obtained that

$$\hat{x}_k = (\prod_{i=1}^{k}((I - \hat{K}_k C)A))x_0 + \sum_{i=1}^{k}((\sum_{j=i}^{k}((I - \hat{K}_j C)A))\hat{K}_i y_i) \tag{3.86}$$

Since $(I - \hat{K}_k C)A$ is stable and x_0, the initial state estimation, is bounded, it can be obtained that there exists a positive number x_0^M, which satisfy $\left\| x_0 \prod_{i=1}^{k}((I - \hat{K}_k C)A) \right\| \le x_0^M$. Introducing a function

$$\phi(K) = (I - KC)A \tag{3.87}$$

Since $\hat{K}_k \to K_{ass}$ as $k \to \infty$, there exist positive numbers K_{min} and K_{max} such that $K_{min} \le \left\| \hat{K}_k \right\| \le K_{max}$, $k = 1, 2, \ldots$. Obviously, for the function $\phi(\cdot)$, there exists a positive number K_M such that $\left\| \phi(\hat{K}_k) \right\| \le K_M$, $k = 1, 2, \ldots$. Since $\hat{K}_k \to K_{ass}$ and y_k is bounded, there exists a positive number C_y such that $\left\| \hat{K}_k y_k \right\| \le C_y$, $k = 1, 2, \ldots$. Under Assumption 3.4, it can be obtained that

$$\begin{aligned}
\left\|\hat{x}_k\right\| &\le \left\| x_0 \prod_{i=1}^{k}((I - \hat{K}_k C)A) \right\| \\
&\quad + \left\| \sum_{i=1}^{k}(\hat{K}_i y_i \sum_{j=i}^{k}((I - \hat{K}_j C)A)) \right\| \\
&\le x_0^M + \left\| \sum_{i=1}^{k}(\hat{K}_i y_i \sum_{j=i}^{k}((I - \hat{K}_j C)A)) \right\| \\
&\le x_0^M + \sum_{i=1}^{k}(\left\| \hat{K}_i y_i \right\| \left\| \sum_{j=i}^{k}((I - \hat{K}_j C)A) \right\|) \\
&\le x_0^M + \sum_{i=1}^{k}(C_y K_M^{(k-i)}) \\
&< x_0^M + \frac{C_y}{1-K_M}
\end{aligned} \tag{3.88}$$

Consequently, we can conclude the boundedness of $\hat{x}_k$. Since C and A are constant matrix, $CA\hat{x}_{k-1}$ is bounded. Then because y_k is bounded, it can be obtained that $y_k + CA\hat{x}_{k-1}$ is bounded. From Theorem 3.4, it can be obtained that $\hat{K}_k - K_k \to 0$ as $k \to \infty$. So $u_k \to 0$ as $k \to \infty$ and $\Phi(k)$ is stable. $\Phi_k \to \Phi$ as $k \to \infty$.

From Lemma 3.2, we have

$$\delta_k \to 0 \quad as \quad k \to \infty \tag{3.89}$$

That is to say, $\hat{x}_k - \hat{x}_k^{op} \to 0$ as $k \to \infty$. □

From Theorems 3.2 to 3.5, it can be obtained that $\text{cov}_k(Bw)$ is convergent to the real covariance matrix $\text{cov}(Bw)$, $\hat{P}_{k,k-1}$ is convergent to the ideal estimating error covariance matrix $P_{k,k-1}$, $\hat{K}_k$ is convergent to the ideal Kalman gain K_k, and $\hat{x}_k$ is convergent to the optimal state estimate $\hat{x}_k^{op}$, where $\text{cov}_k(Bw)$, $\hat{P}_{k,k-1}$, $\hat{K}_k$, and $\hat{x}_k$ are obtained by the KF-RCE algorithm, $P_{k,k-1}$, K_k, and $\hat{x}_k^{op}$ are obtained by the ideal Kalman filter with exact system parameters. In other words, $\text{cov}_k(Bw) - \text{cov}(Bw) \to 0$, $\hat{P}_{k,k-1} - P_{k,k-1} \to 0$, $\hat{K}_k - K_k \to 0$ and $\hat{x}_k - \hat{x}_k^{op} \to 0$ as $k \to \infty$.

3.6 Simulations

In this section, some simulation results, one for a one-dimensional system and two for multidimensional systems, are reported to illustrate the effects of the KF-RCE algorithm.

3.6.1 One-Dimensional Simulation

The discrete-time LTI system is as follows:

$$\begin{cases} x_k = x_{k-1} + w_{k-1} \\ y_k = x_k + v_k \end{cases} \tag{3.90}$$

where $E(w_k) = 0, \mathrm{D}(w_k) = Q = 0.25, E(v_k) = 0, \mathrm{D}(v_k) = 0.25$. $E(\cdot)$ is the exception and $\mathrm{D}(\cdot)$ is the variance.

The simulation results are shown in Fig. 3.3 and Table 3.3.

From the simulation results, it can be seen that $\hat{Q}_k$ is quickly convergent to the real value of Q as $k \to \infty$. The state estimate $\hat{x}_k$ is also quickly convergent to the ideal Kalman filter estimate $\hat{x}_k^{op}$ with exact system parameters.

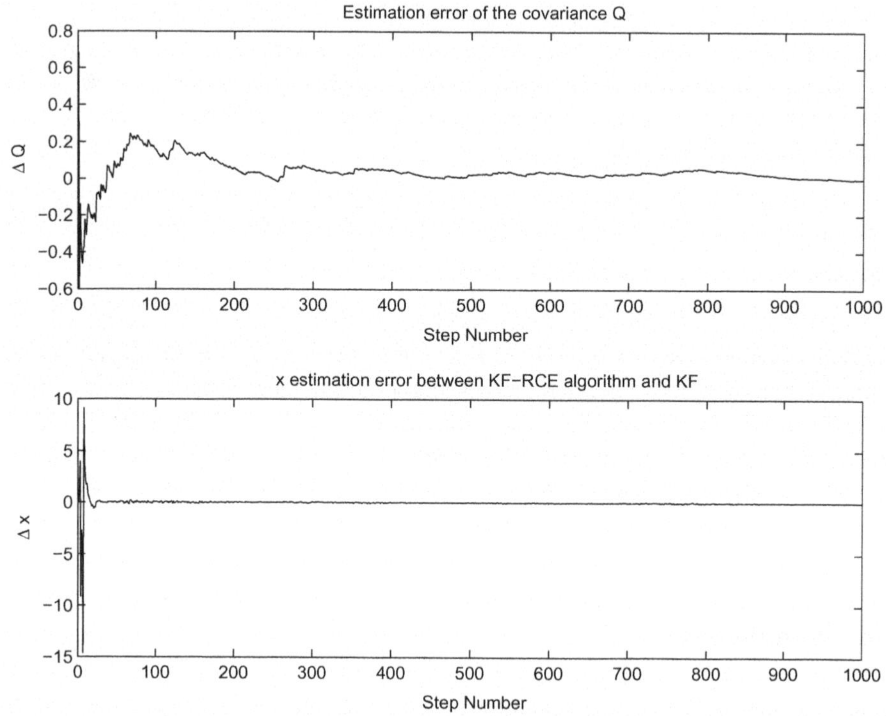

Fig. 3.3 Estimation error of one-dimensional KF-RCE algorithm

3.6.2 *Multidimensional Simulation*

The multidimensional LTI system is as follows:

$$\begin{cases} x_k = \begin{bmatrix} 1 & 0 \\ 0 & 1 \end{bmatrix} x_{k-1} + w_{k-1} \\ \\ y_k = \begin{bmatrix} 1 & 0 \\ 0 & 1 \end{bmatrix} x_k + v_k \end{cases} \tag{3.91}$$

where $E(w_k) = \begin{bmatrix} 0 \\ 0 \end{bmatrix}$, $E(v_k) = \begin{bmatrix} 0 \\ 0 \end{bmatrix}$. $\mathrm{cov}(w_k) = Q = \begin{bmatrix} 0.25 & 0 \\ 0 & 0.25 \end{bmatrix}$, $\mathrm{cov}(v_k) = R = \begin{bmatrix} 0.25 & 0 \\ 0 & 0.25 \end{bmatrix}$. $E(\cdot)$ is the exception and $\mathrm{cov}(\cdot)$ is the covariance matrix.

The simulation results are presented in Figs. 3.4, 3.5, Tables 3.4 and 3.5.

From the simulation results, it can be seen that $\hat{Q}_k$ is quickly convergent to the real value of Q as $k \to \infty$. The state estimate $\hat{x}_k$ is also quickly convergent to the ideal Kalman filter estimate $\hat{x}_k^{op}$, which are consistent with the theory analysis results.

Table 3.3 Estimation error of KF-RCE at different times

Step No.	1	10	100	500	1000
ΔQ	0.6835	−0.2715	0.1596	0.0217	0.0025
$\Delta \hat{x}$	−0.0625	2.1516	0.0887	−0.0061	−0.0018

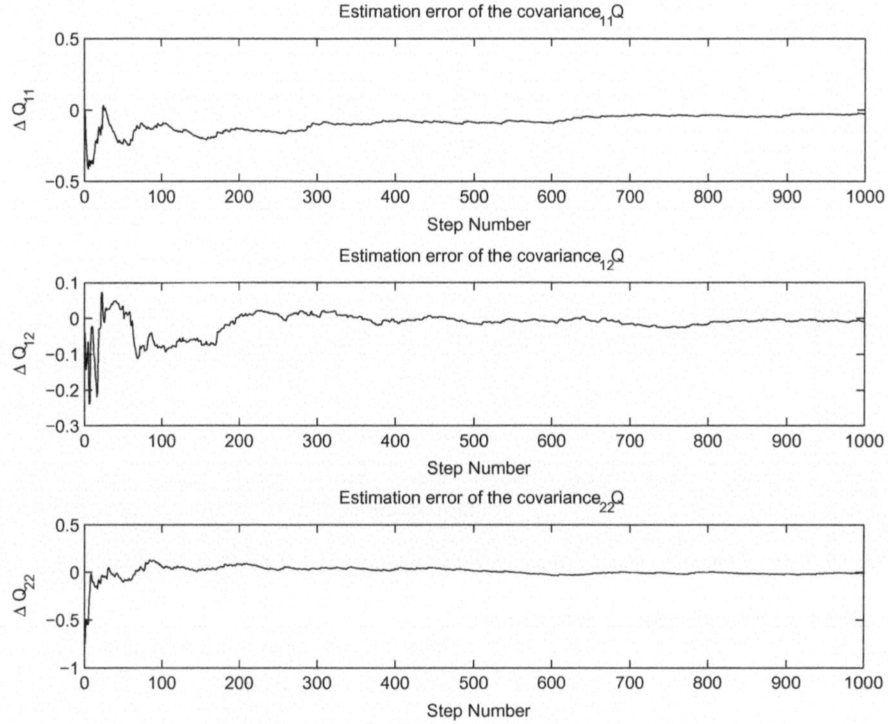

Fig. 3.4 Q estimation error with multidimensional KF-RCE algorithm

3.6.3 *Integrated Navigations Simulation*

The data fusion algorithm plays an important role in the integrated navigation systems. For instance, some positioning systems are always used to be integrated with inertial navigation systems to suppress the navigation error of the inertial navigation systems divergence. The speed and acceleration information of the vehicle could be obtained by the inertial navigation systems, and the position information could be obtained by other aided positioning systems. Then the position information can be used to suppress the inertial navigation error divergence.

For brevity, we adopt a "lineland" (one-dimensional space) system model to demonstrate the application of KF-RCE in integrated navigation systems:

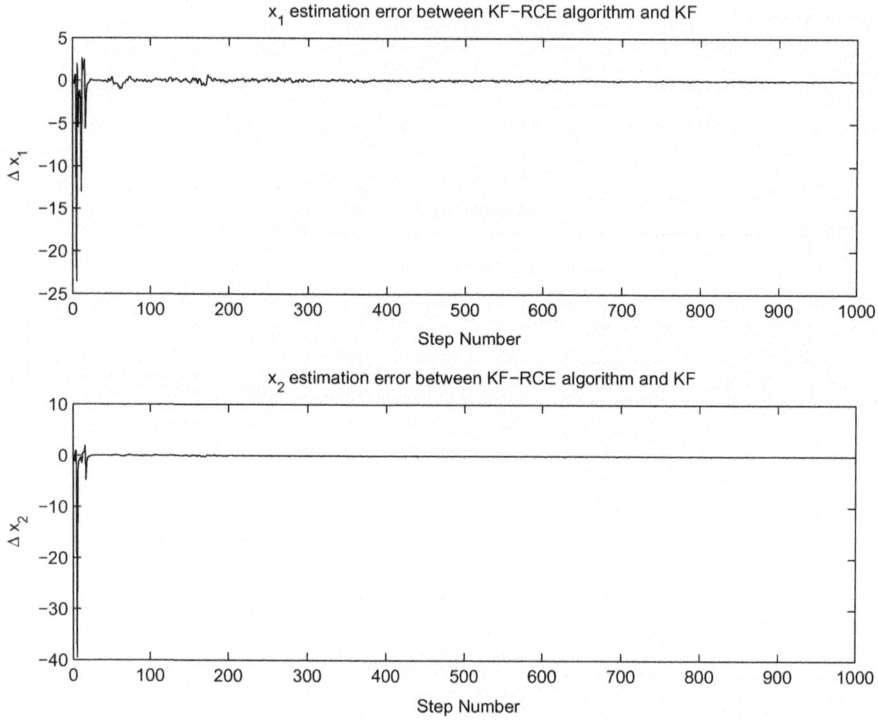

Fig. 3.5 Estimation error of state estimation with multidimensional KF-RCE algorithm

Table 3.4 Q estimation error at different times

Step(k)	1	10	100	500	1000
ΔQ_{11}	0.0029	−0.3504	−0.1022	−0.0856	−0.0261
ΔQ_{12}	−0.0399	−0.0217	−0.0809	−0.0150	−0.0087
ΔQ_{22}	−0.7479	−0.0878	0.0581	0.0143	−0.0064

Table 3.5 State estimation error at different times

Step No.	1	10	100	500	1000
$\Delta\hat{x}_1$	0.0005	−2.1998	−0.1504	0.0194	0.0092
$\Delta\hat{x}_2$	−0.0000	−0.3542	−0.0919	0.0120	0.0051

$$\begin{cases} \begin{bmatrix} x_k \\ \dot{x}_k \\ \ddot{x}_k \end{bmatrix} = \begin{bmatrix} 1 & T & 0.5T^2 \\ 0 & 1 & T \\ 0 & 0 & 1 \end{bmatrix} \begin{bmatrix} x_{k-1} \\ \dot{x}_{k-1} \\ \ddot{x}_{k-1} \end{bmatrix} + w_{k-1} \\ \begin{bmatrix} y_k \\ \dot{y}_k \\ \ddot{y}_k \end{bmatrix} = \begin{bmatrix} 1 & 0 & 0 \\ 0 & 1 & 0 \\ 0 & 0 & 1 \end{bmatrix} \begin{bmatrix} x_k \\ \dot{x}_k \\ \ddot{x}_k \end{bmatrix} + v_k \end{cases} \tag{3.92}$$

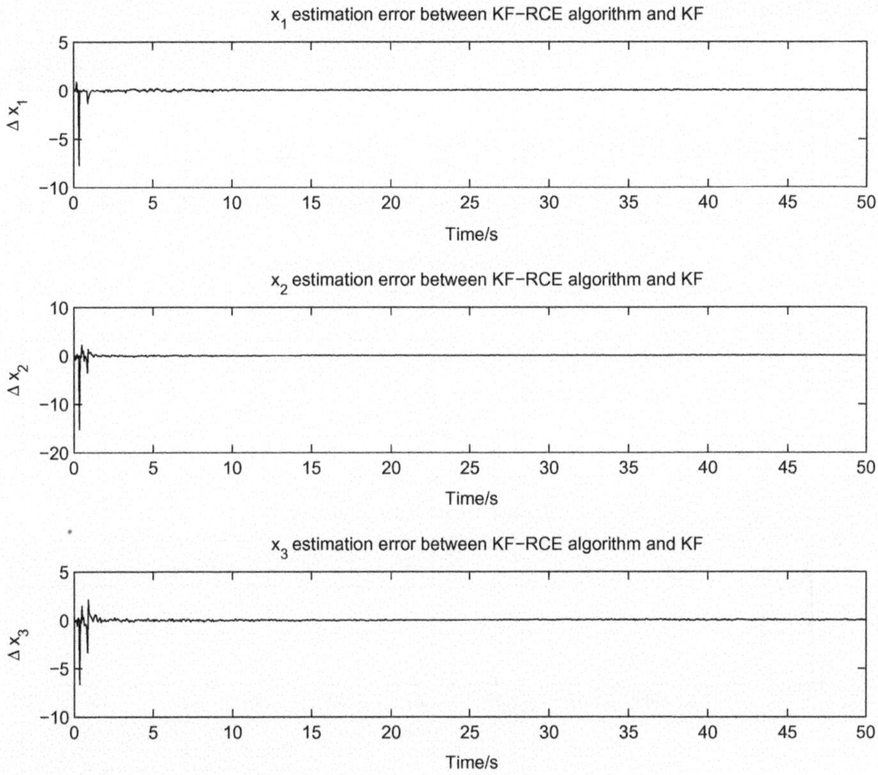

Fig. 3.6 Estimation error of state estimation with KF-RCE algorithm

where T is the sample period, x_k is the vehicle position, $\dot{x}_k$ is the vehicle speed, $\ddot{x}_k$ is the acceleration, and $[y_k \quad \dot{y}_k \quad \ddot{y}_k]^T$ is the measurement signal.

Supposing that the $E(w_k) = E(v_k) = 0$, $T = 0.05s$ and

$$\mathrm{cov}(w_k) = Q = \begin{bmatrix} 0.25 & 0.04 & 0.04 \\ 0.04 & 0.25 & 0.04 \\ 0.04 & 0.04 & 0.25 \end{bmatrix} \tag{3.93}$$

$$\mathrm{cov}(v_k) = R = \begin{bmatrix} 0.25 & 0.04 & 0.04 \\ 0.04 & 0.25 & 0.04 \\ 0.04 & 0.04 & 0.25 \end{bmatrix} \tag{3.94}$$

However, Q is completely unknown in the data fusion of the integrated navigation systems.

The simulation results are presented in Figs. 3.6, 3.7, 3.8, Tables 3.6, 3.7 and 3.8.

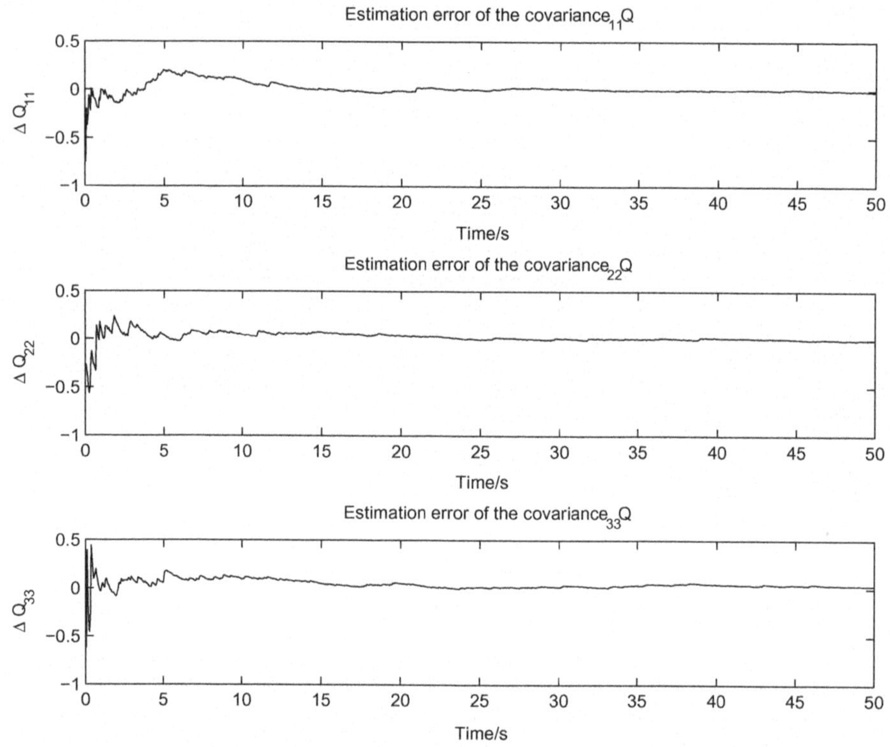

Fig. 3.7 Q estimation error with KF-RCE algorithm

From the simulation results, it can be seen that $\hat{Q}_k$ is quickly convergent to the real value of Q as $k \to \infty$. The state estimate $\hat{x}_k$ is also quickly convergent to the ideal Kalman filter estimate $\hat{x}_k^{op}$, which is consistent with the theory analysis results.

3.7 Conclusion

Kalman filter algorithm is a kind of optimal filters to obtain state estimation in the linear systems with a priori known parameters and noise statistical properties. However, successful applications of standard Kalman filters require that the state-space model in use is accurate enough with exact system matrices or parameters as well as proper a priori information of the process and measurement noises, which in fact request people to make great efforts in the modeling procedure before the real application of Kalman filter. Generally speaking, exact or perfect modeling is impossible or expensive in most practical applications, and inaccurate model often results that the performance of Kalman filter is very poor or completely not acceptable.

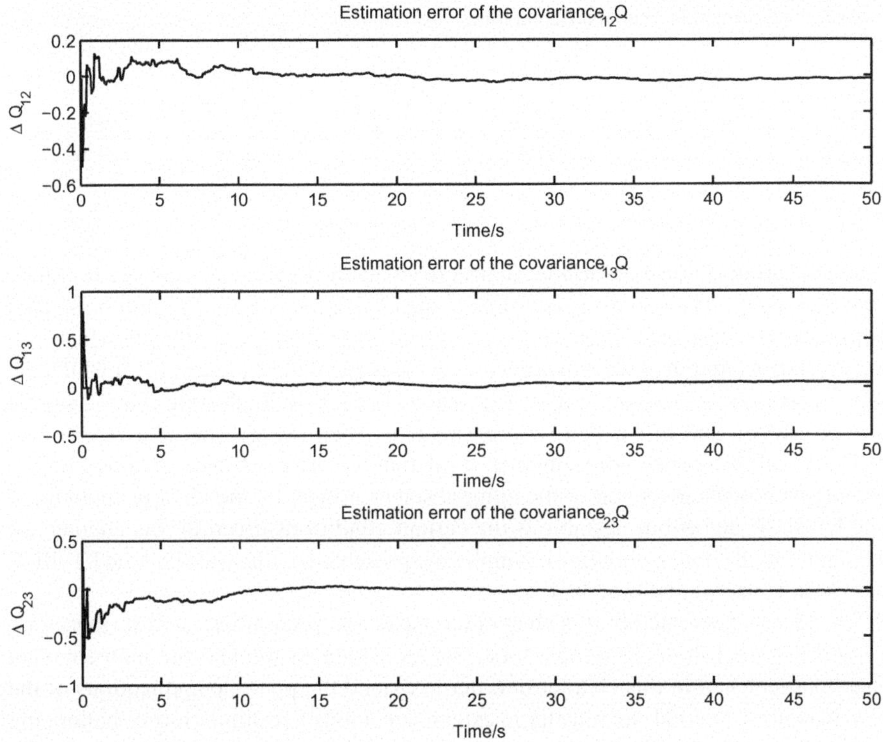

Fig. 3.8 Q estimation error with KF-RCE algorithm

Table 3.6 State estimation error at different times

Time(s)	0.05	0.5	5	25	50
Δx_1	−0.0026	−0.1552	0.0053	−0.0040	0.0156
Δx_2	0.0127	2.3244	−0.0665	−0.0213	−0.0098
Δx_3	0.0045	1.4694	0.0991	0.0035	0.0362

Table 3.7 Q estimation error at different times

Time(s)	0.05	0.5	5	25	50
ΔQ_{11}	−0.7506	−0.0199	0.1904	−0.0061	−0.0152
ΔQ_{22}	−0.2642	−0.2599	0.0375	−0.0005	0.0001
ΔQ_{33}	−0.6226	0.0933	0.0633	0.0136	0.0277

To deal with the completely unknown covariance matrix of the process noise, a new algorithm named as Kalman filter with `RecursiveCovarianceEstimation` algorithm (KF-RCE) is proposed in this chapter. The KF-RCE algorithm tried to

Table 3.8 Q estimation error at different times

Time(s)	0.05	0.5	5	25	50
ΔQ_{12}	−0.0140	0.0344	0.0723	−0.0282	−0.0156
ΔQ_{13}	−0.0011	−0.0516	−0.0344	−0.0051	0.0441
ΔQ_{23}	0.2365	−0.4052	−0.1272	−0.0085	−0.0365

obtain the process noise covariance matrix from the measurement sequence for linear time-invariant (LTI) systems. According to the idea of the law of large numbers taken from statistics, basic `RecursiveCovarianceEstimation` algorithm has been presented first, and then we construct a new measurement sequence, from which the process covariance matrix could be estimated. The KF-RCE algorithm has a simple recursive form, and it is easy to use in practice. Then the stability analysis of KF-RCE algorithm has been presented to show that the state estimate obtained by the new algorithm is convergent to the optimal state estimate by the ideal Kalman filter. The KF-RCE algorithm is stable if the certain conditions given in this chapter are satisfied. Finally, two simulation examples are presented to illustrate that the KF-RCE algorithm is useful and practical.

The results presented in this chapter are basic and preliminary, and furthermore, we address the following issues. First, we are trying to weaken the mathematical assumptions listed in this chapter in order to extend the possible applications of the new algorithm. Second, we may try to extend our stability results to more challenging case in the presence of parametric uncertainties in the system model or give some robustness results. Third, we may make an attempt to explore efficient algorithms to extend our work to the nonlinear systems.

References

1. S. Won, W. Melek, F. Golnaraghi, A Kalman/particle filter-based position and orientation estimation method using a position sensor/inertial measurement unit hybrid system. IEEE Trans. Ind. Electron. **57**(5), 1787–1798 (2010)
2. M.S. Grewal, A.P. Andrews, Applications of Kalman filtering in aerospace 1960 to the present. IEEE Control Syst. Mag. **30**(3), 69–78 (2010)
3. H. Guo, M. Yu, C.W. Zou, W.W. Huang, Kalman filtering for GPS/magnetometer integrated navigation system. Adv. Space Res. **45**(11), 1350–1357 (2010)
4. J. Crassidis, Sigma-point Kalman filtering for integrated GPS and inertial navigation. IEEE Trans. Aerosp. Electron. Syst. **42**(2), 750–756 (2006)
5. Y. Mostofi, R. Murray, To drop or not to drop: design principles for Kalman filtering over wireless fading channels. IEEE Trans. Autom. Control **54**(2), 376–381 (2009)
6. K. Muralidhar, H. Kwok, A low-complexity Kalman approach for channel estimation in doubly-selective OFDM systems. IEEE Signal Process. Lett. **16**(7), 632–635 (2009)
7. W. Wang, H. Zhang, L. Xie, Robust filter design for tracking of time-varying mobile radio channels. Acta Autom. Sinica **34**(5), 523–528 (2008)
8. H. Xu, S. Mannor, A Kalman filter design based on the performance/robustness tradeoff. IEEE Trans. Autom. Control **54**(5), 1171–1175 (2009)

9. R.E. Kalman, A new approach to linear filtering and prediction problems. Trans. ASME - J. Basic Eng. **82**(Series D), 35–45 (1960)
10. R.E. Kalman, R.S. Bucy, New results in linear filtering and prediction theory. Trans. ASME Ser. D, J. Basic Eng. **83**, 95–107 (1961)
11. S.F. Schmidt, The Kalman filter: its recognition and development for aerospace applications. AIAA J. Guid. Control **4**(1), 4–7 (1981)
12. S.J. Lee, Y. Motai, M. Murphy, Respiratory motion estimation with hybrid implementation of extended Kalman filter. IEEE Trans. Ind. Electron. **59**(11), 4421–4432 (2012)
13. Y. Shi, K. Sun, L. Huang, Y. Li, Online identification of permanent magnet flux based on extended Kalman filter for IPMSM drive with position sensorless control. IEEE Trans. Ind. Electron. **59**(11), 4169–4178 (2012)
14. M.A. Khanesar, E. Kayacan, M. Teshnehlab, O. Kaynak, Extended Kalman filter based learning algorithm for type-2 fuzzy logic systems and its experimental evaluation. IEEE Trans. Ind. Electron. **59**(11), 4443–4455 (2012)
15. S. Kluge, K. Reif, M. Brokate, Stochastic stability of the extended Kalman filter with intermittent observations. IEEE Trans. Autom. Control **55**(2), 514–518 (2010)
16. N. Carlson, Fast triangular formulation of the square root filter. AIAA J. **11**(9), 1259–1265 (1973)
17. N.A. Carlson, Federated square root filter for decentralized parallel processors. IEEE Trans. Aerosp. Electron. Syst. **26**(3), 517–525 (1990)
18. S.J. Julier, J.K. Uhlmann, Unscented filtering and nonlinear estimation. Proc. IEEE **92**(3), 401–422 (2004)
19. S. Jafarzadeh, C. Lascu, M.S. Fadali, State estimation of induction motor drives using the unscented Kalman filter. IEEE Trans. Ind. Electron. **59**(11), 4207–4216 (2012)
20. H. Marina, F.J. Pereda, J.M. Giron-Sierra, F. Espinosa, UAV attitude estimation using unscented Kalman filter and TRIAD. IEEE Trans. Ind. Electron. **59**(11), 4465–4474 (2012)
21. S.A. Pasha, H.D. Tuan, V. Ba-Ngu, Nonlinear bayesian filtering using the unscented linear fractional transformation model. IEEE Trans. Signal Process. **58**(2), 477–489 (2010)
22. I.R. Petersen, A.V. Savkin, *Robust Kalman Filtering for Signals and Systems with Large Uncertainties* (Birkhause, Boston, 1999)
23. B. Hassibi, A.H. Sayed, T. Kailath, H_∞ optimality of the LMS algorithm. IEEE Trans. Signal Process. **44**(2), 267–280 (1996)
24. X. Kai, C. Wei, L. Liu, Robust extended Kalman filtering for nonlinear systems with stochastic uncertainties. IEEE Trans. Syst. Man Cybern. Part A: Syst. Hum. **40**(2), 399–405 (2010)
25. R.K. Mehra, On the identification of variances and adaptive Kalman filtering. IEEE Trans. Autom. Control **15**(2), 175–184 (1970)
26. X. Gao, D. You, S. Katayama, Seam tracking monitoring based on adaptive Kalman filter embedded elman neural network during high-power fiber laser welding. IEEE Trans. Ind. Electron. **59**(11), 4315–4325 (2012)
27. X. Xiao, B. Feng, B. Wang, On-line realization of SVM Kalman filter for MEMS gyro, in *Proceedings of the 3rd International Conference on Measuring Technology and Mechatronics Automation* , pp. 768–770
28. M. Karasalo, X. Hu, An optimization approach to adaptive Kalman filtering. Automatica **47**(8), 1785–1793 (2011)
29. Y. Yang, W. Gao, An optimal adaptive Kalman filter. J. Geod. **80**(4), 177–183 (2006)
30. K. Myers, B. Tapley, Adaptive sequential estimation with unknown noise statistics. IEEE Trans. Autom. Control **21**(4), 520–523 (1976)
31. Z.L. Deng, Y. Gao, C. Li, G. Hao, Self-tuning decoupled information fusion Wiener state component filters and their convergence. Automatica **44**(3), 685–695 (2008)
32. Y. Gao, W.J. Jia, X.J. Sun, Z. Deng, Self-tuning multisensor weighted measurement fusion Kalman filter. IEEE Trans. Aerosp. Electron. Syst. **45**(1), 179–191 (2009)
33. C. Ran, G. Tao, J. Liu, Z. Deng, Self-tuning decoupled fusion Kalman predictor and its convergence analysis. IEEE Sens. J. **9**(12), 2024–2032 (2009)

Chapter 4
Kalman Filter with Recursive Covariance Estimation Revisited with Technical Conditions Reduced

Kalman filter algorithm is an optimal state estimator in the sense of minimum mean squared errors and maximum likelihood estimation. However, standard Kalman filter is highly restricted by strict a priori requirements on the information of dynamic model and statistical information of process noises. In practice, the covariance matrix of process noise is almost impossible to be directly determined a priori due to the intrinsic coupling of process noise and system state dynamics. Considering such background of Kalman filter algorithm, one algorithm, named as recursive covariance estimation (RCE), is introduced to estimate the covariance matrix when there is no information of process noise covariance matrix in the linear time-invariant (LTI) systems. Based on the framework of standard Kalman filter, a new algorithm named as Kalman filter with recursive covariance estimation (KF-RCE) is introduced to resolve this challenging problem of state estimation without the covariance matrix of process noise. Then the rigorous stability analysis is presented to show that the algorithm is optimal and the covariance matrix and state estimations are asymptotically consistent with the ideal Kalman filter using the exact covariance matrix of process noise. Finally, some simulation results of navigation system are given to examine theory analysis theorem and algorithm optimality.

4.1 Introduction

Nowadays, researchers pay more and more attention to intelligent systems and autonomous systems, which seldom need intervention of human beings, with modern information technology. Cybernetics or control theory, plays an important role in the intelligent systems, especially the feedback control is almost an essential mean for the modern high-accuracy equipments and devices. In practice, feedback control can be only implemented based on collecting data via various sensors, which usually provide corrupted measurement with accuracy to certain extent. Considering the fact that the measurement data are always noisy and the true states of a system are often not available for use directly, filtering technology is treated as a standard method to

H. Ma et al., *Kalman Filtering and Information Fusion*,
https://doi.org/10.1007/978-981-15-0806-6_4

improve the accuracy of system measurement and controlling output or state precision. For the modern control theory based on the state-space model, constructing essential system state estimation is necessary for the application of various modern control techniques. With the rapid development of digital computer, digital filtering techniques have been widely used as the mainstream in the intelligent systems, especially Kalman filtering algorithm [1, 2] due in large part to advances in digital computing.

For the linear state-space model, the most commonly used model or approximation model of nonlinear systems in practice, Kalman filter algorithm [3–5] is an optimal state estimating method in the sense of minimum mean squared errors and maximum likelihood estimation. However, it is well known that standard Kalman filter is highly restricted by the strict requirements on a priori information of the dynamic model and statistics information of the process and measurement noises. Following Kalman's basic work, many extensions and generalizations are expanded to deal with the case where the conditions of standard Kalman filter cannot be strictly satisfied [6, 7]. For example, in 1990, L. Guo presented a useful method to estimate time-varying parameter by Kalman filter-based algorithm [8] and in 1991 J. F. Zhang presented a detail analysis for the L_p−stability of tracking error when Kalman filter is used to track unknown time-varying parameters [9]. However, in most existing work, the objective is to obtain good denoising and estimating effects when system model has small uncertainty or the statistical properties of the process and observation noises are uncertain in a small range [10]. There are little research to focus on the problem that there is no a priori information for the covariance matrix for process noise sequence in the system dynamic model.

Previously, there have been attempts to estimate the unknown covariance matrices, for example, in 1970, Mehra proposed an adaptive Kalman filter to identify the covariance matrices with an intuitive way [11]; later in 1976, Myers and Tapley provided an improved version with limited memory by introducing the optimization approach [12]. However, these efforts did not address the mathematical proofs of the proposed algorithms; in fact, due to the intrinsically introduced adaptation in the Kalman filter loop, the closed-loop systems become very complex and the closed-loop theoretical analysis remained *unresolved* problem for *several decades* since 1970s. Unlike these existing works, our new approach is based on a solid theoretical foundation of ergodicity of a constructed i. i. d. time series, which makes the asymptotic consistent covariance estimation possible for a class of discrete-time linear time-invariant (LTI) systems, where the covariance matrix of process noise is completely unknown, and greatly simplifies the convergence analysis for the closed-loop system. Based on this idea, a new Kalman filtering algorithm named as *Kalman filter with recursive covariance estimation (KF-RCE)* is presented to resolve this challenging problem of state estimation without the covariance matrix of process noise, and the rigorous stability analysis is given to show that this algorithm is optimal in sense that the covariance matrix and state estimations are asymptotically consistent with the ideal Kalman filter when the exact covariance matrix of process noise is completely known a priori. Extensive simulation studies have also verified the theoretical results and the effectiveness of the proposed algorithm.

This chapter is an extension to our recent work [13] published in IEEE Transactions on Industrial Electronics, where the KF-RCE algorithm is first introduced and investigated under several technical conditions, among which two key technical conditions imposed on the closed-loop signals in [13] are now completely removed in this work. Compared with our previous work, in this chapter, the first removed technical assumption is about the stability of a linear time-varying autonomous system which is determined by the filtering gains in the KF-RCE algorithm. Such a condition is reasonable yet usually not easy to be established directly. In this version, we are able to show that this autonomous system is exponentially stable if the system model meets the conditions of complete controllability and observability, hence, the technical assumption on this autonomous system is in fact intrinsically implied in other conditions. The second removed technical assumption in this chapter is the condition of the bounded measurement, which is the essential assumption for the stability analysis in our previous work [13]. In this chapter, the bounded estimating covariance matrix is used to prove the stability of the state estimating based on the asymptotic stability of estimating covariance matrix, which makes the condition on the boundedness of the measurements unnecessary. In this way, the closed-loop stability and optimality can now be presented elegantly.

The rest of this chapter will be organized as follows: the problem will be reformulated which the KF-RCE is designed for in Sect. 4.2; then for easy to be discussed later, a simple statistical algorithm named *recursive covariance estimation (RCE)* is presented to deal with the case that how to obtain the covariance matrix of a vector from the sample sequence corrupted with a noise in real time in Sect. 4.3, for the discrete-time LTI systems, KF-RCE algorithm will be presented to illustrate that how to obtain the state estimating sequence when there is no statistical information of covariance matrix of process noise; to guarantee the theory stability of KF-RCE algorithm, the results of stability analysis will be formulated to prove that the state estimating sequence is statistical consistent with the estimation obtained by the ideal Kalman filter with exact system parameters; In Sect. 4.5, a practical simulation experiment will be presented to show the effectiveness of the KF-RCE algorithm; finally, in Sect. 4.6 some concluding remarks will be presented to summarize KF-RCE algorithm and some feasible research method will be proposed to improve our algorithm in the future.

4.2 Problem Formulation

Kalman filter algorithm is a kind of standard signal processing method for the linear system, while the system model in the state space has exact system parameters and known noise statistical properties. In this chapter, a new filtering algorithm based on Kalman filter algorithm named KF-RCE algorithm will be presented for certain linear time-invariant systems and in order to be convenient the standard Kalman filter will be reviewed for the linear time-invariant systems. Kalman filter algorithm is proposed to deal with the state x_k estimation for the dynamic system model as follows:

$$\begin{cases} x_k = Ax_{k-1} + w_{k-1} \\ y_k = Cx_k + v_k \end{cases} \tag{4.1}$$

where matrix A is the coefficient matrix of state transfer, C is the observation coefficient matrix, and y_k is observable measurements at the time t_k.

For linear state-space model, standard Kalman filter is the optimal state model estimating algorithm when exact a priori knowledge for the system structure (i.e., coefficient matrices A, C) and the statistical properties, mainly covariance matrices of Q_k and R_k, for the process and measurement noise sequences are exactly known [14, 15]. However, such ideal conditions are hardly completely satisfied in the engineering systems due to different factors such as modeling error and nonlinear characteristics. Especially, the exact statistical properties of the noise are seldom obtained from the sample data since there are no such methods to directly isolate the noise sequence from measurement sequence. Indeed, the mechanism of the noise cannot be completely comprehended in most cases. However, previous research works [16, 17] have gained the theory results that it may degrade the performance of state estimation or even make the filter not work at all without the exact covariance matrices of Q_k and R_k in Kalman filter. To befittingly apply the traditional Kalman filter in practice, it is often necessary to make much effort to obtain the exact statistical properties of the corrupted noise or alternatively, taking the upper bound Q of covariance matrix Q_k as the unknown covariance matrix Q_k, which in theory reduce the performance of state estimating.

From the above all issues, it can be obtained that it is essential to using the correct covariance matrix Q_k for Kalman filter in various applications. To reduce the side effects of covariance matrix Q_k, in 1970 R. K. Mehra first proposed the original architecture of adaptive Kalman filter (AKF) in his research paper [11], where an innovation sequence is constructed according to the system dynamic model in order to guarantee the estimating optimality of covariance matrix based on the state estimation and measurement sequence, which makes theory analysis of closed-loop stability complex since the covariance estimation is intrinsically coupled with the state estimation. Without the information of noise statistics, Kenneth Myers and Byron Tapley proposed a new adaptive sequential estimation algorithm, named adaptive limited memory filter (ALMF) in 1976 [12]. Under the framework of Kalman filter algorithm, it is a novel attempt to treat the parameter uncertainties as the noise statistics and from the simulation experiment it can be obtained that in the application of an orbit determination problem ALMF algorithm was very effective for a near-earth satellite with distinct modeling errors [12]. However, the estimating process of covariance matrix was coupled with the state estimating, which may increase the mathematical difficulty of stability analysis. Besides, it is one more question that the computation is little complex for most AKF algorithm and its complexity is computational load to satisfy the real-time demand in practice, which is one of the basic requirements in most state estimating systems.

From current approaches proposed in the research works, in our point of view, we believe it is essential to explore new approach to revisit the state estimating problem without statistical information of covariance matrix Q_k in the data fusion field, especially when there is no a priori knowledge for process noise except that it is zero-mean mathematical expectation and unrelated with measurement noise sequence. In other words, is it possible to construct a new algorithm for state estimation when there is no range of Q_k or nominal values of Q_k while it can cope with large uncertainty problem of covariance matrix of process noise and guarantee state estimating accuracy and stability?

Mathematically speaking, this fundamental problem will be investigated under the following mathematical conditions in this chapter:

Assumption 4.1 For the dynamic system model as in Eq. (4.1), supposing that state transfer matrix A and measurement matrix C are obtained, constant coefficient matrices and the system satisfy controllable and observable requirements.

Remark 4.1 To guarantee the stability of Kalman filter, the system must satisfy the conditions of controllability and observability. For easy to be formulated, the system parameter matrices, A and C, are supposed known and constant for the dynamic model.

Assumption 4.2 Suppose that the coefficient matrix C, measurement matrix, exists left inverse matrix.

Remark 4.2 It is a technical assumption for measurement matrix C, and this condition will be used to prove the stability of estimating algorithm. This condition is specialized for this chapter, which may be greatly weakened in our future research results.

Assumption 4.3 Suppose that the covariance matrix of process noise, $Q_k = Q$, is constant matrix yet completely unknown. The process noise sequence is uncorrelated with measurement noise and the covariance matrix of measurement noise, R, has been exactly obtained.

Remark 4.3 In practice, LTI dynamic models could be used to approximatively describe most engineering systems. For easy to be designed, the covariance matrices of process and measurement noise are always supposed to be constant matrix since it is hard to be determined if the covariance matrix is time-varying.

Problem 4.1 *Under the mathematical Assumptions 4.1–4.3, how to construct a filter algorithm which could solve the estimating problem with completely unknown covariance matrix of process noise?*

Remark 4.4 In our previous work [13], the state estimation problem has been investigated under the uncertainty of the covariance matrix of process noise for the following linear time-invariant system:

$$\begin{aligned} x_{k+1} &= Ax_k + Bw_k \\ y_k &= Cx_k + v_k \end{aligned} \tag{4.2}$$

where the covariance $Q = \mathrm{Cov}(w_k)$ of the process noise is completely unknown while the covariance $R = \mathrm{Cov}(v_k)$ of the measurement noise is supposed to be known a priori. In this chapter, we will try to get rid of two technical conditions among five assumptions presented in the previous work [13], and some essential new results and techniques of closed-loop stability analysis will be presented in this contribution.

Generally speaking, the covariance matrix Q_k should be estimated from sample sequence of process noise. However, there are at least two practical difficulties: (a) There is no way to obtain the sample sequence from the measurement sequence based on the dynamic models. (b) Even if there is any method to isolating the process noise from the measurement sequence, according to the law of large number in statistics, it is essential to obtain enough sample data to guarantee the reliability of covariance estimation. Considering such mathematical background, it is more effective approach to estimate the covariance matrix based on the measurement data sequence in real time rather than obtain the estimated covariance matrix of process noise from complex offline statistical experiments.

4.3 Kalman Filter with Recursive Covariance Estimation

4.3.1 Basic Method: Covariance Matrix Estimation

For easy to be formulated later, the basic idea is first introduced that how to estimate covariance matrix for a random vector, which is the statistical basis of parameter estimating the method but plays an important role in this chapter.

First, consider how to deal with the simple statistical yet nontrivial estimating problem as follows:

Problem 4.2 *There is a random vector w, which is supposed to satisfy that*

$$\xi = w + v \tag{4.3}$$

where $\mathrm{E}(w) = 0$, $\mathrm{E}(v) = 0$ *and* $cov(v) = R$.

$\mathrm{E}(\cdot)$ *stands for the mathematical expectation and* $cov(\cdot)$ *indicates the covariance matrix.*

Provided that it can be obtained from an experiment, an i.i.d sample sequence $\{\xi_i, i = 1, 2, \ldots\}$ *of* ξ, *how to obtain the estimation of covariance matrix of random vector* w *from the i.i.d sample sequence* ξ *in real time?*

Theorem 4.1 *Provided that* $\xi \in R^{m\times 1}$ *is a random vector and* $E(\xi) = 0$, *the estimation of covariance matrix can be calculated from its sample sequence as follows:*

$$cov(\xi) = E(\xi\xi^T) = \lim_{n\to\infty} cov_n(\xi) = \lim_{n\to\infty} \frac{1}{n}\sum_{i=1}^{n} \xi_i\xi_i^T \tag{4.4}$$

then the estimation of covariance matrix of random vector w *could be calculated as follows:*

$$cov(w) = \lim_{n\to\infty} \frac{1}{n} \sum_{i=1}^{n} \xi_i \xi_i^T - R \tag{4.5}$$

where the parameter matrix R is covariance matrix of random vector v.

However, the above estimating approach is not reliable for real-time matrix calculation since there is no way to obtain all of sample sequence at one moment. If it is necessary to obtain the parameter matrix estimation in a real time, we need to develop the method to restructure the estimating algorithm and obtain a recursive calculation form.

Provided that the covariance matrix estimation $\mathrm{cov}_n(\xi)$ is obtained from the sample sequence up to and including the time n, when there is a new sample data, the $(n+1)$th data should be added to modify the estimation $\mathrm{cov}_n(\xi)$ and obtain new covariance matrix $\mathrm{cov}_{n+1}(\xi)$.

From Eq. (4.4), restructuring the calculating process

$$\begin{aligned}
\mathrm{cov}_{n+1}(\xi) &= \tfrac{1}{n+1} \sum_{i=1}^{n+1} \xi_i \xi_i^T \\
&= \tfrac{1}{n+1} \left(\sum_{i=1}^{n} \xi_i \xi_i^T + \xi_{n+1} \xi_{n+1}^T \right) \\
&= \mathrm{cov}_n(\xi) + \tfrac{1}{n+1} \left(\xi_{n+1} \xi_{n+1}^T - \mathrm{cov}_n(\xi) \right)
\end{aligned} \tag{4.6}$$

Using the method as in Eq. (4.6), the estimating covariance matrix of random vector ξ could be obtained as the arrival of new sample data in real time. From the law of large number in statistics, it can be obtained that $\mathrm{cov}_n(\xi) \to \mathrm{cov}(\xi)$ as $n \to \infty$, where covariance matrix $\mathrm{cov}(\xi)$ is the true value of covariance matrix of random vector ξ. Obviously, the estimation of covariance matrix of random vector w could be obtained as follows:

$$\mathrm{cov}_k(w) = \mathrm{cov}_k(\xi) - R \tag{4.7}$$

Then it can be obtained that Algorithm 3 could be adapted to solve Problem 4.2.

Algorithm 3 Recursive Covariance Estimation($\{\xi_i\}$, R)

Initialization: $\mathrm{cov}_0(w) = 0$, $\mathrm{cov}_0(\xi) = 0$, R
Input: the sequence of sample data $\{\xi_k\}$ of random vector ξ
Output: the sequence of estimated covariance matrix $\{\mathrm{cov}_k(w)\}$ of random vector w
1: **for** $k = 1$ to n **do**
2: Using the measurement data ξ_k to update the estimation of covariance matrix $\mathrm{cov}_k(\xi)$ as in Eq. (4.6)
3: Using the approach as in Eq. (4.7) to obtain the estimation of covariance matrix $\mathrm{cov}_k(w)$
4: **end for**
5: Return $\{\mathrm{cov}_k(w)\}$

4.3.2 KF-RCE Algorithm for LTI Systems

In this section, the basic approach in Algorithm 3 will be applied to cope with parameter matrix estimating problem in the discrete-time linear time-invariant (LTI) system dynamic model, which could be described as follows:

$$\begin{cases} x_k = Ax_{k-1} + w_{k-1} \\ y_k = Cx_k + v_k \end{cases} \tag{4.8}$$

where x_k is real value of the state at time k, y_k is the measurement vector at time k, w_k stands for process noise vector, and v_k is measurement noise vector at time k. Besides, $A \in \boldsymbol{R}^{n\times n}$ stands for state transfer coefficient matrix, $C \in \boldsymbol{R}^{m\times n}$ stands for the measurement coefficient matrix, $E(w_k) = 0$, $E(v_k) = 0$, $\mathrm{cov}(w_k) = Q$, $\mathrm{cov}(v_k) = R$, w_k and v_k are uncorrelated with each other.

From Eq. (4.8), it can be obtained that

$$\begin{aligned} y_k &= Cx_k + v_k \\ &= C[Ax_{k-1} + w_{k-1}] + v_k \\ &= CAx_{k-1} + Cw_{k-1} + v_k \end{aligned} \tag{4.9}$$

and

$$y_{k-1} = Cx_{k-1} + v_{k-1} \tag{4.10}$$

Then we have

$$Cx_{k-1} = y_{k-1} - v_{k-1} \tag{4.11}$$

From Assumption 4.2, we have

$$[C^T C]^{-1} C^T C = I \tag{4.12}$$

where I is the unit matrix.

Since the measurement matrix C has left inverse matrix, $[C^T C]^{-1} C^T$, from Eq. (4.11) it can be obtained that

$$x_{k-1} = [C^T C]^{-1} C^T y_{k-1} - [C^T C]^{-1} C^T v_{k-1} \tag{4.13}$$

Then take the x_{k-1} as in Eq. (4.13) into Eq. (4.9), we have

$$y_k = CAMy_{k-1} - CAMv_{k-1} + Cw_{k-1} + v_k \tag{4.14}$$

where $M = [C^T C]^{-1} C^T$.

Multiplying M to Eq. (4.14), we obtain

$$My_k - AMy_{k-1} = w_{k-1} + Mv_k - AMv_{k-1} \tag{4.15}$$

Introduce the notations

$$\xi_k = My_k - AMy_{k-1} \tag{4.16}$$

$$V_k = Mv_k - AMv_{k-1} \tag{4.17}$$

Since the covariance matrix Q is constant, from Eq. (4.15), we have

$$\xi_k = w_{k-1} + V_k \tag{4.18}$$

Since the parameter matrices A, C, and R have been known a priori, the covariance matrix $\mathrm{cov}(V)$ could be obtained. Considering the definition of random vector ξ_k as in Eq. (4.16), the sample sequence ξ_k could be obtained as the measurement sequence y_k goes on in real time. From the sequence of sample data ξ_k, based on Algorithm 3, the estimation of covariance matrix $\hat{Q}_k$ could be obtained as follows:

$$\hat{Q}_k = \mathrm{cov}_k(\xi) - \mathrm{cov}(V) \tag{4.19}$$

From the process above, a reliable approach has been obtained by calculating the estimation of covariance matrix sequence $\hat{Q}_k$ in real time, which can be treated as previous work for applying traditional Kalman filter to obtain state estimation in linear dynamic models.

Algorithm 4 Kalman Filter with Recursive Covariance Estimation($\{y_k\}$,R,A,C)

Initialization: R,A,C, $\hat{x}_0$ the initial value of state estimate,
P_0 the initial error covariance matrix,
$\mathrm{cov}_0(\xi) = 0$, $\hat{Q}_0 = 0$
Obtain the covariance matrix $\mathrm{cov}(V)$ of the notation defined in Eq. (4.17) based on A, C and R
Input: measurement data sequence $\{y_k\}$
Output: state estimate sequence $\{\hat{x}_k\}$, the covariance matrix of estimation error $\{P_k\}$

```
for k = 1 to n do
    Calculate ξ(k) as in Eq. (4.16) with the measurement y_k
    Update the estimated covariance matrix cov_k(ξ) as in Eq. (4.6) with the sample ξ(k)
    Obtain the estimated process noise matrix Q̂_k as in Eq. (4.19) with cov_k(ξ) and cov(V)
    x̂_{k,k-1} = A x̂_{k-1}
    P̂_{k,k-1} = A P_{k-1} A^T + Q̂_k
    K̂_k = P̂_{k,k-1} C^T (C P̂_{k,k-1} C^T + R)^{-1}
    x̂_k = x̂_{k,k-1} + K̂_k (y_k - C x̂_{k,k-1})
    P_k = (I - K̂_k C) P̂_{k,k-1}
end for
Return {(Q̂)_k}, {x̂_k} and {P_k}
```

4.4 Stability Analysis

In this subsection, the theory results of stability analysis for KF-RCE algorithm will be presented to illustrate that under Assumption 4.1–4.3 KF-RCE algorithm is effective and the state estimating sequence from KF-RCE algorithm is consistent with Kalman filter with exact covariance matrix of process noise in the statistical aspects.

Lemma 4.1 (The bounded input-to-bounded output stability and the infinitesimal input-to-infinitesimal output stability) *Considering the dynamic model of error system*

$$\delta_k = F_k \delta_{k-1} + u_k \tag{4.20}$$

where $k = 1, 2, \ldots, \delta_k \in \boldsymbol{R}^m$ *is the measurement sequence, and* $u_k \in \boldsymbol{R}^m$ *is the driven signal. Suppose that* $F_k \to F$ *as* $k \to \infty$, *and the matrix F is stable. If the amplitude of driven signal* u_k *is bounded, then the amplitude of* δ_k *is bounded. If* $u_k \to 0$, *as* $k \to \infty$, $\delta_k \to 0$, *as* $k \to \infty$ *[18–20].*

Proof The proof can be found in the literature [13].

Theorem 4.2 *Under the conditions as in Assumptions 4.1–4.3 for the discrete-time LTI system, based on Algorithm 3, the estimated sequence of covariance matrix of process noise* $\{\hat{Q}_k\}$ *is convergent to the true value of covariance matrix Q as* $k \to \infty$, *i.e.,*

$$\hat{Q}_k - Q \to 0 \quad as\ k \to \infty \quad a.s. \tag{4.21}$$

where $\dot{Q}_k$ *is the estimated sequence of covariance matrix from Eq. (4.19) and Q is the true value of covariance matrix.*

Proof The proof can be found in the literature [13].

Theorem 4.3 *Under the conditions as in Assumptions 4.1–4.3 for the discrete-time LTI system, based on Algorithm 4,* $\limsup_{k\to\infty} \left\| \hat{P}_{k,k-1} \right\| < \infty$, *and* $\hat{P}_{k,k-1} - P_{k,k-1} \to 0$ *as* $k \to \infty$, *where* $\hat{P}_{k,k-1}$ *is the* priori *covariance matrix of state estimate from the KF-RCE algorithm and* $P_{k,k-1}$ *is the* priori *covariance matrix of state estimate from the ideal Kalman filter with exact covariance matrix of process noise.*

Proof The proof can be found in the literature [13].

Theorem 4.4 *Under the conditions as in Assumptions 4.1–4.3 for the discrete-time LTI system, based on Algorithm 4, the Kalman gain error* $\hat{K}_k - K_k \to 0$ *as* $k \to \infty$, *where* $\hat{K}_k$ *is the Kalman gain from KF-RCE algorithm and* K_k *from the standard Kalman filter with exact covariance matrix of process noise.*

Proof The proof can be found in the literature [13].

Lemma 4.2 *For the discrete-time LTI system (4.8), suppose that Assumption 4.1 holds, then the matrix*

$$\bar{A} = (I - K_{ss}C)A \tag{4.22}$$

is Schur stable, where K_{ss} is the steady Kalman gain of the the ideal Kalman filter with exact system parameters.

Lemma 4.3 *1. The dynamics of state estimation error $\tilde{x}_k^{op} \triangleq x_k - \hat{x}_k^{op}$ in the standard Kalman filter with exact parameters is governed by one autonomous linear time-varying system*

$$\eta_{k+1} = (I - K_k C)A\eta_k \tag{4.23}$$

where K_k is the Kalman gain in the ideal Kalman filter with exact parameters, and $\hat{x}_k^{op}$ denotes the optimal state estimation obtained by the ideal Kalman filter with exact parameters. If the LTI system (4.8) is controllable and observable, then system (4.23) must be asymptotically exponentially stable.

2. The dynamics of state estimation error $\tilde{x}_k \triangleq x_k - \hat{x}_k$ in the KF-RCE algorithm is governed by one autonomous linear time-varying system

$$\eta_{k+1} = (I - \hat{K}_k C)A\eta_k \tag{4.24}$$

where $\hat{K}_k$ is the Kalman gain matrix from KF-RCE algorithm. If the LTI system (4.8) is controllable and observable, then system (4.24) must be asymptotically exponentially stable.

Proof (1) First consider the ideal Kalman filter with exact parameters, which generate optimal state estimates $\hat{x}_k^{op}$ at time step k. It is easy to see that

$$\begin{aligned}
& x_{k+1} - \hat{x}_{k+1}^{op} \\
&= [Ax_k + Bw_{k+1}] - [\hat{x}_{k+1,k}^{op} + K_{k+1}(y_{k+1} - C\hat{x}_{k+1,k}^{op})] \\
&= [Ax_k + Bw_{k+1}] \\
&\quad -[A\hat{x}_k^{op} + K_{k+1}(Cx_{k+1} + v_{k+1} - CA\hat{x}_k^{op})] \\
&= [Ax_k + Bw_{k+1}] \\
&\quad -[A\hat{x}_k^{op} + K_{k+1}(CAx_k + CBw_{k+1} + v_{k+1} - CA\hat{x}_k^{op})] \\
&= [A - K_{k+1}CA](x_k - \hat{x}_k^{op}) \\
&\quad +[Bw_{k+1} - K_{k+1}CBw_{k+1} - K_{k+1}v_{k+1}]
\end{aligned} \tag{4.25}$$

whose homogeneous part is exactly

$$\eta_{k+1} = (I - K_k C)A\eta_k \tag{4.26}$$

Let

$$\bar{A} = (I - K_{ss}C)A \tag{4.27}$$

$$\Delta_k = (K_{ss} - K_k)CA \tag{4.28}$$

where K_{ss} is the steady Kalman gain matrix of standard Kalman filter with exact covariance matrix of process noise, and K_k from the ideal Kalman filter with exact system parameters. Then the autonomous system (4.26) becomes

$$\eta_{k+1} = (\bar{A} + \Delta_k)\eta_k \tag{4.29}$$

For the ideal Kalman filter with exact system parameters, from Assumption 4.1, matrix $\bar{A}$ is Schur stable; hence, for any positive definite matrix $\bar{Q}$, there exists a positive definite matrix $\bar{P}$ such that the Lyapunov equation holds

$$\bar{A}^T \bar{P} \bar{A} - \bar{P} = -\bar{Q} \tag{4.30}$$

Take a Lyapunov function $V(k) = \eta_k^T \bar{P} \eta_k$. Therefore,

$$\begin{aligned} \Delta V(k) &\overset{\Delta}{=} V(k+1) - V(k) \\ &= \eta_{k+1}^T \bar{P} \eta_{k+1} - \eta_k^T \bar{P} \eta_k \end{aligned} \tag{4.31}$$

Substituting Eq. (4.29) into Eq. (4.31) yields

$$\begin{aligned} \Delta V(k) &= \eta_k^T [(\bar{A} + \Delta_k)^T \bar{P} (\bar{A} + \Delta_k) - \bar{P}] \eta_k \\ &= -\eta_k^T \bar{Q} \eta_k + \eta_k^T E_k \eta_k \end{aligned} \tag{4.32}$$

where

$$E_k = \Delta_k^T \bar{P} \bar{A} + \bar{A}^T \bar{P} \Delta_k + \Delta_k^T \Delta_k \tag{4.33}$$

According to properties of the ideal Kalman filter with exact system parameters, we know that $K_k \to K_{ss}$ exponentially as $k \to \infty$, hence $\Delta_k = (K_{ss} - K_k)CA \to 0$ as $k \to \infty$. Consequently, $E_k \to 0$ as $k \to \infty$. Thus, for arbitrarily small $\epsilon > 0$ such that $\epsilon < \lambda_{\min}(\bar{Q})$, there exists an integer k_0 such that for any $k > k_0$,

$$\begin{aligned} \Delta V(k) &\le -\lambda_{\min}(\bar{Q})\eta_k^T \eta_k + \epsilon \eta_k^T \eta_k \\ &= -(\lambda_{\min}(\bar{Q}) - \epsilon)\|\eta_k\|^2 \\ &\le -\tfrac{\lambda_{\min}(\bar{Q}) - \epsilon}{\lambda_{\max}(\bar{P})} \eta_k^T \bar{P} \eta_k \\ &= -C_1 V(k) \end{aligned} \tag{4.34}$$

where

$$C_1 = \frac{\lambda_{\min}(\bar{Q}) - \epsilon}{\lambda_{\max}(\bar{P})} > 0 \tag{4.35}$$

with $\lambda_{\min}(\cdot)$ and $\lambda_{\max}(\cdot)$ being the minimal and the maximal eigenvalues of a matrix, respectively. Noting also

$$\bar{P} - \bar{Q} = \bar{A}^T \bar{P} \bar{A} > 0 \tag{4.36}$$

one can conclude that $\bar{P} > \bar{Q}$ hence

$$\lambda_{\max}(\bar{P}) > \lambda_{\max}(\bar{Q}) \geq \lambda_{\min}(\bar{Q}) \tag{4.37}$$

which implies that $0 < C_1 < 1$. Together with Eq. (4.34), immediately we can conclude that time-varying system (4.23) is asymptotically exponentially stable at a rate $(1 - C_1)^t$.

(2) Next consider the KF-RCE algorithm, i.e., the Kalman filter with recursive covariance estimation, for LTI system (4.8). Similar ideas can be adopted except that K_k should be replaced with $\hat{K}_k$, and $\hat{x}_k^{op}$, $\hat{x}_{k+1,k}^{op}$ need to be replaced with $\hat{x}_k$, $\hat{x}_{k+1,k}$, respectively. To establish the asymptotic exponential stability of the autonomous system (4.24), one needs only to notice that $\hat{K}_k - K_k \to 0$ as $k \to \infty$ by Theorem 4.4 or $\hat{K}_k - K_{ss} \to 0$ as $k \to \infty$ by Theorem 4.4. Consequently, the definition of Δ_k in above proof should be replaced with

$$\Delta_k = (K_{ss} - \hat{K}_k)CA \tag{4.38}$$

which is also convergent to zero as $k \to \infty$. Hence, the rest proof details can be similarly repeated, which complete the proof of this key lemma. □

Theorem 4.5 *Under the conditions as in Assumptions 4.1–4.3 for the discrete-time LTI system, based on Algorithm 4, the sequence of state estimates $\{\hat{x}_k\}$ asymptotically approaches to the sequence of $\{\hat{x}_k^{op}\}$ as $k \to \infty$ if $\{y_k\}$ is bounded, where $\hat{x}_k$ is the state estimate from KF-RCE algorithm and $\hat{x}_k^{op}$ is the optimal state estimate from standard Kalman filter with exact covariance matrix of process noise.*

Proof The estimation of system state could be obtained from Kalman filter with $\hat{Q}_k$ as follows:

$$\hat{x}_k = A\hat{x}_{k-1} + \hat{K}_k[y_k - CA\hat{x}_{k-1}] \tag{4.39}$$

State estimating process of Kalman filter with Q is as follows:

$$\hat{x}_k^{op} = A\hat{x}_{k-1}^{op} + K_k[y_k - CA\hat{x}_{k-1}^{op}] \tag{4.40}$$

Then from Eqs. (4.39)–(4.40), we have

$$\begin{aligned} \hat{x}_k - \hat{x}_k^{op} = &(I - K_kC)A(\hat{x}_{k-1} - \hat{x}_{k-1}^{op}) \\ &+(\hat{K}_k - K_k)(y_k - CA\hat{x}_{k-1}) \end{aligned} \tag{4.41}$$

Introduce the notations

$$\delta_k = \hat{x}_k - \hat{x}_k^{op} \tag{4.42}$$

$$\Phi_k = (I - K_k C)A \tag{4.43}$$

$$\Phi = (I - K_{ass} C)A \tag{4.44}$$

$$u_k = (\hat{K}_k - K_k)(y_k - CA\hat{x}_{k-1}) \tag{4.45}$$

Then we can obtain

$$\delta_k = \Phi_k \delta_{k-1} + u_k \tag{4.46}$$

Since the LTI system (4.8) is controllable and observable, Φ_k is always a stable matrix.

From Algorithm 4, we have that $\hat{x}_{k-1}$ is the unbiased estimate for the state x_{k-1} at the $(k-1)$th step if $\mathrm{E}(x_0) = x_0$, the initial state estimation. Then it could be obtained that

$$\mathrm{E}(y_k - CA\hat{x}_{k-1}) = 0 \tag{4.47}$$

From Theorem 4.3, it can be obtained that $\hat{P}_{k,k-1} - P_{k,k-1} \to 0$ and $\hat{P}_{k,k-1} - P_{ass} \to 0$ as $k \to \infty$. So there exists a covariance matrix P_{sup} which satisfied that

$$\mathrm{cov}(y_k - CA\hat{x}_{k-1}) \le P_{sup} \tag{4.48}$$

From Algorithm 4, we know that $\hat{x}_{k-1}$ is the unbiased estimate of the system state x_{k-1} and it can be obtained that

$$\mathrm{E}(u_k) = 0 \tag{4.49}$$

and

$$\mathrm{cov} u_k \le P_{sup}(\hat{K}_k - K_k)^2 \tag{4.50}$$

From Theorem 4.4, we have

$$\hat{K}_k - K_k \to \mathbf{0} \quad as \quad k \to \infty \tag{4.51}$$

So $u_k \to \mathbf{0}$ as $k \to \infty$ and the transfer matrix $\Phi(k)$ is stable. $\Phi_k \to \Phi$ as $k \to \infty$. From *Lemma* 4.1, it could be obtained that

$$\delta_k \to \mathbf{0} \quad as \quad k \to \infty \tag{4.52}$$

In one word, $\hat{x}_k - \hat{x}_k^{op} \to \mathbf{0}$ as $k \to \infty$. □

From Theorems 4.2–4.5, it can be obtained that $\hat{x}_k$ is convergent to the optimal state estimate $\hat{x}_k^{op}$, where $\hat{x}_k$ is obtained by the KF-RCE algorithm, $\hat{x}_k^{op}$ is obtained by the ideal Kalman filter with exact system parameters. In other words, $\hat{x}_k - \hat{x}_k^{op} \to 0$ as $k \to \infty$. □

Remark 4.5 From the proof, we can see that the condition of boundedness of $\{y_k\}$ is not very necessary because one key to establish the convergence or boundedness of $\hat{x}_k - \hat{x}_k^{op}$ lies in the convergence or boundedness of u_k, which is in turn determined by the product of $\hat{K}_k - K_k$ and $y_k - CA\hat{x}_{k-1}$.

4.5 Simulation Experiments

In this part, the results of simulation experiments will be presented to show the effects of the estimating algorithm proposed in this chapter.

Data fusion approach attracts many interests of research for various integrated positioning systems. For instance, some subsystem could obtain the positioning information; other subsystems can be used to sense speed-related information, while another system could offer the acceleration of different vehicles. In order to obtain full-dimensional space information, it is essential to combine different subsystems with each other.

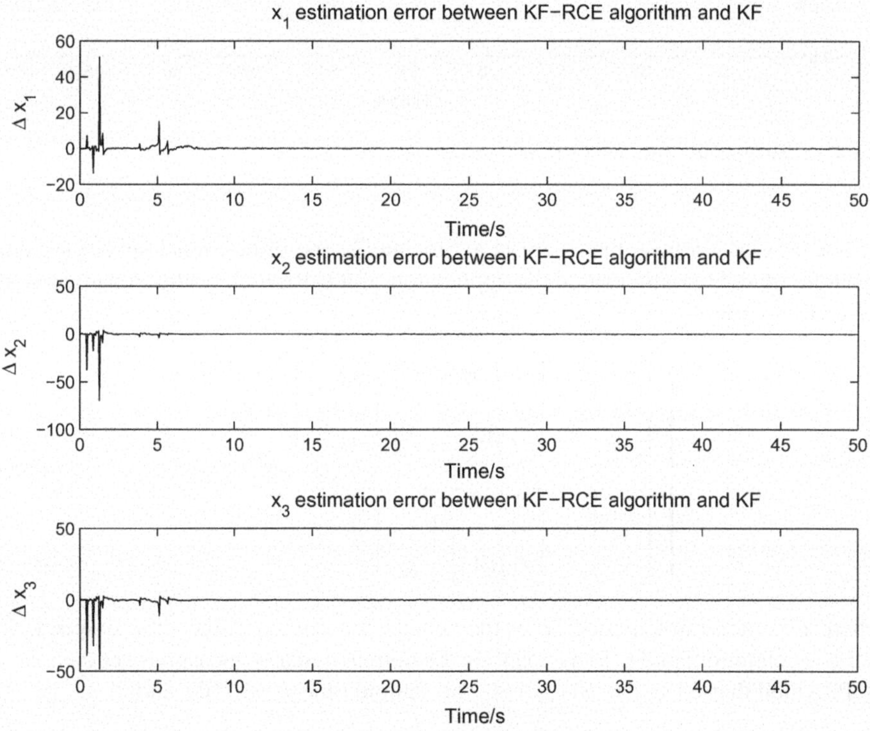

Fig. 4.1 Estimation error of state estimation with KF-RCE algorithm

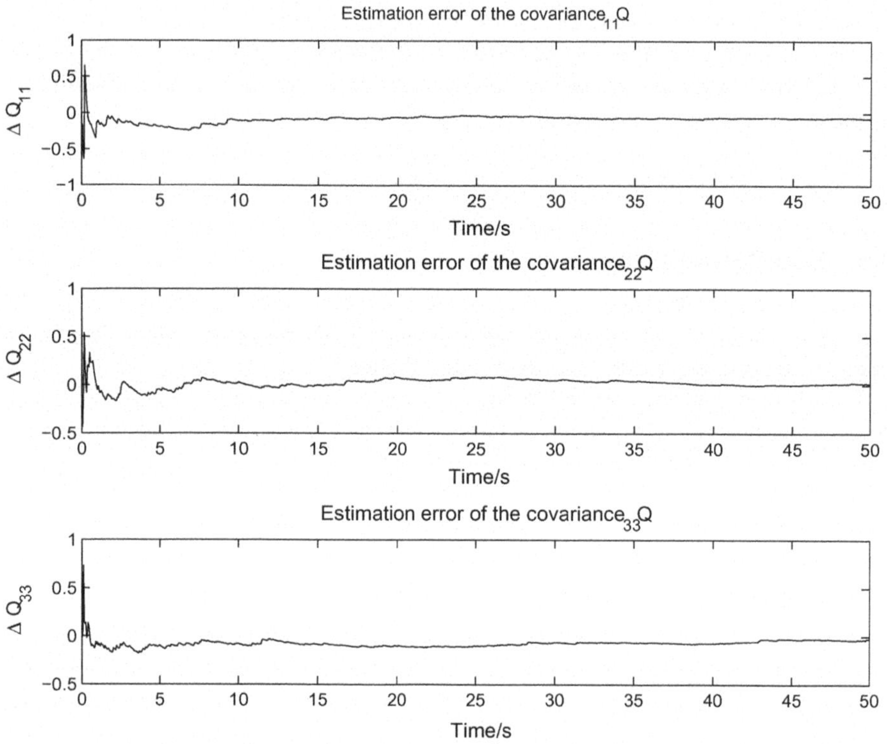

Fig. 4.2 Q estimation error with KF-RCE algorithm

For brevity, suppose that there is a "lineland" (one-dimensional space) system dynamic model to demonstrate the application of KF-RCE algorithm for integrated navigation systems:

$$\begin{cases} \begin{bmatrix} x_k \\ \dot{x}_k \\ \ddot{x}_k \end{bmatrix} = \begin{bmatrix} 1 & T & 0.5T^2 \\ 0 & 1 & T \\ 0 & 0 & 1 \end{bmatrix} \begin{bmatrix} x_{k-1} \\ \dot{x}_{k-1} \\ \ddot{x}_{k-1} \end{bmatrix} + w_{k-1} \\ \begin{bmatrix} y_k \\ \dot{y}_k \\ \ddot{y}_k \end{bmatrix} = \begin{bmatrix} 1 & 0 & 0 \\ 0 & 1 & 0 \\ 0 & 0 & 1 \end{bmatrix} \begin{bmatrix} x_k \\ \dot{x}_k \\ \ddot{x}_k \end{bmatrix} + v_k \end{cases} \tag{4.53}$$

where T is the sample period, x_k is the vehicle position, $\dot{x}_k$ is the vehicle speed, $\ddot{x}_k$ is the acceleration, and $[y_k \quad \dot{y}_k \quad \ddot{y}_k]^T$ is the output signal of measurement data.

Provided that $E(w_k) = E(v_k) = 0$, the sample time $T = 0.05s$ and

$$\text{cov}(w_k) = Q = \begin{bmatrix} 0.25 & 0.04 & 0.04 \\ 0.04 & 0.25 & 0.04 \\ 0.04 & 0.04 & 0.25 \end{bmatrix} \tag{4.54}$$

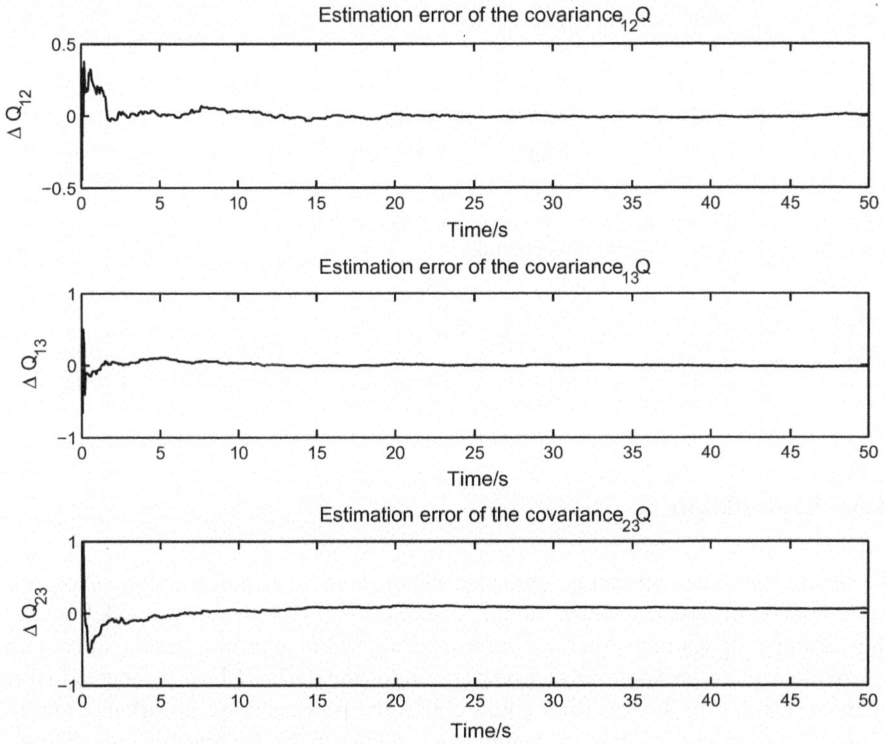

Fig. 4.3 Q estimation error with KF-RCE algorithm

Table 4.1 State estimation error at different times

Time(s)	0.05	0.5	5	25	50
Δx_1	−0.0482	0.1431	2.4564	0.0194	0.0390
Δx_2	0.0350	−0.5259	−0.6203	−0.0222	−0.0762
Δx_3	0.0534	−0.2916	−1.7390	0.0951	0.0694

$$\text{cov}(v_k) = R = \begin{bmatrix} 0.25 & 0.04 & 0.04 \\ 0.04 & 0.25 & 0.04 \\ 0.04 & 0.04 & 0.25 \end{bmatrix} \tag{4.55}$$

However, Q is completely unknown in the data fusion of the integrated navigation systems.

The simulation results are presented in Figs. 4.1, 4.2, 4.3, Tables 4.1, 4.2, and 4.3.

From the simulation results, it can be seen that $\hat{Q}_k$ is quickly convergent to the true value of covariance matrix of process noise Q as $k \to \infty$. The estimating sequence of system state $\hat{x}_k$ is also consistent with standard Kalman filter estimate $\hat{x}_k^{op}$ in the statistical aspect, which is coincident with the theoretic results.

Table 4.2 Q estimation error at different times

Time(s)	0.05	0.5	5	25	50
ΔQ_{11}	−0.1566	−0.1266	−0.1937	−0.0388	−0.0755
ΔQ_{22}	−0.4334	0.3387	−0.0505	0.0820	0.0171
ΔQ_{33}	−0.0113	0.0095	−0.1009	−0.1036	−0.0278

Table 4.3 Q estimation error at different times

Time(s)	0.05	0.5	5	25	50
ΔQ_{12}	−0.4466	0.3000	0.0004	−0.0072	0.0079
ΔQ_{13}	−0.6627	−0.1467	0.1047	0.0040	−0.0186
ΔQ_{23}	0.4716	−0.5395	−0.0574	0.0888	0.0594

4.6 Conclusion

To obtain the state estimating sequence when there is no information of covariance matrix of process noise in the discrete-time LTI dynamic model, under the structure of Kalman filter, an improved algorithm named Kalman filter with `RecursiveCovarianceEstimation` algorithm (KF-RCE) is presented in this chapter. The KF-RCE algorithm could obtain the process noise covariance matrix estimating sequence from the measurement sequence in the discrete-time linear time-invariant (LTI) systems. Based on the idea of the law of large numbers, simple `RecursiveCovarianceEstimation` algorithm has been presented first, and then a new measurement sequence is constructed based on the system dynamic model, from which the process covariance matrix could be estimated.

Since the KF-RCE algorithm has a simple recursive form, it is actually easy to be used in practice. Then, the stability analysis of KF-RCE algorithm has been introduced to guarantee that the state estimation obtained by the new algorithm is convergent to the optimal state estimation by the ideal Kalman filter. The KF-RCE algorithm is stable and reliable when the certain conditions given in this chapter are satisfied. Finally, the results of simulation experiment are presented to illustrate that the KF-RCE algorithm is effective and practical.

References

1. Y. Zhang, G.Y. Wang, J.T. Xu, Z.F. Shi, T. Feng, D.X. Dong, G.Q. Chi, A method of eliminating the signal-dependent random noise from the raw CMOS image sensor data based on Kalman filter. Signal Process. **104**, 401–406 (2014)
2. S. Ghandour-Haidar, L. Ros, J.M. Brossier, On the use of first-order autoregressive modeling for Rayleigh flat fading channel estimation with Kalman filter. Signal Process. **92**(2), 601–606 (2012)

3. Q.B. Ge, C. Wen, T.L. Xu, Networked Kalman filtering with combined constraints of bandwidth and random delay. Sci. China Inf. Sci. **55**(10), 2343–2357 (2012)
4. J. Xu, J.X. Li, S. Xu, Data fusion for target tracking in wireless sensor networks using quantized innovations and Kalman filtering. Sci. China Inf. Sci. **55**(3), 530–544 (2012)
5. X.J. Shen, Y.T. Luo, Y.M. Zhu, E.B. Song, Globally optimal distributed Kalman filtering fusion. Sci. China Inf. Sci. **55**(3), 512–529 (2012)
6. Z. Deng, S. Sun, Wiener state estimatiors based on Kalman filtering. Acta Automatica Sinica **30**(1), 126–130 (2004)
7. B.S. Koh, J. Junkins, Kalman filter for linear fractional order systems. J. Guid. Control Dyn. **35**(6), 1816–1827 (2012)
8. L. Guo, Estimating time-varying parameters by the Kalman filter based algorithm: stability and convergence. IEEE Trans. Autom. Control **35**(2), 141–147 (1990)
9. J.F. Zhang, L. Guo, H.F. Chen, l_p-stability of estimation errors of Kalman filter for tracking time-varying parameters. Int. J. Adap. Control Signal Process. **5**(3), 155–174 (1991)
10. X. Xiao, B. Feng, B. Wang, On-line realization of SVM Kalman filter for MEMS gyro, in *Proceedings of the 3rd International Conference on Measuring Technology and Mechatronics Automation*, pp. 768–770
11. R.K. Mehra, On the identification of variances and adaptive Kalman filtering. IEEE Trans. Autom. Control **15**(2), 175–184 (1970)
12. K. Myers, B. Tapley, Adaptive sequential estimation with unknown noise statistics. IEEE Trans. Autom. Control **21**(4), 520–523 (1976)
13. B. Feng, M.Y. Fu, H.B. Ma, Y. Xia, Kalman filter with recursive covariance estimation - sequentially estimating process noise covariance. IEEE Trans. Ind. Electron. **61**(11), 6253–6263 (2014)
14. R.E. Kalman, A new approach to linear filtering and prediction problems. Trans. ASME - J. Basic Eng. **82**(Series D), 35–45 (1960)
15. R.E. Kalman, R.S. Bucy, New results in linear filtering and prediction theory. Trans. ASME Ser. D, J. Basic Eng. **83**, 95–107 (1961)
16. M. Karasalo, X. Hu, An optimization approach to adaptive Kalman filtering. Automatica **47**(8), 1785–1793 (2011)
17. Y. Yang, W. Gao, An optimal adaptive Kalman filter. J. Geod. **80**(4), 177–183 (2006)
18. Z.L. Deng, Y. Gao, C. Li, G. Hao, Self-tuning decoupled information fusion Wiener state component filters and their convergence. Automatica **44**(3), 685–695 (2008)
19. Y. Gao, W.J. Jia, X.J. Sun, Z. Deng, Self-tuning multisensor weighted measurement fusion Kalman filter. IEEE Trans. Aerosp. Electron. Syst. **45**(1), 179–191 (2009)
20. C. Ran, G. Tao, J. Liu, Z. Deng, Self-tuning decoupled fusion Kalman predictor and its convergence analysis. IEEE Sens. J. **9**(12), 2024–2032 (2009)

Chapter 5
Modified Kalman Filter with Recursive Covariance Estimation for Gyroscope Denoising

The precision of inertial navigation system (INS) mainly depends on the precision of inertial devices, among which a kernel device, gyroscope, is of fundamental importance. In this chapter, a physical dynamic model is introduced to represent the dynamic behavior of gyroscope measurement output, yet without a priori knowledge on the covariance matrices of process and measurement noises. For such a model, standard Kalman filter is not applicable to denoise gyroscope signal due to the lack of statistics of the noises. To resolve this challenging problem, a modified Kalman filter with recursive covariance estimation (MKF-RCE) is presented to cope with the general case when the covariance matrices of process and measurement noises in the physical model cannot be obtained a priori and the rigorous stability analysis is given to show that this algorithm is optimal in sense that the covariance matrix and state estimations are asymptotically consistent with the ideal Kalman filter using the exact covariance matrices. By applying MKF-RCE algorithm, the experiments for one MEMS gyroscope have demonstrated satisfactory denoising results, which verified the effectiveness of MKF-RCE algorithm in state estimation problem in the presence of large uncertainty of process and measurement noise covariance matrices.

5.1 Introduction

Inertial navigation system (INS) [1] is a kind of navigation system which is suitable for use in the autonomous navigation applications considering such following advantages: (1) It does not require any external information from other devices unlike other navigation techniques such as global positioning system (GPS); (2) It has no very strict requirements on the weather conditions; in other words, it can work well under different weather conditions; and (3) It has certain feature of anti-inference with the help of inertial devices [2]. Hence, inertial navigation systems are widely equipped [3] on modern vehicles, ships, submarines, and aircraft [4, 5].

H. Ma et al., *Kalman Filtering and Information Fusion*,
https://doi.org/10.1007/978-981-15-0806-6_5

Inertial navigation systems could be divided into gimbaled inertial navigation systems and strapdown navigation systems according to different system structures. Recently, there are more and more strapdown inertial navigation systems (SINS), since SINS do not require a fixed gimbaled gyro-stabilized platform and SINS reduces the cost, eliminates gimbal lock, removes some calibration need, and increases the reliability by eliminating some of the shifting parts. However, for a gyroscope in strapdown inertial navigation systems, it must have a larger measurement range yet with more uncertainty in the input angular velocity due to high-speed data updating [6]. Since gyroscope working conditions are much different between platform inertial navigation systems and strapdown inertial navigation systems, in most time only using analog circuits to deal with the measurement signal may not effectively satisfy the demand of signal processing for the gyroscope output measurement in the strapdown inertial navigation systems [7].

In practice, one main approach is the so-called Kalman filtering [8], which is proposed in R. E. Kalman's pioneering work [9, 10], where he formulated one general state estimating problem and investigated linear filtering and prediction problems [11]. Standard Kalman filter is optimal in the sense of minimum mean squared errors and maximum likelihood estimation [12, 13], providing that the system model is linear and precisely known a priori, and the process and measurement noises are jointly Gaussian with known covariance matrices [4, 14–16]. In order to use Kalman filter in the gyros signal denoising problem, a linear system model should be known with exact parameters and the process and measurement noises are independent with each other, zero mathematical expectation, Gaussian with known covariance matrices [17–20].

To apply Kalman filter for the purpose of filtering gyroscope signals in inertial navigation systems, the auto regressive moving average (ARMA) [21] model was often used to describe the performance of gyroscope output signal and the parameters in the ARMA model were determined by the offline sample data a priori [22]. However, in the strapdown inertial navigation systems, a priori system model may not be much suitable for the dynamic performance of gyroscope measurement since the working conditions may not be consistent as the expectation a priori.

In this chapter, one physical model is established to mathematically describe gyroscope measurement signal. Considering the dynamic and repeatability uncertainty, suppose that the covariance matrices of process and measurement noises are completely unknown a priori. A new algorithm named as modified Kalman filtering with recursive covariance estimation (MKF-RCE) will be presented to deal with the case when the covariance matrices of process and measurement noises cannot be determined a priori for the physical system model. Then, a simulation example will be presented to illustrate that MKF-RCE algorithm can be applied to deal with the problem that the covariance matrices of process and measurement noises are completely unknown a priori and state estimation obtained by the new algorithm is convergent to the optimal estimation by the standard Kalman filter with exact system parameters. Finally, experiment tests will be presented to show that the method proposed in this chapter can effectively deal with MEMS gyroscope noise signal and improve the precision of the measurement data.

In our previous work [23], preliminary research on the estimating problem had been presented when the covariance matrix of process noise is completely unknown a priori and the stability analysis of Kalman filter with recursive covariance estimation had been presented under some technical assumptions. Compared with the previous work [23], a different estimation problem will be dealt with when the process and measurement covariance matrices are completely unknown a priori. The stability analysis of MKF-RCE algorithm is more complex than KF-RCE algorithm in [23] since both process noise and measurement noise are with unknown covariance matrix, and more efforts have been made to weaken the mathematical assumptions for a more complete stability analysis. Considering the background of MEMS gyroscope denoising application, one simple yet nontrivial physical system model will be presented to describe the dynamic behavior of signal output with some uncertainty.

The rest of this chapter is organized as follows: in Sect. 5.2, the standard Kalman filter algorithm will be reviewed with mathematical equations for convenience of later discussions; then the problem will be formulated which is dealt with in this chapter; in Sect. 5.3, to figure out how to obtain the covariance matrix of a vector from the sample sequence corrupted with a noise in real time, a new algorithm named recursive covariance estimation will be presented, and then a new algorithm named MKF-RCE will be introduced to show how to cope with the case when the covariance matrices of process and measurement noises are completely unknown a priori and a simulation example will be presented to illustrate that the new algorithm is practically useful and effective; in Sect. 5.4, testing experiments will be presented to illustrate the effectiveness of the MKF-RCE algorithm for denoising the measurement signal of MEMS130 gyroscope; finally, in Sect. 5.5 concluding remarks are presented to summarize the denoising problem of MEMS gyroscope measurement and the MKF-RCE algorithm.

5.2 Problem Formulation

5.2.1 Kalman Filter

In 1960, R. E. Kalman published his most famous paper describing a recursive solution to the discrete-time linear filtering problem [9, 11]. Since then, due in large part to advantages in recursive digital computing, the Kalman filter has been the subject of extensive research and application, particularly in the area of autonomous or assisted navigation [24].

It is the program core to design Kalman filter's observation update and state update. Suppose that the state x_k to be estimated is governed by the following dynamics:

$$\begin{cases} x_k = \phi_{k,k-1}x_{k-1} + \Gamma_{k-1}w_{k-1} \\ y_k = C_k x_k + v_k \end{cases} \tag{5.1}$$

where $\phi_{k,k-1}$ is the transfer matrix, C_k is the observation matrix, w_k is the process noise sequence, v_k is the measurement noise sequence, Γ_k is the noise input matrix, and y_k represents observable measurements at time t_k.

If the system noise and observation noise are zero mean, Gaussian, and uncorrelated, $\hat{x}_k$, the estimate of x_k, can be recursively obtained by the following equations [9]:

$$\begin{cases} \hat{x}_{k,k-1} = \phi_{k,k-1}\hat{x}_{k-1} \\ \hat{x}_k = \hat{x}_{k,k-1} + K_k(y_k - C_k\hat{x}_{k,k-1}) \\ P_{k,k-1} = \phi_{k,k-1}P_{k-1}\phi_{k,k-1}^T + \Gamma_{k-1}Q_{k-1}\Gamma_{k-1}^T \\ K_k = P_{k,k-1}C_k^T(C_kP_{k,k-1}C_k^T + R_k)^{-1} \\ P_k = (I - K_kC_k)P_{k,k-1} \end{cases} \tag{5.2}$$

5.2.2 Problem to Be Resolved

Considering the real-time demand of signal processing, ARMA model is always used to describe the gyroscope measurement dynamic in the inertial navigation systems. The parameters of ARMA model were determined a priori based on the offline sample data from gyroscope output inspired by designed input angular rate. However, the gyroscope working conditions are always complex and hardly determined a priori, which may cause that a priori ARMA model is not consistent with online gyroscope dynamic and make the signal processing unsuccessful.

The basic demand of signal processing for gyroscope measurement is real-time computation in inertial navigation systems. To satisfy the real-time demand and meet high precision for inertial navigation systems, it should make a trade-off between computation complexity and signal processing accuracy. Considering the real-time demand, a simple yet nontrivial physical system model is presented to describe the dynamic performance of the gyroscope for inertial navigation systems in this chapter as follows:

$$\begin{cases} \begin{bmatrix} x_k \\ s_k \end{bmatrix} = \begin{bmatrix} 1 & T \\ 0 & 1 \end{bmatrix}\begin{bmatrix} x_{k-1} \\ s_{k-1} \end{bmatrix} + \begin{bmatrix} 0 \\ 1 \end{bmatrix} w_{k-1} \\ y_k = \begin{bmatrix} 1 & 0 \end{bmatrix}\begin{bmatrix} x_k \\ s_k \end{bmatrix} + v_k \end{cases} \tag{5.3}$$

where T is the sample period, x_k is the angular velocity measured by the gyroscope at the kth step, s_k is the rate of angular velocity at the kth step, y_k is the gyroscope output signal at the kth step, v_k is the measurement noise at the kth step, and w_k is the process noise which stands for the uncertainty in the angular velocity.

Remark 5.1 The gyroscope output signal can be classified as two types, angular velocity or angular increment. No matter what type the signal is of, angular velocity or increment, it can be described by the system model (5.3). From physical meanings of v_k and w_k, it is natural that they are two independent random signals which cannot be observed. Hence, using model (5.3) to describe one arbitrary gyroscope signal, the biggest problem is that it has no any idea on the statistics information of the noise sequences $\{v_k\}$ and $\{w_k\}$, which motivates the preliminary case studies in this chapter and may be a starting point for investigating more general and challenging estimation problem for any uncertain linear systems with completely unknown statistics of both the process noise and measurement noise.

Remark 5.2 In our previous work [23], the state estimation problem has been investigated under the uncertainty of the covariance matrix of process noise only for the following linear time-invariant system:

$$\begin{aligned} x_{k+1} &= Ax_k + Bw_k \\ y_k &= Cx_k + v_k \end{aligned} \tag{5.4}$$

where the covariance $Q = \text{Cov}(w_k)$ of the process noise is completely unknown, while the covariance $R = \text{Cov}(v_k)$ of the measurement noise is supposed to be known a priori. Besides, the KF-RCE algorithm proposed in [23] has one technical condition that the matrix C should be of full column rank. In this work, it is obvious that system model (5.3) cannot satisfy the technical condition given in our previous work [23]; hence, it is necessary to explore new algorithm to cope with the difficulties encountered in the practical problem of gyroscope denoising with model (5.3).

In this chapter, the system model (5.3) will be used to describe the gyroscope measurement dynamic and a new algorithm will be presented to deal with the case where the covariance matrices of process and measurement noise are completely unknown a priori.

For clarity to be formulated, the problem to be resolved in this chapter is based on the assumptions as follows:

Assumption 5.1 For the system model (5.3), suppose that w_k and v_k are uncorrelated zero-mean Gaussian random sequences.

Assumption 5.2 For the system model (5.3), suppose that the covariance matrices of process and measurement noise are constant such as $\text{cov}(w_k) = Q$ and $\text{cov}(v_k) = R$. But Q and R are completely unknown a priori.

Problem 5.1 Under Assumptions 5.1–5.2, for the gyroscope system model (5.3), how to design a filter which is able to cope with the state estimating problem when the covariance matrices of process and measurement noise are completely unknown a priori?

Generally speaking, such statistical properties should be obtained via many samples of the process and measurement noises. However, three difficulties may be encountered: (a) It is almost impossible to isolate the process noise and measurement noise from the output data generated by the system dynamics since the real system states cannot be accurately obtained. (b) Even if the process and measurement noise can be isolated from the output data, according to statistical knowledge, many samples of output data are needed so as to generate enough samples for the purpose of estimating statistical properties. (c) The estimated statistical properties based on offline sample data may be not consistent with the statistical properties of real dynamic noises. Facing these challenges, it is hoped to use only the measurement sequence to estimate covariance matrices of process and measurement noises online rather than doing many experiments offline for the purpose of obtaining statistical properties.

5.3 Modified Kalman Filter with Recursive Covariance Matrix

5.3.1 Basic Idea: Estimating Covariance Matrix

For the convenience of later discussions, it is needed to introduce first the basic idea of estimating covariance matrix for a random variable, which is the basis of constructing filter algorithm proposed in this chapter.

The following simple yet nontrivial problem will be considered:

Problem 5.2 For a random variable ξ given an sample sequence $\{\xi_i, i = 1, 2, 3, \ldots\}$ of ξ, it is supposed that $\mathrm{E}(\xi) = 0$ and $\mathrm{E}(\xi_k \xi_{k-p}^T) = 0$ when $p \geq k_m$. Based on the sample sequence $\{\xi_i, i = 1, 2, 3, \ldots\}$ of ξ, how to obtain the optimal estimated variance and covariance of the random variable ξ in real time?

To answer this basic question, with the law of large numbers, the following theorem is easily obtained:

Theorem 5.1 *Suppose that $\xi \in \boldsymbol{R}^{m\times 1}$ be a random variable and $E(\xi) = 0$, the covariance matrix of the random variable ξ can be obtained from its samples as follows:*

$$\mathrm{E}(\xi_k \xi_{k-i}^T) = \lim_{k\to\infty} \frac{1}{k} \sum_{j=1}^{k} \xi_j \xi_{j-i}^T \tag{5.5}$$

where $i = 1, 2, 3, \ldots, k_m$

However, the calculation method indicated above is not suitable for online parameter estimating since it is not possible to collect all the vector samples at a certain time. Therefore, it is desirable for real-time computation if a recursive form of the above estimation algorithm can be implemented.

Supposing that $\hat{e}_n(\xi_k\xi_{k-i}^T)$ is the estimated covariance matrix from the samples up to and including the time n, with the arrival of a new sample, the $(n+1)$th sample should be used to modify $\hat{e}_n(\xi_k\xi_{k-i}^T)$ in order to obtain new estimation $\hat{e}_{n+1}(\xi_k\xi_{k-i}^T)$. As in Eq. (5.5), it can be obtained that

$$\begin{aligned}
&\hat{e}_{n+1}(\xi_k\xi_{k-i}^T)\\
&=\frac{1}{n+1}\sum_{j=1}^{n+1}\xi_j\xi_{j-i}^T\\
&=\hat{e}_n(\xi_k\xi_{k-i}^T)+\frac{1}{n+1}(\xi_{n+1}\xi_{n+1-i}^T-\hat{e}_n(\xi_k\xi_{k-i}^T))
\end{aligned} \tag{5.6}$$

Therefore, Algorithm 5 can be used to resolve Problem 5.2.

Algorithm 5 RCE($\{\xi_j\}$)

Initialization: $\hat{e}_0(\xi_k\xi_{k-i}^T)=0, i=1,2,\ldots,k_m$
Input: the sample sequence $\{\xi_j\}$ of random variable ξ
Output: estimated covariance matrix $\hat{e}_n(\xi_k\xi_{k-i}^T)$ of the random variable ξ
1: **for** $j=1$ to n **do**
2: Update the estimated covariance matrix $\hat{e}_n(\xi_k\xi_{k-i}^T)$ just as in Eq. (5.6) with the sample data ξ_j
3: **end for**
4: Return $\hat{e}_n(\xi_k\xi_{k-i}^T)$

5.3.2 *Modified Kalman Filter with Recursive Covariance Matrix*

In the subsection, the basic idea presented in Algorithm 5 will be applied to the discrete-time linear time-invariant (LTI) system model (5.7).

$$\begin{cases}\begin{bmatrix}x_k\\ s_k\end{bmatrix}=\begin{bmatrix}1 & T\\ 0 & 1\end{bmatrix}\begin{bmatrix}x_{k-1}\\ s_{k-1}\end{bmatrix}+\begin{bmatrix}0\\ 1\end{bmatrix}w_{k-1}\\ y_k=\begin{bmatrix}1 & 0\end{bmatrix}\begin{bmatrix}x_k\\ s_k\end{bmatrix}+v_k\end{cases} \tag{5.7}$$

where T is the sample period, x_k is the angular velocity measured by the gyroscope at the kth step, s_k is the equivalent rate of angular velocity at the kth step, y_k is the gyroscope output signal at the kth step, v_k is the measurement noise at the kth step, and w_k is the process noise which stands for the uncertainty in the angular velocity. Here $\mathrm{E}(w_k)=0$, $\mathrm{E}(v_k)=0$, $\mathrm{cov}(w_k)=Q$, $\mathrm{cov}(v_k)=R$, w_k and v_k are uncorrelated but Q and R are completely unknown a priori. Introducing the backward shift operator q^{-1}, $q^{-1}x_k=x_{k-1}$.

From Eq. (5.7), it can be obtained that

$$x_k = \frac{q^{-1}T}{1-q^{-1}} s_k \tag{5.8}$$

$$s_k = \frac{1}{1-q^{-1}} w_{k-1} \tag{5.9}$$

$$y_k = x_k + v_k \tag{5.10}$$

Then, it can be obtained that

$$(1-q^{-1})^2 y_k = q^{-1} T w_{k-1} + (1-q^{-1})^2 v_k \tag{5.11}$$

Introduce the notations

$$\xi_k = y_k - 2y_{k-1} + y_{k-2} \tag{5.12}$$

$$W_k = q^{-1} T w_{k-1} = T w_{k-2} \tag{5.13}$$

$$V_k = v_k - 2v_{k-1} + v_{k-2} \tag{5.14}$$

From Eq. (5.11), it can be obtained that

$$\mathrm{E}(\xi_k) = 0 \tag{5.15}$$

$$\mathrm{E}(\xi_k \xi_k) = T^2 Q + 6R \tag{5.16}$$

$$\mathrm{E}(\xi_k \xi_{k-1}) = -4R \tag{5.17}$$

From the definition as in Eq. (5.12), one more sample ξ_k can be obtained with the arrival of measurement y_k. Then using the new sample, the estimated covariance matrix $\hat{\mathrm{e}}_k(\xi_k \xi_k)$ and $\hat{\mathrm{e}}_k(\xi_k \xi_{k-1})$ can be obtained as in Algorithm 5. From the relation as in Eqs. (5.16) and (5.17), it can be obtained that

$$\begin{bmatrix} \hat{Q}_k \\ \hat{R}_k \end{bmatrix} = \begin{bmatrix} T^2 & 6 \\ 0 & -4 \end{bmatrix}^{-1} \begin{bmatrix} \hat{\mathrm{e}}_k(\xi_k \xi_k) \\ \hat{\mathrm{e}}_k(\xi_k \xi_{k-1}) \end{bmatrix} \tag{5.18}$$

Using this method, the estimated covariance matrices $\hat{Q}_k$ and $\hat{R}_k$ can be obtained in real time, which can serve as a basis for applying standard Kalman filter to generate system state estimates.

Algorithm 6 MKF-RCE($\{y_k\}$,T)

Initialization: P_0 the initial state error covariance matrix,
  $\hat{e}_0(\xi_k\xi_k)=0$, $\hat{e}_0(\xi_k\xi_{k-1})=0$,$\hat{Q}_0=0$,$\hat{R}_0=0$
Input: measurement sequence $\{y_k\}$
Output: real-time system state estimate $\{\hat{x}_k\}$, state estimation error covariance matrix $\{P_k\}$

for $k=1$ to n **do**
  Calculate ξ_k as in Eq. (5.12) with the measurement y_k
  Update the estimated covariance matrix $\hat{e}_k(\xi_k\xi_k)$ and $\hat{e}_k(\xi_k\xi_{k-1})$as in Algorithm 5 with the sample ξ_k to
  Calculate the estimated $\hat{Q}_k$ and $\hat{R}_k$ as in Eq. (5.18) with $\hat{e}_k(\xi_k\xi_k)$ and $\hat{e}_k(\xi_k\xi_{k-1})$
  $\hat{x}_{k,k-1}=A\hat{x}_{k-1}$
  $\hat{P}_{k,k-1}=AP_{k-1}A^T+\hat{Q}_k$
  $\hat{K}_k=\hat{P}_{k,k-1}(\hat{P}_{k,k-1}+\hat{R}_k)^{-1}$
  $\hat{x}_k=\hat{x}_{k,k-1}+\hat{K}_k(y_k-\hat{x}_{k,k-1})$
  $P_k=(I-\hat{K}_k)\hat{P}_{k,k-1}$
end for
Return $\{\hat{Q}_k\}$, $\{\hat{R}_k\}$,$\{\hat{x}_k\}$ and $\{P_k\}$

Obviously, for Problem 5.1, Algorithm 6 can be used to deal with the problem that the covariance matrices of process and measurement noises are completely unknown a priori for the gyroscope signal system model (5.3).

Remark 5.3 The new MKF-RCE Algorithm 6 benefits from the special structure of the gyroscope signal system model (5.3), which makes the deduction in Eqs. (5.8)–(5.18) successful. Such an idea was not adopted in our previous work [23]; hence, the technical requirement on measurement gain matrix C is completely removed in this contribution. It is possible to extend the MKF-RCE algorithm to more general uncertain linear systems; however, the technical conditions to guarantee the convergence and stability may still need further investigation. The ideas in this preliminary work may shed a light on the more challenging state estimation problems for general linear systems or even nonlinear systems under more uncertainties of the process noise and the measurement noise.

5.3.3 Stability Analysis

In this subsection, some theoretical results of MKF-RCE algorithm will be presented to show that MKF-RCE algorithm is stable under Assumption 5.1–5.2 and the sequence of state estimate obtained by MKF-RCE algorithm will be convergent to the ideal Kalman filter with exact system parameters.

Remark 5.4 Due to the new problem formulation, the stability analysis for MKF-RCE algorithm is different from that for KF-RCE algorithm in our previous work [23], despite that the essential idea of recursive covariance estimation is common in both algorithms. Besides, owing to the special structure of system model (5.3), in this contribution, the technical condition on the boundedness of $\{y_k\}$ is not necessary for ensuring the convergence and stability.

Theorem 5.2 *Under Assumptions 5.1–5.2 for the discrete-time LTI system* (5.7), *using Algorithm 5 and Algorithm 6, the sequence of estimated process noise covariance matrix* $\{\hat{Q}_k\}$ *is convergent to the real covariance matrix* Q *as* $k \to \infty$, *the sequence of estimated measurement noise covariance matrix* $\{\hat{R}_k\}$ *is convergent to the real covariance matrix* R *as* $k \to \infty$, *i.e.,*

$$\hat{Q}_k - Q \to 0 \quad as \quad k \to \infty \quad a.s. \tag{5.19}$$

$$\hat{R}_k - R \to 0 \quad as \quad k \to \infty \quad a.s. \tag{5.20}$$

where $\hat{Q}_k$ *and* $\hat{R}_k$ *are the estimated covariance matrix obtained by Eq.* (5.18) *and* Q *and* R *are the real covariance matrices.*

Proof From Algorithm 5, it can be obtained that

$$\hat{\mathrm{e}}_k(\xi_k\xi_k) - \mathrm{E}(\xi_k\xi_k) \to 0 \ \text{ as } \ k \to \infty \quad a.s. \tag{5.21}$$

$$\hat{\mathrm{e}}_k(\xi_k\xi_{k-1}) - \mathrm{E}(\xi_k\xi_{k-1}) \to 0 \ \text{ as } \ k \to \infty \quad a.s. \tag{5.22}$$

Based on the relationship as in Eqs. (5.16) and (5.17), it can be obtained that

$$\hat{Q}_k - Q \to 0 \quad \text{as } k \to \infty \quad a.s. \tag{5.23}$$

$$\hat{R}_k - R \to 0 \quad \text{as } k \to \infty \quad a.s. \tag{5.24}$$

□

Theorem 5.3 *Under Assumptions 5.1–5.2 for the discrete-time LTI system* (5.7), *using Algorithm 6, it can be obtained that* $\limsup_{k\to\infty} \left\| \hat{P}_{k,k-1} \right\| < \infty$, *and* $\hat{P}_{k,k-1} - P_{k,k-1} \to 0$ *as* $k \to \infty$, *where* $\hat{P}_{k,k-1}$ *is the priori state estimating covariance matrix in the MKF-RCE algorithm and* $P_{k,k-1}$ *is the priori state estimating covariance matrix in the ideal Kalman filter with exact system parameters.*

Proof Introducing a notation

$$\begin{aligned} \mathcal{G}(P, R) &= APA^T - APC^T(CPC^T + R)^{-1}CPA^T \\ &= A(I + PC^T R^{-1}C)^{-1}PA^T \end{aligned} \tag{5.25}$$

For arbitrary positive matrices P_1 and P_2, supposing that $\mathcal{G}(P_1, R) > \mathcal{G}(P_2, R)$, it can be obtained that

$$A(I + P_1 C^T R^{-1} C)^{-1} P_1 A^T > A(I + P_2 C^T R^{-1} C)^{-1} P_2 A^T \tag{5.26}$$

Obviously,

$$(I + P_1 C^T R^{-1} C)^{-1} P_1 > (I + P_2 C^T R^{-1} C)^{-1} P_2 \tag{5.27}$$

Calculating the inverse matrix at the both sides of Eq. (5.27), it can be obtained that

$$P_1^{-1}(I + P_1 C^T R^{-1} C) < P_2^{-1}(I + P_2 C^T R^{-1} C) \tag{5.28}$$

Then $P_1^{-1} < P_2^{-1}$. Therefore, $P_1 > P_2$ Thus, for arbitrary positive matrices P_1 and P_2, $\mathcal{G}(P_1, R) > \mathcal{G}(P_2, R)$ can be obtained if and only if $P_1 > P_2$.

For arbitrary positive matrices R_1 and R_2, provided $\mathcal{G}(P, R_1) > \mathcal{G}(P, R_2)$, it can be obtained that

$$P^{-1}(I + PC^T R_1^{-1} C) < P^{-1}(I + PC^T R_2^{-1} C) \tag{5.29}$$

Therefore, it can be obtained that

$$R_1 > R_2 \tag{5.30}$$

For arbitrary positive matrices R_1 and R_2, $\mathcal{G}(P, R_1) > \mathcal{G}(P, R_2)$ can be obtained if and only if $R_1 > R_2$.

From Theorem 5.2, it can be obtained that

$$\hat{R}_k - R \rightarrow \mathbf{0} \quad as \quad k \rightarrow \infty \tag{5.31}$$

$$\hat{Q}_k - Q \rightarrow \mathbf{0} \quad as \quad k \rightarrow \infty \tag{5.32}$$

Since $R, Q > \mathbf{0}$. There exists a positive matrix R_M and Q_M, which satisfies that

$$\hat{R}_k \leq R_M \tag{5.33}$$

$$\hat{Q}_k \leq Q_M \tag{5.34}$$

Suppose there exists a sequence $\{\hat{P}_{k,k-1}, k = 1, 2, \ldots\}$, which is generated from the initial value P_0 with the following equation:

$$\hat{P}_{k+1,k} = \mathcal{G}(\hat{P}_{k,k-1}, \text{cov}_k(v)) + B\hat{Q}_k B^T \tag{5.35}$$

Consider also a sequence $\{\hat{P}^M_{k,k-1}, k = 1, 2, \ldots\}$ which is generated from the initial value P_0 with the following equation:

$$\hat{P}^M_{k+1,k} = \mathcal{G}(\hat{P}^M_{k,k-1}, R_M) + B\hat{Q}_M B^T \tag{5.36}$$

Since the sequence $\{\hat{P}_{k,k-1}\}$ and $\{\hat{P}^M_{k,k-1}\}$ have the same initial value P_0 and the function $\mathcal{G}(\cdot)$ is monotonically increasing respect to P and R, it can be obtained that

$$\hat{P}_{k,k-1} \le \hat{P}^M_{k,k-1} \quad k = 1, 2, \ldots \tag{5.37}$$

Since the system (5.1) is controllable and observable, the sequence $\{\hat{P}^M_{k,k-1}, k = 1, 2, \ldots\}$ is a convergent sequence. Since $\hat{P}_{k,k-1} > \mathbf{0}$ and $\hat{P}_{k,k-1} \le \hat{P}^M_{k,k-1}$, it can be obtained that the sequence $\{\hat{P}_{k,k-1}, k = 1, 2, \ldots\}$ is a convergent sequence.

That is to say, $\lim\limits_{k\to\infty} \hat{P}_{k+1,k}$ exists. Then it can be obtained that

$$\begin{aligned} \lim_{k\to\infty} \hat{P}_{k+1,k} &= \lim_{k\to\infty} (\mathcal{G}(\hat{P}_{k,k-1}, \mathrm{cov}_k(v)) + B\hat{Q}_k B^T) \\ &= \lim_{k\to\infty} (\mathcal{G}(\hat{P}_{k,k-1}, R) + B\hat{Q}_k B^T) \end{aligned} \tag{5.38}$$

Since the LTI system (5.1) is controllable and observable, there exists P_{ass} which satisfies the Riccati equation

$$P_{ass} = \mathcal{G}(P_{ass}, R) + BQB^T \tag{5.39}$$

Supposing that

$$\lim_{k\to\infty} \hat{P}_{k+1,k} = P'_{ass} \tag{5.40}$$

from Eq. (5.38), it can be obtained that

$$P'_{ass} = \mathcal{G}(P'_{ass}, R) + BQB^T \tag{5.41}$$

Since the solution of Riccati equation (5.39) is unique, it can be obtained that

$$P'_{ass} = P_{ass} \tag{5.42}$$

Obviously, it can be obtained that

$$\lim_{k\to\infty} (\hat{P}_{k+1,k} - P_{ass}) = \mathbf{0} \tag{5.43}$$

From Eq. (5.43), it can be obtained that

$$\limsup_{k\to\infty} \left\| \hat{P}_{k,k-1} \right\| < \infty \tag{5.44}$$

and

$$\hat{P}_{k,k-1} - P_{ass} \to \mathbf{0} \quad as \quad k \to \infty \tag{5.45}$$

Since

$$\hat{P}_{k,k-1} - P_{k,k-1} = (\hat{P}_{k,k-1} - P_{ass}) - (P_{k,k-1} - P_{ass}) \tag{5.46}$$

it can be obtained that

$$\hat{P}_{k,k-1} - P_{k,k-1} \to \mathbf{0} \quad as \quad k \to \infty \tag{5.47}$$

□

Theorem 5.4 *Under Assumptions 5.1–5.2 for the discrete-time LTI system* (5.7), *using Algorithm 6, the Kalman gain $\hat{K}_k - K_k \to 0$ as $k \to \infty$, where $\hat{K}_k$ is the Kalman gain obtained from the MKF-RCE algorithm and K_k from the ideal Kalman filter with exact system parameters.*

Proof Introducing the notations

$$K_{ass} = P_{ass} C^T (C P_{ass} C^T + R)^{-1} \tag{5.48}$$

where P_{ass} is the solution of the Riccati equation (5.39). From the improved Kalman filter with recursive covariance estimation algorithm, it can be obtained that

$$\hat{K}_k = \hat{P}_{k,k-1} C^T (C \hat{P}_{k,k-1} C^T + R)^{-1} \tag{5.49}$$

From the ideal Kalman filter algorithm, it can be obtained that

$$K_k = P_{k,k-1} C^T (C P_{k,k-1} C^T + R)^{-1} \tag{5.50}$$

Since $\hat{P}_{k,k-1} - P_{ass} \to \mathbf{0}$ and $P_{k,k-1} - P_{ass} \to \mathbf{0}$ as $k \to \infty$, it can be obtained that

$$\hat{K}_k - K_{ass} \to \mathbf{0},\ K_k - K_{ass} \to \mathbf{0} \tag{5.51}$$

as $k \to \infty$.

Therefore,

$$\hat{K}_k - K_k \to \mathbf{0} \quad as \quad k \to \infty \tag{5.52}$$

□

Lemma 5.1 ([23]) *Consider a dynamic error system*

$$\delta_k = F_k \delta_{k-1} + u_k \tag{5.53}$$

where $k = 1, 2, \ldots$, $\delta_k \in \boldsymbol{R}^m$ is the output, $u_k \in \boldsymbol{R}^m$ is the input. Assume that $F_k \to F$ as $k \to \infty$, where F is a stable matrix and F_k are stable. If u_k is bounded, then δ_k is bounded. If $u_k \to 0$, as $k \to \infty$, then $\delta_k \to 0$, as $k \to \infty$ [23].

Theorem 5.5 *Under Assumptions 5.1–5.2 for the discrete-time LTI system* (5.7), *using Algorithm 6, the sequence of state estimates* $\{\hat{x}_k\}$ *asymptotically approaches to the sequence of* $\{\hat{x}_k^{op}\}$ *as* $k \to \infty$, *where* $\hat{x}_k$ *is the state estimate from MKF-RCE algorithm and* $\hat{x}_k^{op}$ *is the optimal state estimate from the ideal Kalman filter with exact system parameters.*

Proof For the LTI system (5.7), real-time state estimate can be obtained by MKF-RCE algorithm. State estimate by MKF-RCE is as follows:

$$\hat{x}_k = A\hat{x}_{k-1} + \hat{K}_k[y_k - CA\hat{x}_{k-1}] \tag{5.54}$$

The optimal state estimate given by Kalman filter with R is as follows:

$$\hat{x}_k^{op} = A\hat{x}_{k-1}^{op} + K_k[y_k - CA\hat{x}_{k-1}^{op}] \tag{5.55}$$

Then from Eqs. (5.54) and (5.55), it can be obtained that

$$\begin{aligned} \hat{x}_k - \hat{x}_k^{op} = & (I - K_k C)A(\hat{x}_{k-1} - \hat{x}_{k-1}^{op}) \\ & +(\hat{K}_k - K_k)(y_k - CA\hat{x}_{k-1}) \end{aligned} \tag{5.56}$$

Introducing the notation

$$\delta_k = \hat{x}_k - \hat{x}_k^{op} \tag{5.57}$$

$$\Phi_k = (I - K_k C)A \tag{5.58}$$

$$u_k = (\hat{K}_k - K_k)(y_k - CA\hat{x}_{k-1}) \tag{5.59}$$

Then

$$\delta_k = \Phi_k \delta_{k-1} + u_k \tag{5.60}$$

Since the LTI system (5.7) is controllable and observable, Φ_k is always a stable matrix.

From Algorithm 6, it can be obtained that $\hat{x}_{k-1}$ is the unbiased estimate for the state x_{k-1} at the $(k-1)$th step if $\mathrm{E}(x_0) = x_0$, the initial state estimation. Then

$$\mathrm{E}(y_k - CA\hat{x}_{k-1}) = 0 \tag{5.61}$$

From Theorem 5.3, it can be obtained that $\hat{P}_{k,k-1} - P_{k,k-1} \to 0$ and $\hat{P}_{k,k-1} - P_{ass} \to 0$ as $k \to \infty$. So there exists a covariance matrix P_{sup} which satisfied that

$$\mathrm{cov}(y_k - CA\hat{x}_{k-1}) \le P_{sup} \tag{5.62}$$

From Eqs. (5.61) and (5.62), it can be obtained that

$$\mathrm{E}(u_k) = 0 \tag{5.63}$$

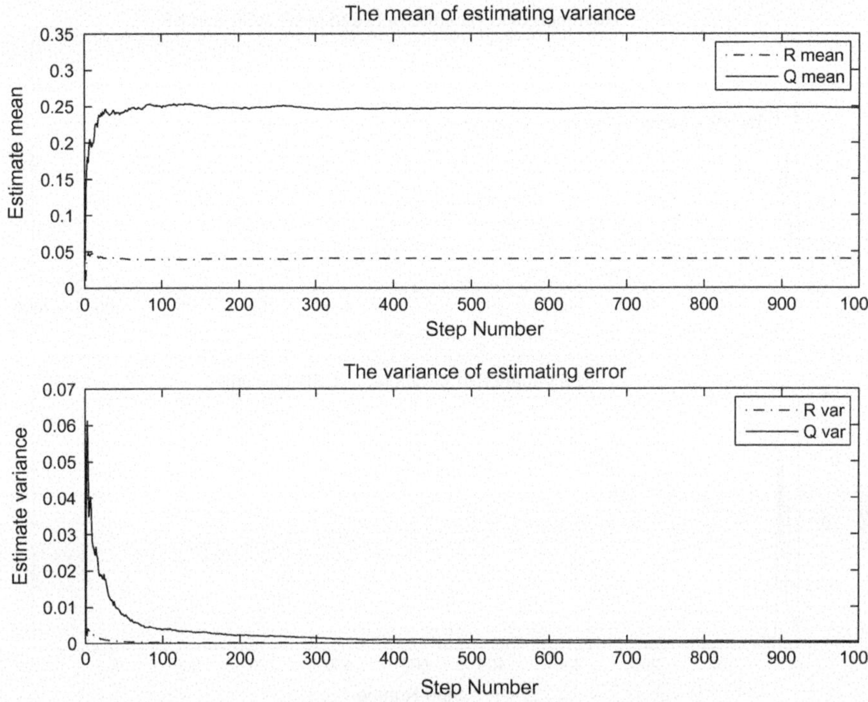

Fig. 5.1 The mean of estimating error for covariance matrix with MKF-RCE

and

$$\text{cov}u_k \leq P_{sup}(\hat{K}_k - K_k)^2 \tag{5.64}$$

From Theorem 5.4, it can be obtained that

$$\hat{K}_k - K_k \to \mathbf{0} \quad as \quad k \to \infty \tag{5.65}$$

So $u_k \to \mathbf{0}$ as $k \to \infty$ and $\Phi(k)$ is stable. And $\Phi_k \to \Phi$ as $k \to \infty$.
From *Lemma* 5.1, it can be obtained that

$$\delta_k \to \mathbf{0} \quad as \quad k \to \infty \tag{5.66}$$

That is to say, $\hat{x}_k - \hat{x}_k^{op} \to \mathbf{0}$ as $k \to \infty$. □

From Theorems 5.2–5.5, it can be obtained that $\hat{x}_k$ is convergent to the optimal state estimate $\hat{x}_k^{op}$, where $\hat{x}_k$ is obtained by the KF-RCE algorithm, and $\hat{x}_k^{op}$ is obtained by the ideal Kalman filter with exact system parameters. In other words, $\hat{x}_k - \hat{x}_k^{op} \to 0$ as $k \to \infty$.

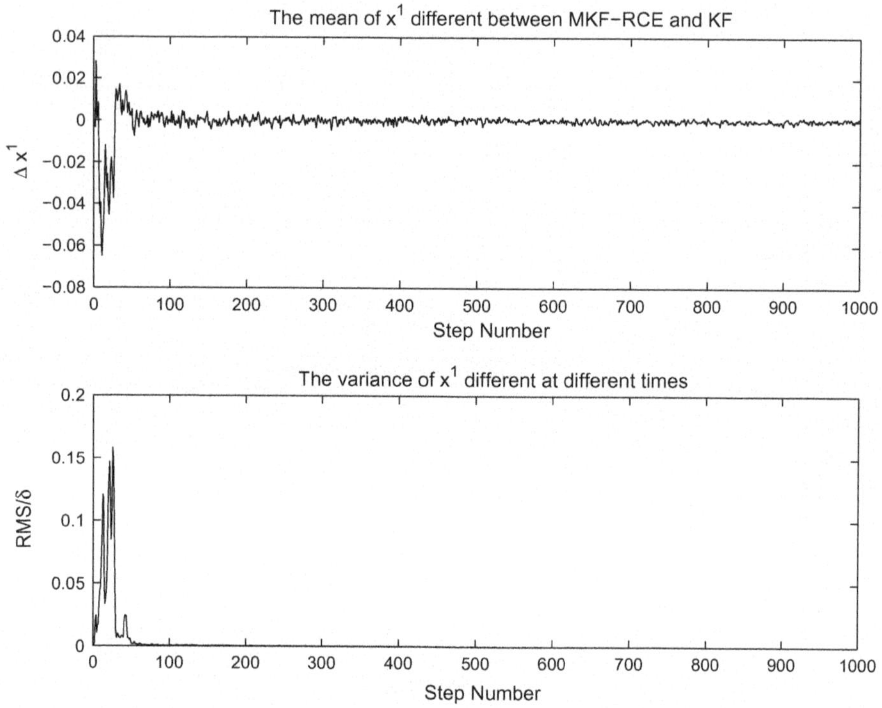

Fig. 5.2 The mean and variance of estimating error for $\hat{x}_1$

5.3.4 *Simulation Study*

In this subsection, simulation results of the gyroscope system model proposed in this chapter are reported to illustrate the effects of the MKF-RCE algorithm in the gyroscope denoising problem when the process and measurement covariance matrices are completely unknown a priori. The system model used in this simulation experiment is as follows:

$$\begin{cases} \begin{bmatrix} x_k \\ s_k \end{bmatrix} = \begin{bmatrix} 1 & T \\ 0 & 1 \end{bmatrix} \begin{bmatrix} x_{k-1} \\ s_{k-1} \end{bmatrix} + \begin{bmatrix} 0 \\ 1 \end{bmatrix} w_{k-1} \\ y_k = \begin{bmatrix} 1 & 0 \end{bmatrix} \begin{bmatrix} x_k \\ s_k \end{bmatrix} + v_k \end{cases} \tag{5.67}$$

where $T = 1s$, $\mathrm{E}(w_k) = 0$, $Q = 0.25$, $\mathrm{E}(v_k) = 0$, $R = 0.04$, w_k is unrelated with v_k. Here $\mathrm{E}(\cdot)$ is the expectation. The simulations suppose that Q and R are completely unknown a priori.

The simulation experiment is carried out based on Monte Carlo method, and the Monte Carlo number is 100. The simulation results are shown in Figs. 5.1, 5.2, and 5.3.

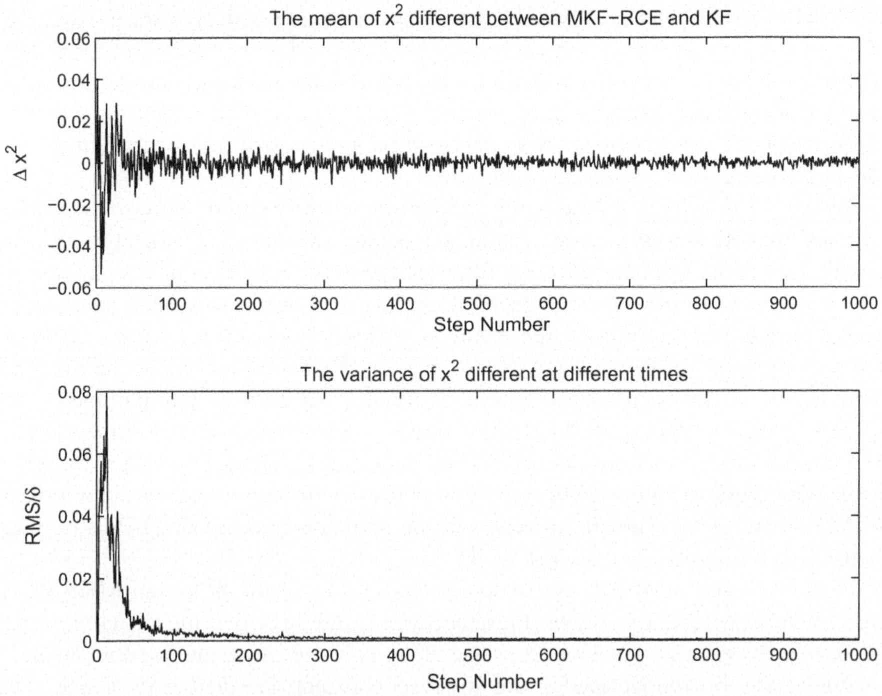

Fig. 5.3 The mean and variance of estimating error for $\hat{x}_2$

In Figs. 5.1, 5.2, and 5.3, estimating mean means the average value of Monte Carlo simulation at each step number and estimating variance means the variance of Monte Carlo sample results at each step number. From the simulation results, it can be seen that $\hat{Q}_k$ and $\hat{R}_k$ are quickly convergent to the real value of Q and R as step number $k \to \infty$. The state estimate $\hat{x}_k$ obtained with MKF-RCE is also quickly convergent to the ideal Kalman filter estimate $\hat{x}_k^{op}$ with exact system parameters, which is consistent with the stability analysis results.

5.4 Experimental Tests

In the inertial navigation systems, inertial devices play an important role as the kernel. Generally speaking, the precision of inertial navigation system mainly depends on the measurement accuracy of inertial devices. Compared with the accelerometer, the gyroscope is more important in the inertial navigation systems considering that accelerometer measurement accuracy is generally higher than the gyroscope in most inertial navigation system.

MEMS gyroscope is usually low-cost, small-size, and relatively easy to make for mass production; thus, in recent years there are more and more engineering systems, such as mobile phones, using MEMS gyroscope to obtain the angular rate signal in recent years. However, the one thing which cannot be ignored is that the measurement precision of MEMS gyroscope may not be good enough and the accuracy may not satisfy the demand of various applications.

Recently, digital signal processing technology is always used to improve the stability of the MEMS gyroscope output. Especially, Kalman filtering algorithm is usually treated as a standard digital filter for the MEMS gyroscope output signal. Before using the Kalman filtering technology, a system model with fixed parameters was often adopted to describe the behavior of MEMS gyroscope output. In most engineering applications, linear ARMA model is often chosen to represent the performance of MEMS gyroscope measurement output. But the system parameters of ARMA model should be absolutely determined a priori based on the offline sample data obtained by certain driven signal. If the real-time behavior of the MEMS gyroscope measurement signal is not consistent with the assumptions a priori, the certain ARMA model is not suitable to represent the dynamic process of MEMS output, then Kalman filtering cannot work well.

In this section, a MEMS gyroscope is chosen to test the proposed MKF-RCE algorithm. For MEMS gyroscope, the existence of random error is the most important factor to influence the precision. Because of the nature of uncertain random error, it cannot be simply compensated as unknown yet constant zero drift error; hence, it is the main error to limit measurement precision. Therefore, it is particularly important to reduce random error to improve measurement accuracy of MEMS gyroscope. In our experiments, MEMS130 gyroscope in the condition of 25°C is adopted to test the practical performance of MKF-RCE algorithm. Shaft angle rate information of the measurement output data of MEMS130 gyroscope is used as original data for MKF-RCE algorithm. The sampling period is $T = 5$ ms in all experiments.

Fig. 5.4 MEMS130 gyroscope and the turntable designed for the experiment tests

Fig. 5.5 The controlling box of the turntable designed for the experiment tests

The MEMS130 gyroscope and turntable are assembled as in Figs. 5.4 and 5.5. In order to obtain the acceptable experiments results, MEMS130 gyroscope has been fixed on the turntable with a working fixture.

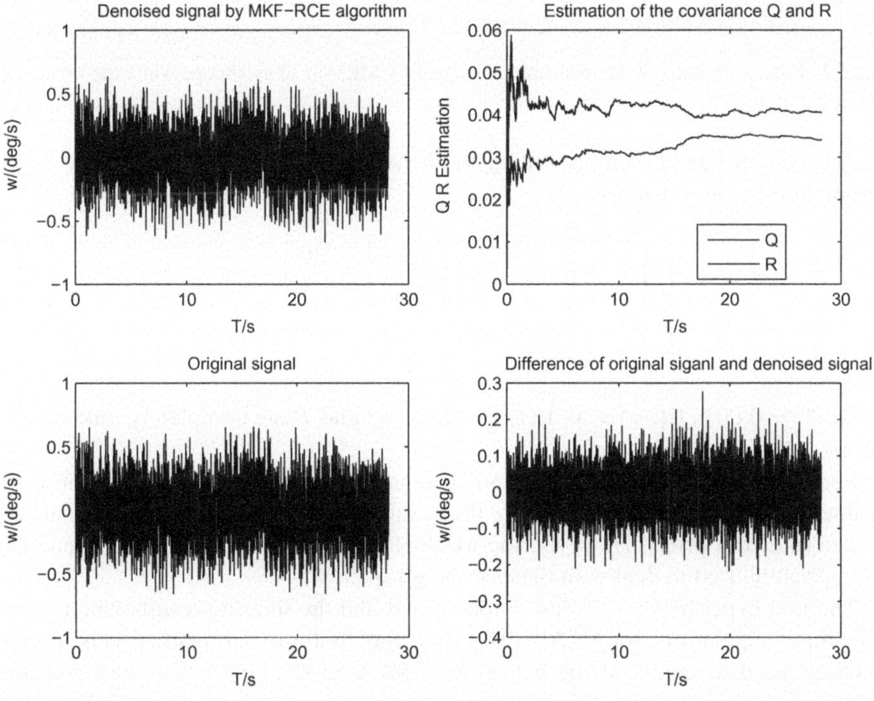

Fig. 5.6 Filtering results of the measurement signal by MEMS130 gyroscope while there is no input angular rate

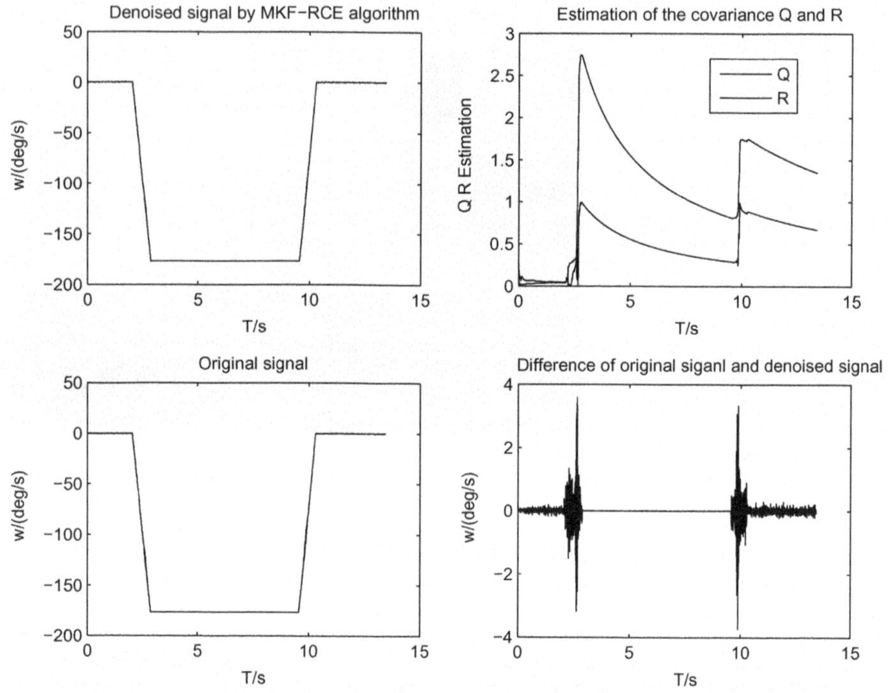

Fig. 5.7 Filtering results of the measurement signal by MEMS130 gyroscope while the input is a constant angular rate

The system model of representing the behavior of MEMS130 gyroscope measurement is given as follows:

$$\begin{cases} \begin{bmatrix} x_k \\ s_k \end{bmatrix} = \begin{bmatrix} 1 & T \\ 0 & 1 \end{bmatrix} \begin{bmatrix} x_{k-1} \\ s_{k-1} \end{bmatrix} + \begin{bmatrix} 0 \\ 1 \end{bmatrix} w_{k-1} \\ y_k = \begin{bmatrix} 1 & 0 \end{bmatrix} \begin{bmatrix} x_k \\ s_k \end{bmatrix} + v_k \end{cases} \tag{5.68}$$

where $T = 0.005\text{s}$ $\mathrm{E}(w_k) = 0$, $\mathrm{E}(v_k) = 0$, but Q and R are completely unknown a priori.

For the system (5.68), the standard Kalman filter is not suitable any more to deal with the denoising problem since the covariance matrix of process noise cannot be determined a priori. However, the MKF-RCE presented in this chapter can be appropriately used to deal with the case as in Eq. (5.68).

The first experiment is testing output signal and the filtering results when there is no input angular rate for MEMS130 gyroscope. In this experiment, system model (5.68) is used to represent the behavior of MEMS130 gyroscope output and the covariance matrices of process and measurement noises are completely unknown a

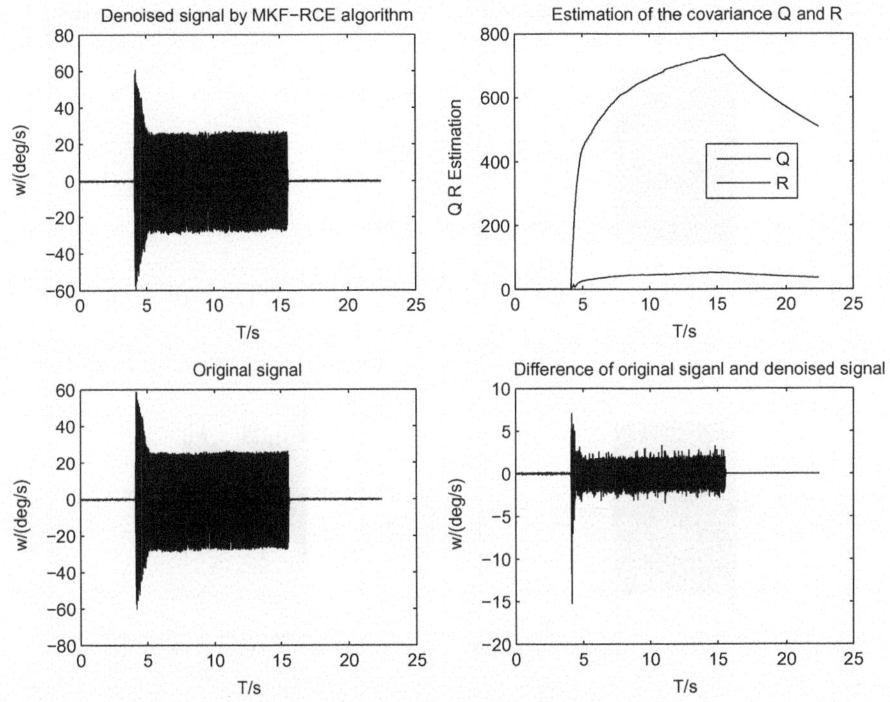

Fig. 5.8 Filtering results of the measurement signal by MEMS130 gyroscope while the input is a swing angular rate with 20 Hz frequency of 0° amplitude

priori. The measurement signal filtering results are depicted in Fig. 5.6 for the case when there is no input angular rate.

From the experimental results, it can be seen obviously that when there is no input angular rate or there is not much change in the dynamic uncertainty of system state, the estimation of covariance matrices of process and measurement noises will be convergent to a certain value and MKF-RCE algorithm can be effectively used to improve the measurement accuracy of MEMS130 gyroscope output signal. The system model (5.68) and MKF-RCE algorithm could be used effectively to decrease the random error of gyroscope angular velocity measurement.

Remark 5.5 In all experiments, the constant zero draft of MEMS130 gyroscope has been compensated before the filtering algorithm is applied. In other words, the effect caused by the constant zero draft of MEMS130 gyroscope can be ignored in all experiments.

The second experiment is testing the physical model (5.68) and MKF-RCE algorithm while the input is a constant angular rate for MEMS130 gyroscope. The filtering results are shown in Fig. 5.7.

The estimation results can follow the tend of input angular rate, and MKF-RCE algorithm could improve the measurement accuracy of MEMS130 gyroscope. The

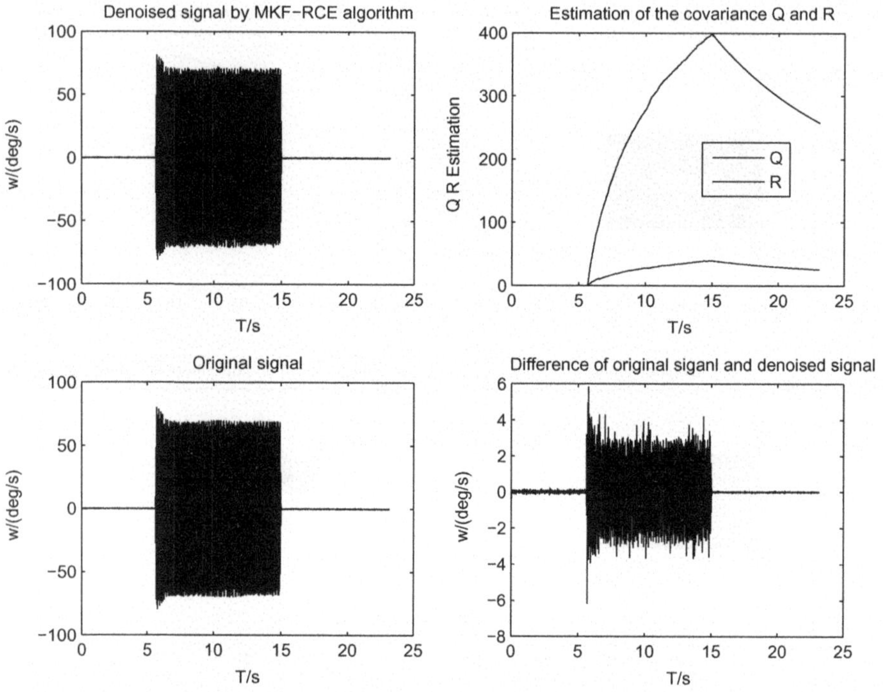

Fig. 5.9 Filtering results of the measurement signal by MEMS130 gyroscope while the input is a swing angular rate with 10 Hz frequency of 1° amplitude

system model (5.68) and MKF-RCE algorithm are suitable for tracking the trend of angular velocity in the gyroscope measurements.

The third experiment is designed to test the filtering effect for MEMS130 gyroscope while the input signal is a swing angular rate with 20 Hz frequency of 0.3° amplitude or 10 Hz frequency of 1° amplitude. The estimating results are shown in Figs. 5.8 and 5.9.

From Figs. 5.8 and 5.9, one conclusion could be obviously obtained that the system (5.68) is suitable to represent the behavior of MEMS130 gyroscope output and MKF-RCE algorithm can calculate the proper estimation of noise covariance matrices as the dynamic uncertainty is changing with the input angular rate. The filtering results of MKF-RCE algorithm can follow the trend of input angular rate and improve the measurement accuracy of MEMS130 gyroscope.

From the testing experiments above, it can be obviously obtained that the physical system model (5.68) is suitable to describe the behavior of MEMS130 gyroscope output and MKF-RCE algorithm is effectively used to deal with the case when the covariance matrices of process and measurement noises are completely unknown a priori for the physical system model.

5.5 Conclusion

Kalman filter is a widely used algorithm in the inertial navigation systems, especially for the purpose of improving the precision of inertial devices. However, in order to use standard Kalman filter to deal with the denoising problem, the exact system model and parameters must be determined a priori. Considering the real-time demands, a simple physical system model instead of classic ARMA model has been proposed to describe the behavior of the gyroscope measurement data. But for this physical system model, the covariance matrix of process noise is completely unknown a priori, which result in difficulties and challenges in applying standard Kalman filter.

To resolve such challenging practical problems, which can also be seen in many other applications requiring Kalman filtering technology, a modified Kalman filter named MKF-RCE, Kalman filter with recursive covariance estimation, has been presented to cope with the estimation problem for the physical model, and one typical simulation example has been presented to illustrate the effectiveness of the MKF-RCE algorithm for denoising the measurement signal with the proposed physical model. Finally, three testing experiments using MEMS130 gyroscope under different conditions were presented to illustrate that the MKF-RCE algorithm is useful and practically effective for the denoising problem when the covariance matrices of process and measurement noises are completely unknown in advance.

References

1. L. Zhao, Z.F. He, An in-coordinate interval adaptive Kalman filtering algorithm for INS/GPS/SMNS, in *Proceedings of 2012 10th IEEE International Conference on Industrial Informatics*, pp. 41–44
2. J.L. Li, F. Jiao, J.C. Fang, Y.H. Ma, Integrated calibration method for dithered RLG POS using a hybrid analytic/Kalman filter approach. IEEE Trans. Instrum. Meas. **62**(12), 3333–3342 (2013)
3. J. Vasconcelos, C. Silvestre, P. Oliveira, B. Guerreiro, Embedded UAV model and LASER aiding techniques for inertial navigation systems. Control Eng. Pract. **18**(3), 262–278 (2010)
4. J. Bijker, W. Steyn, Kalman filter configurations for a low-cost loosely integrated inertial navigation system on an airship. Control Eng. Pract. **16**(12), 1509–1518 (2008)
5. D. Berdjaga, J. Cieslak, A. Zolghadri, Fault diagnosis and monitoring of oscillatory failure case in aircraft inertial system. Control Eng. Pract. **20**(12), 1337–1352 (2012)
6. B. Feng, Q. Li, P.P. Yang, G.D. Fu, On-line realization of multi-scale soft-threshold de-noising of MEMS gyro by wavelet, in *Proceedings of the 2009 9th International Conference on Electronic Measurement and Instruments*, pp. 2442–2445
7. A. Vaccarella, E.D. Momi, A. Enquobahrie, G. Ferrigno, Unscented Kalman filter based sensor fusion for robust optical and electromagnetic tracking in surgical navigation. IEEE Trans. Instrum. Meas. **62**(7), 2067–2081 (2013)
8. B.S. Koh, J. Junkins, Kalman filter for linear fractional order systems. J. Guid., Control Dyn. **35**(6), 1816–1827 (2012)
9. R.E. Kalman, A new approach to linear filtering and prediction problems. Trans. ASME-J. Basic Eng. **82**(Series D), 35–45 (1960)
10. A.H. Jazwinski, *Stochastic Processes and Filtering Theory* (Academic Press, New York, 1970)
11. R.E. Kalman, R.S. Bucy, New results in linear filtering and prediction theory. Trans. ASME Ser. D, J. Basic Eng. **83**, 95–107 (1961)

12. L. Ljung, *System Identification, Theory for the User*, 2nd edn. (Prentice-Hall, New York, 1999)
13. A. Gelb, *Applied Optimal Estimation* (M.I.T Press, Cambridge, 2001)
14. O. Rosén, A. Medvedev, T. Wigren, Parallelization of the Kalman filter on multicore computational platforms. Control. Eng. Pract. **21**(9), 1188–1194 (2013)
15. Y.Q. Xia, J.J. Yan, J.Z. Shang, M.Y. Fu, B. Liu, Stabilization of quantized systems based on Kalman filter. Control. Eng. Pract. **20**(10), 954–962 (2012)
16. C. Guardiola, B. Pla, D. Blanco-Rodriguez, L. Eriksson, A computationally efficient Kalman filter based estimator for updating look-up tables applied to no_x estimation in diesel engines. Control. Eng. Pract. **21**(11), 1455–1468 (2013)
17. J.D. Searcy, H.J. Pernicka, Magnetometer-only attitude determination using novel two-step Kalman filter approach. AIAA J. Guid. Control Dyn. **35**(6), 1693–1701 (2012)
18. R. Zanetti, R.H. Bishop, Kalman filters with uncompensated biases. AIAA J. Guid. Control Dyn. **35**(1), 327–330 (2012)
19. M. Joerger, B. Pervan, Kalman filter-based integrity monitoring against sensor faults. AIAA J. Guid. Control Dyn. **36**(2), 349–361 (2013)
20. S.M. Persson, I. Sharf, Invariant trapezoidal Kalman filter for application to attitude estimation. AIAA J. Guid. Control Dyn. **36**(3), 721–733 (2013)
21. Q.H. Chi, Z.S. Fei, Z. Zhao, L. Zhao, J. Liang, A model predictive control approach with relevant identification in dynamic PLS framework. Control Eng. Pract. **22**, 181–193 (2014)
22. B. Feng, H.B. Ma, M.Y. Fu, S.T. Wang, A framework of finite-model Kalman filter with case study: MVDP-FMKF algorithm. Acta Autom. Sin. **39**(8), 1246–1256 (2013)
23. B. Feng, M.Y. Fu, H.B. Ma, Y. Xia, Kalman filter with recursive covariance estimation–sequentially estimating process noise covariance. IEEE Trans. Ind. Electron. **61**(11), 6253–6263 (2014)
24. X. Xiao, B. Feng, B. Wang, On-line realization of SVM Kalman filter for MEMS gyro, in *Proceedings of the 3rd International Conference on Measuring Technology and Mechatronics Automation*, pp. 768–770

Chapter 6
Real-Time State Estimator Without Noise Covariance Matrices Knowledge

The digital filtering technology has been widely applied in a majority of signal processing applications. For the linear systems with state-space model, Kalman filter provides optimal state estimates in the sense of minimum mean squared errors and maximum likelihood estimation. But only with accurate system parameters and noise statistical properties, the estimation obtained by standard Kalman filter is the optimal state estimate. Most of time, the exact noise statistical properties could not be obtained as a priori information or even wrong statistical properties may be captured by the offline method. This may lead to a poor performance (even divergence) of Kalman filtering algorithm. In this chapter, a novel real-time filter, named as fast minimum norm filtering algorithm, has been proposed to deal with the case when the covariance matrices of the process and measurement noises were unknown in the linear time-invariant systems with state-space model. Tests have been performed on numerical examples to illustrate that the fast minimum norm filtering algorithm could be used to obtain acceptable precision state estimation in comparison with the standard Kalman filter for the discrete-time linear time-invariant systems.

6.1 Introduction

State estimation is a key part in most applications of model-based control techniques since feedback control is usually designed based on accurate state information in most existing control approaches such as model predictive control (MPC), linear-quadratic regulator (LQR), and so on. For example, in [1] a high-gain observer is adopted to cope with the problem of estimating partial state information for the purpose of synchronized tracking control, and it is well known that linear state observers can be designed to generate asymptotically accurate state estimates for a deterministic linear system. For stochastic linear systems with state-space model, Kalman filtering algorithm is known as a common optimal state estimating method and has been widely used in many applications, such as navigation [2], communication [3], and fault diagnosis [4]. In Kalman's pioneering work [5, 6], state estimate and prediction

H. Ma et al., *Kalman Filtering and Information Fusion*,
https://doi.org/10.1007/978-981-15-0806-6_6

problems were described in a different version; he also proposed a method for the optimal solution of this general problem when system model is linear, precisely known and the statistical properties of process and measurement noises are obtained precisely in advance.

However, the requirements of Kalman filtering algorithm can be seldom completely satisfied in the practical engineering systems for a variety of reasons. In order to use Kalman filtering algorithm to deal with different state estimating problems, a series of modified Kalman filter algorithms were proposed to resolve certain practical issues. Stanley F. Schmidt proposed extend Kalman filter (EKF) [7, 8] algorithm to solve filtering and prediction problems of nonlinear systems encountered in Apollo program in the 1960s. Based on the idea of Taylor expansion, EKF is designed to approximate nonlinear system by linear model and then use standard Kalman filter to estimate system states for the linearized system model in real time. Such a modified filtering algorithm was proved to be very useful in many applications, and hence EKF algorithm was ever regarded as a kind of standard filtering method for nonlinear systems for a long time and has attracted huge amount of research interests [9, 10]. Then Carlson [11] presented federal Kalman filter (FKF) [12] algorithm to deal with multisystem optimal data fusion. Based on the covariance matrix of the estimation error, different estimations from various measurement systems were used to obtain more precise real-time state estimate for multisystem with state-space model.

Since the estimation obtained by EKF algorithm has unacceptable precision when system model is strongly nonlinear, unscented Kalman filter (UKF) [13] was introduced by S. J. Julier and J. K. Uhlmann in 1997, and later applied in various areas such as induction motor drives [14], UAV attitude estimation [15], and so on. In recent years, some variants of UKF were proposed in the literature [16, 17]. When the system model has uncertainty in the parameters, the modified Kalman filter algorithms above were not suitable to be used to obtain state estimation in real time. Robust Kalman filter (RKF) [18–21] was proposed to solve the problems when the transfer and/or measurement matrices have uncertainty error with a bounded norm in a linear system with state-space model. Adaptive Kalman filter (AKF) [22, 23] was used to deal with the cases where process and measurement noise statistical properties are uncertain.

In this chapter, we will present a new algorithm to solve the estimating problem for the discrete-time linear time-invariant systems when the covariance matrices of both the process and measurement noises were completely unknown. Since the process and measurement noise covariance matrices could not be obtained in advance, the standard Kalman filter and adaptive Kalman filter became not suitable to be used to solve the estimating problem in such cases any more. Because most modified filtering algorithms are in fact based on Kalman filter, the results from these modified filters are thus always suboptimal. Referring to the framework of standard Kalman filter, the so-called fast minimum norm filtering algorithm proposed in this chapter tries to design a new framework to solve the filtering problem when the covariance matrices of process and measurement noises were absent in the state-space model. Finally, numerical simulation examples were presented to illustrate that the fast minimum norm filtering algorithm can effectively deal with the problem of state estimation when the process and measurement noise covariance matrices were completely unknown for the discrete-time linear time-invariant systems.

The rest of this chapter is organized as follows: in Sect. 6.2, problem formulation is performed for the fast minimum norm filtering algorithm and some assumptions are made for convenience of further analysis. In Sect. 6.3, the fast minimum norm filtering algorithm is presented to obtain state estimation of discrete-time linear time-invariant systems in the absence of process and measurement covariance matrices. For the convenience of discussion, the problem has been classified into four different cases; In Sect. 6.4, numerical simulation studies on a number examples will be carried out to illustrate the effectiveness of the fast minimum norm filtering algorithm; finally, in Sect. 6.5 concluding remarks are presented to summarize the fast minimum norm filtering algorithm and our prospective further work to improve the proposed algorithm.

6.2 Problem Formulation

A linear state-space model is often used in practice to approximate true plants. Consider the following typical state-space model

$$\begin{cases} x_k = A_k x_{k-1} + B_k u_{k-1} + w_{k-1} \\ y_k = C_k x_k + D_k u_{k-1} + v_k \end{cases} \tag{6.1}$$

where x_k is the system state vector, u_k represents the system input, A_k stands for the state transfer matrix, C_k is the measurement matrix, y_k is the system measurement, w_k is the process noise, and v_k is the measurement noise.

For linear plant (6.1), the control signal u_k is often designed as a linear controller, e.g., $u_k = K_k x_k$, thus we may obtain the resulting closed-loop system without explicit control inputs. Hence, without loss of generality, we assume that $u_k \equiv 0$. The system model should be simplified as follows:

$$\begin{cases} x_k = A_k x_{k-1} + w_{k-1} \\ y_k = C_k x_k + v_k \end{cases} \tag{6.2}$$

The linear filtering problem is trying to obtain system state estimates in real time according to the measurement sequence based on the system model. Kalman filter could obtain the optimal state estimation based on the measurement sequence when the state-space model of the linear system and the statistical properties of both the process and measurement noises have been precisely obtained. However, the a priori requirements of Kalman filter can be hardly completely satisfied in practice. This may degrade the performance of the Kalman filter or even cause the estimation error divergent with the incorrect system parameters. In general, state transfer matrix A_k could be often obtained by the relationship of the system states or kinetic relationship among each variable of the state. The measurement matrix C_k is on behalf of the mathematical relationship between the system state and measurement signal. The system process noise sequence $\{w_k\}$ represents dynamic uncertainty. Since the filter deals with the measurement signal in real time, the statistical properties of the random variable w_k can be seldom determined in advance. The measurement noise sequence $\{v_k\}$ represents the credibility of the measurement signal. When the measurement signal is directly obtained from the sensors, the statistical properties of measurement noise could be determined by the accuracy of measurement physical sensors. In other cases, the statistical properties of the random variable v_k were hardly to be determined in practice. However, these statistical properties play an important role in the linear filtering algorithm, such as Kalman filter. Kalman filter is not suitable for obtaining the state estimates without the statistical properties of w_k and v_k.

In this chapter, we are trying to deal with the linear system state estimating problem when the statistical properties of w_k and w_k were completely unknown for the state-space system model.

Mathematically speaking, in this chapter we are going to investigate this fundamental problem under the following technical assumptions:

Assumption 6.1 For simplicity, supposing that the discrete-time linear system is time-invariant. Equivalently, the system could be described as follows:

$$\begin{cases} x_k = Ax_{k-1} + w_{k-1} \\ y_k = Cx_k + v_k \end{cases} \tag{6.3}$$

where x_k is the system state, A is the state transfer matrix, C is the measurement matrix, y_k is the system measurement, w_k is the process noise, and v_k is the measurement noise.

Assumption 6.2 For the discrete-time linear time-invariant (LTI) system (6.3), supposing that the transfer matrix A and measurement matrix C are constant and known, and the controllability (regarding the process noise as control input) and observability matrices

$$\mathcal{Q}_c = [e, Ae, \ldots, A^{p-1}e] \tag{6.4}$$

$$\mathcal{Q}_o = [C; CA; \ldots; CA^{p-1}] \tag{6.5}$$

are of full rank, where $e = [1, 1, \ldots, 1]^T$ and p is the least number such that the $\mathcal{Q}_c$ and $\mathcal{Q}_o$ are of full rank.

Assumption 6.3 The mathematical expectation of the variable w_k and v_k is equal to 0. The state transfer matrix A and measurement matrix C could be exactly obtained.

Problem 6.1 *In the absence of the statistical covariance matrices of processing and measurement noises, how to design a real-time filtering algorithm for the discrete-time linear time-invariant systems with state-space model as described in Eq. (6.3)?*

For the discrete-time linear time-invariant systems, the state estimation by Kalman filter is a kind of optimal estimation in the sense of least mean square and maximum likelihood when the system parameters and noise statistical properties could be determined a priori. However, the statistical properties of the process and measurement noises were hardly determined or even cannot be obtained in advance in most engineering systems. Then various modified filtering algorithms based on Kalman filter had been presented in order to use Kalman filter to deal with real-time practical state estimating problem. In engineering practice, one frequently used method is to use the upper bound of noise covariance matrix replacing the correct one in the Kalman filter. The side effect of this method is that the state estimation from the filter is not optimal any more, and this method may even cause the filter divergent if the upper value is not suitable for the system model. For the reasons above, the idea of adaptive control was introduced in state estimation problem to deal with the case where the noise statistical properties could not be completely determined a priori. According to the measurement sequence, adaptive regulating principle was designed a priori to adjust the noise covariance matrix based on the idea of adaptive control. The above two methods were not suitable to be used to obtain the state estimation if the process and measurement noise covariance matrices were completely unknown. Besides, all of these modified algorithms were based on the framework of standard Kalman filter and the results from these filters were always suboptimal and the stability is hard to be established in the mathematical principle.

In this chapter, we will focus on the filtering problem when the process and measurement covariance matrices were completely unknown and propose a new algorithm, named as fast minimum norm filtering algorithm, which is based on a new framework compared with the standard Kalman filter, and two theorems will be presented to illustrate that the covariance matrix of estimation error of the fast minimum norm filtering algorithm has a certain upper bound.

6.3 The Fast Minimum Norm Filtering Algorithm

In this section, a new algorithm will be presented to cope with the problem described in the previous section. When the mean of the noise random variable is equal to 0 and the covariance matrix is completely unknown in advance, we try to formulate a new framework which is different from Kalman filer and present a new algorithm to deal with the estimating problem without the knowledge noise on covariance matrices for the discrete-time linear time-invariant systems with state-space model.

For simplicity yet without loss of generality, the following linear time-invariant system is considered:

$$\begin{cases} x_k = Ax_{k-1} + w_{k-1} \\ y_k = Cx_k + v_k \end{cases} \tag{6.6}$$

where $w_k \in \boldsymbol{R}^{n\times 1}$, $x_k \in \boldsymbol{R}^{n\times 1}$, $v_k \in \boldsymbol{R}^{m\times 1}$, $y_k \in \boldsymbol{R}^{m\times 1}$, $A \in \boldsymbol{R}^{n\times n}$, $C \in \boldsymbol{R}^{m\times n}$ are process noise, state, measurement noise, output, transfer matrix, and output matrix, respectively. Supposing that $E(w_k) = 0$, $E(w_k) = 0$ and $\mathrm{Cov}(w_k) = Q$, $\mathrm{Cov}(v_k)=R$

but Q and R are completely unknown a priori, where $E(\cdot)$ is the mathematical expectation and $\mathrm{Cov}(\cdot)$ is the covariance matrix. We hope to estimate the unknown states $\{x_k\}$ with the only available output data.

To resolve Problem 6.1, a new algorithm named as fast minimum norm filtering algorithm will be presented in this section and the algorithm can be divided into two steps: (i) computing $\hat{x}_{k,k-p}$ based on previous measurements where $\hat{x}_{k,k-p}$ is the prediction of the system state x_k at the kth time step; (ii) using the new measurement to modify the prediction $\hat{x}_{k,k-p}$ and obtain the real-time estimation $\hat{x}_k$ at the kth time step based on the measurement up to and including the kth measurement data.

6.3.1 Time Update

From Assumption 6.2, it can be obtained that the system observation matrix is

$$Q_o = [C; CA; \ldots; CA^{p-1}] \tag{6.7}$$

and $\mathrm{Rank}(Q_o) = n$, where n is the dimension of the system state.

Introducing the notation

$$Q_{ob} \triangleq [C; CA; \ldots; CA^p] \tag{6.8}$$

it can be obtained that $\mathrm{Rank}(Q_{ob}) = n$.

Then the least square estimation of the variable x_{k-p} could be obtained as follows:

$$\hat{x}^{ls}_{k-p} = [Q^T_{ob} Q_{ob}]^{-1} Q^T_{ob} Y_k \tag{6.9}$$

where $Y_k = [y_{k-p}, y_{k-p+1}, \ldots, y_k]^T$.

The estimation of x_k at $(k-p)$-th step can be obtained as

$$\hat{x}_{k,k-p} = A^p \hat{x}^{ls}_{k-p} \tag{6.10}$$

From Assumption 6.3, we have $E[w_k] = 0$ and $E[v_k] = 0$. Then it can be obtained that $\hat{x}^{ls}_{k-p}$ is a kind of unbiased estimation of the variable x_{k-p}, and $\hat{x}_{k,k-p}$ is an unbiased estimation of x_k, according to the discrete-time linear time-invariant systems with state-space model as in Eq. (6.6).

6.3.2 Measurement Update

It is well known that Kalman filtering algorithm is a general method of using the measurement to modify the time prediction of state estimate as follows:

$$\begin{aligned}\hat{x}_k &= \hat{x}_{k,k-p} + \Delta\hat{x}_k \\ &= \hat{x}_{k,k-p} + K_k(y_k - C\hat{x}_{k,k-p})\end{aligned} \tag{6.11}$$

where K_k is the Kalman filter gain, i.e., the optimal gain under the mean of least square with exact system parameter. According to the formulation of Kalman filter, K_k is coupled with the noise covariance matrices Q and R. However, the optimal K_k is hard to obtain when the noise covariance matrices are completely unknown. For such cases, we are trying to present some effective choice of modified gain K_k to make the incremental correction $\Delta\hat{x}_k$ minimum norm and easy to be formulated in this chapter.

To this end, the measurement updating part is divided into four different cases, and we design different measurement updating methods separately for each case for the discrete-time linear time-invariant system (6.6).

Case I: $m \geq n$ and Rank$(C) = n$

For this case, it can be obtained that the measurement matrix is of full column rank. Referring to the framework of Kalman filter algorithm, time updating and measurement updating parts are redesigned for Problem (6.1).

In this case, since the measurement matrix C is of full column rank, the filtering gain can be obtained as follows:

$$K_k = [C^T C]^{-1} C^T \tag{6.12}$$

So the state estimation at the kth time step could be obtained as

$$\begin{aligned}\hat{x}_k &= \hat{x}_{k,k-1} + \Delta\hat{x}_k \\ &= \hat{x}_{k,k-p} + K_k(y_k - C\hat{x}_{k,k-p})\end{aligned} \tag{6.13}$$

Remark 6.1 To intuitively understand the gain in Eq. (6.12), a simple example would be presented in this remark when the measurement matrix $C = [1 \quad 1 \quad 1]^T$. From Eq. (6.12), we have that $K_k = [\frac{1}{3} \quad \frac{1}{3} \quad \frac{1}{3}]^T$, which means that

$$\Delta\hat{x}_k = [\frac{1}{3} \quad \frac{1}{3} \quad \frac{1}{3}]^T (y_k - C\hat{x}_{k,k-p}) \tag{6.14}$$

since the measurement noise covariance matrix is completely unknown. In such case, the K_k choice is consistent with the commonly used method of taking mean of sensor measurements.

Case II: $m \geq n$ and Rank$(c) = r < n$

It can be seen that this case is different from Case I and the solution for Case I is not suitable any more.

Similar to Case I, it is hard to construct a filtering gain. Hence, we try to modify the measurement equation in order to find suitable filtering gain K_k. Since Rank$(C) = r < n$, the singular value decomposition of measurement matrix C can be given by

$$C = U \begin{bmatrix} \varepsilon_{r\times r} & 0' \\ 0 & 0_{(m-r)\times(n-r)} \end{bmatrix} V^T \tag{6.15}$$

where U and V is $m \times m$ unitary matrix, $\varepsilon_{r\times r} = \text{diag}(\delta_1, \delta_2, \ldots, \delta_r)$, and δ_i is the singular value of the matrix C .

Introducing the notation

$$\Delta z_k \triangleq (U^T(y_k - \hat{y}_k))_r \tag{6.16}$$

where $(U^T(y_k - \hat{y}_k))_r$ denotes the first r rows of the matrix $U^T(y_k - \hat{y}_k)$.

$$C_{SVD} \triangleq [\varepsilon_{r\times r} \quad 0]V^T \tag{6.17}$$

$$v_k' \triangleq (U^T v_k)_r \tag{6.18}$$

Then it can be obtained that

$$\Delta z_k = C_{SVD}\Delta x_k + v_k' \tag{6.19}$$

where $\Delta x_k = x_k - \hat{x}_{k,k-p}$.

Since Rank$(C_{SVD}) = r$ and $C_{SVD} \in \boldsymbol{R}^{r\times n}$, $C_{SVD}C_{SVD}^T$ is invertible, therefore the minimum norm solution of the variable Δx_k could be obtained as follows:

$$\Delta\hat{x}_k = C_{SVD}^T(C_{SVD}C_{SVD}^T)^{-1}\Delta z_k \tag{6.20}$$

Then the state estimation at the kth time step could be obtained as

$$\hat{x}_k = \hat{x}_{k,k-p} + C_{SVD}^T(C_{SVD}C_{SVD}^T)^{-1}\Delta z_k \tag{6.21}$$

Case III: $m < n$ and Rank$(C) = m$

From Eq. (6.6), it can be obtained that

$$y_k - \hat{y}_k = C[x_k - \hat{x}_{k,k-p}] + v_k \tag{6.22}$$

where $\hat{y}_k = C\hat{x}_{k,k-p}$.

Introducing the notation, $\Delta x_k \triangleq x_k - \hat{x}_{k,k-p}$.

Since $\mathrm{Rank}(C) = m$, the minimum norm solution of the variable Δx_k could be obtained by

$$K_k = C^T(CC^T)^{-1} \tag{6.23}$$

$$\Delta\hat{x}_k = K_k(y_k - \hat{y}_k) \tag{6.24}$$

Then the state estimation could be obtained as follows:

$$\hat{x}_k = \hat{x}_{k,k-p} + K_k(y_k - \hat{y}_k) \tag{6.25}$$

Remark 6.2 For clarity of understanding, a simple example is presented under such case when the measurement matrix $C = [1 \quad 0 \quad 0]$. From Eq. (6.23), we have that $K_k = [1 \quad 0 \quad 0]^T$. Obviously, there is no correct information for the second and third elements of state time prediction in the $\Delta\hat{x}_k$ since it can be obtained that there is no information of the second and third elements of system state in the measurement and the measurement cannot be used to update the state prediction. In such case, the K_k choice is also consistent with the intuitive physical method.

Case IV: $m < n$ and $\mathrm{Rank}(C) = r < m$

From Eq. (6.6), it can be obtained that

$$y_k - \hat{y}_k = C[x_k - \hat{x}_{k,k-p}] + v_k \tag{6.26}$$

where $y_k = C\hat{x}_{k,k-p}$.

Since $\mathrm{Rank}(C) = r < n$, the singular value decomposition of measurement matrix C is given by

$$C = U\begin{bmatrix} \varepsilon_{r\times r} & 0 \\ 0 & 0_{(m-r)\times(n-r)} \end{bmatrix} V^T \tag{6.27}$$

where U and V is $m \times m$ unitary matrix, $\varepsilon_{r\times r} = \mathrm{diag}(\delta_1, \delta_2, \ldots, \delta_r)$, and δ_i is the singular value of the matrix C .

Introducing the notation

$$\Delta z_k \triangleq (U^T(y_k - \hat{y}_k))_r \tag{6.28}$$

where $(U^T(y_k - \hat{y}_k))_r$ represents the first r rows of the matrix $(U^T(y_k - \hat{y}_k))$. Let

$$C_{SVD} \triangleq [\varepsilon_{r\times r} \quad 0]V^T \tag{6.29}$$

$$v_k^{'} \triangleq (U^T v_k)_r \tag{6.30}$$

Then it can be obtained that

$$\Delta z_k = C_{SVD} \Delta x_k + v_k^{'} \tag{6.31}$$

where $\Delta x_k = x_k - \hat{x}_{k,k-p}$.

Noting that $\mathrm{Rank}(C_{SVD}) = r$ and $C_{SVD} \in \boldsymbol{R}^{r \times n}$, the minimum norm solution of the variable Δx_k could be obtained as follows:

$$\Delta \hat{x}_k = C_{SVD}^T (C_{SVD} C_{SVD}^T)^{-1} \Delta z_k \tag{6.32}$$

The state estimation $\hat{x}_k$ at the kth step could be obtained as

$$\hat{x}_k = \hat{x}_{k,k-p} + C_{SVD}^T (C_{SVD} C_{SVD}^T)^{-1} \Delta z_k \tag{6.33}$$

For Problem 6.1, it can be divided into four different cases according to the measurement matrix and the estimating results could be obtained by different means presented in this section.

Algorithm 7 FMNF($\{y_k\}, A, B, C$)

Initialization: $x_0 = 0$
Input: the measurement sequence $\{y_k\}$
Output: state estimated sequence $\{\hat{x}_k\}$
Calculating the matrix Q_{ob} as in Eq. (6.8)
Calculating the matrix $Q_t = [Q_{ob}^T Q_{ob}]^{-1} Q_{ob}^T$
Switching cases by judging the measurement matrix C
  Case I: $K_k = [C^T C]^{-1} C^T$
  Case II: $K_k = C_{SVD}^T (C_{SVD} C_{SVD}^T)^{-1}$
  Case III: $K_k = C^T (C C^T)^{-1}$
  Case VI: $K_k = C_{SVD}^T (C_{SVD} C_{SVD}^T)^{-1}$
for $k = 1$ to N **do**
  Calculating the state prediction $\hat{x}_{k,k-p}$ as Eq. (6.9)
  Switching cases by judging the measurement matrix C
  Case I: calculating the state estimate $\hat{x}_k$ as in Eq. (6.13)
  Case II: calculating the Δz as in Eq. (6.16), then obtained the state estimate $\hat{x}_k$ as in Eq. (6.21)
  Case III: calculating the state estimate $\hat{x}_k$ as in Eq. (6.25)
  Case VI: calculating the Δz as in Eq. (6.28), then obtained the state estimate $\hat{x}_k$ as in Eq. (6.32)
end for
Return $\{\hat{x}_k\}$

Two mathematical theorems were presented in order to illustrate the closed-loop properties of the fast minimum norm filtering algorithm.

Theorem 6.1 *Under Assumptions 6.1–6.3, for Problem 6.1, the estimation $\hat{x}_k$ obtained by Eqs. (6.8)–(6.33) for different four cases is an unbiased estimate of the system state x_k at the kth step in the discrete-time linear time-invariant system (6.6).*

Proof It can be seen that the estimated state $\hat{x}_k$ could be obtained as follows:

$$\hat{x}_k = \hat{x}_{k,k-p} + K_k[y_k - C\hat{x}_{k,k-p}] \tag{6.34}$$

where $\hat{x}_{k,k-p}$ is the prediction of x_k based on the measurement sequence and K_k is the filtering gain.

Introducing the notations

$$\varepsilon_{k,pre} \triangleq x_k - \hat{x}_{k,k-p} \tag{6.35}$$

$$\varepsilon_k \triangleq x_k - \hat{x}_k \tag{6.36}$$

From Eq. (6.34), it can be obtained that

$$\begin{aligned} E[\varepsilon_k] &= E[x_k - \hat{x}_k] \\ &= E[x_k - \hat{x}_{k,k-p} - K_k[y_k - C\hat{x}_{k,k-p}]] \\ &= E[[I - K_kC][x_k - \hat{x}_{k,k-p}] - K_k v_k] \\ &= [I - K_kC]E[\varepsilon_{k,pre}] + K_k E[v_k] \end{aligned} \tag{6.37}$$

From Assumption 6.3, we have $E[v_k] = 0$. Then $E[\varepsilon_k] = 0$ if $E[\varepsilon_{k,k-p}] = 0$. In other words, the estimation $\hat{x}_k$ obtained by Eq. (6.34) is unbiased if $E[\varepsilon_{k,pre}] = 0$.

From the time updating process of the fast minimum norm filtering algorithm, it can be obtained that $\hat{x}_{k,k-p}$ is a kind of unbiased estimation for the state x_k and then the estimation obtained by the fast minimum norm filtering algorithm will be unbiased. □

Remark 6.3 Obviously, it can be seen that the initial state estimation $\hat{x}_0$ is not necessary for the new algorithm from the process of time update. We just need p measurements as a "built-in" parameter for the fast minimum norm filtering algorithm. The estimation error of the initial state has no effect on the unbiased property of the new algorithm.

Theorem 6.2 *Under Assumptions 6.1–6.3, for Problem 6.1, the estimate $\hat{x}_k$ obtained by Eqs. (6.8)–(6.33) for different four cases has a certain covariance matrix, which is determined by the measurement matrix and covariance matrix of the measurement noise in the discrete-time linear time-invariant system (6.6).*

Proof From the framework of the fast minimum norm filtering algorithm, it can be obtained that

$$\begin{aligned} E[\varepsilon_k\varepsilon_k^T] &= E[[x_k - \hat{x}_k][x_k - \hat{x}_k]^T] \\ &= [I - K_kC]E[\varepsilon_{k,k-p}\varepsilon_{k,k-p}^T][I - K_kC]^T + K_kRK_k^T \end{aligned} \tag{6.38}$$

For Case I, it can be obtained that

$$K_k = (C^TC)^{-1}C^T \tag{6.39}$$

Substituting Eq. (6.39) into Eq. (6.38), we obtain

$$E[\varepsilon_k \varepsilon_k^T] = (C^T C)^{-1} C^T R((C^T C)^{-1} C^T)^T \tag{6.40}$$

Therefore, for Case I, the covariance matrix of estimation error is determined by the measurement matrix C and measurement noise covariance matrix R.

For Case III, we have

$$K_k = C^T (C C^T)^{-1} \tag{6.41}$$

Substituting Eq. (6.41) into Eq. (6.38), then multiplying matrix C, and multiplying C^T to Eq. (6.38), it can be obtained that

$$C E[\varepsilon_k \varepsilon_k^T] C^T = R \tag{6.42}$$

So for Case III the covariance matrix of estimation error is constrained by Eq. (6.42), which is governed by the measurement matrix C and measurement noise covariance matrix R.

For Cases II and IV, they can be treated as Case III with different measurement equations obtained by the singular value decomposition method. Therefore, the estimation error covariance matrix is determined by the equivalent measurement equation.

Finally, we can conclude that the covariance matrix of the estimation error will always be determined by C and R. □

Remark 6.4 For the linear time-invariant systems, compared with the standard Kalman filter algorithm, the fast minimum norm filtering algorithm works well when the covariance matrices of the process and measurement noises were completely unknown. Besides, the fast minimum norm filtering algorithm can deal with the case where the initial state estimation cannot be obtained in advance. However, the estimation precision of the fast minimum norm filtering algorithm was relevant with measurement matrix and measurement noise covariance matrix though the covariance matrix of the measurement noise is not necessary in the process of the fast minimum norm filtering algorithm.

6.4 Numerical Examples

In this section, three numerical examples were presented to illustrate the filtering performance of the fast minimum norm filtering algorithm. In order to test the estimating results, the standard Kalman filter was chosen to compare with the fast minimum norm filtering algorithm.

6.4.1 Example I: Measurement Feedback Simulation

This simulation example is for the case where matrix C is a "fat" matrix (columns greater than rows). The discrete-time linear time-invariant system model is given as follows:

$$\begin{cases} x_k = \begin{bmatrix} 1 & 1 \\ 0 & 1 \end{bmatrix} x_{k-1} + w_{k-1} \\ y_k = \begin{bmatrix} 1 & 0 \end{bmatrix} x_k + v_k \end{cases} \tag{6.43}$$

where $x_0 = [0 \quad 0]^T$, $E(w_k) = 0$, $E(v_k) = 0$, $\mathrm{Cov}(w_k) = \mathrm{diag}(0.02, 0.02)$ and $\mathrm{Cov}(v_k) = 0.01$. Here $E(\cdot)$ is the mathematical expectation and $\mathrm{Cov}(\cdot)$ stands for the covariance matrix.

Different from the standard Kalman filter algorithm, it is supposed that the process and measurement noise covariance matrices were completely unknown in the fast minimum norm filtering algorithm. From Algorithm 7, it can be obtained that

$$K_k = [C^T C]^{-1} C^T = [1 \quad 0]^T \tag{6.44}$$

$$Q_t = [Q_{ob}^T Q_{ob}]^{-1} Q_{ob}^T = \begin{bmatrix} 0.8333 & 0.3333 & -0.1667 \\ -0.5000 & 0 & 0.5000 \end{bmatrix} \tag{6.45}$$

Then the real-time state estimate sequence could be obtained by Algorithm 7. The simulation experiment is carried out based on Monte Carlo method with 100 random simulations. The filtering results are shown in Figs. 6.1 and 6.2.

From Figs. 6.1 and 6.2, it can be seen that the filtering results illustrate that the fast minimum norm filtering algorithm could be used to track the state signal trend in such case and the estimation has an acceptable performance compared with the standard Kalman filter with exact system parameters. This simulation illustrates the acceptable performance of the fast minimum norm filtering algorithm in the case where matrix C is a "fat" matrix (columns greater than rows).

6.4.2 Example II: Data Fusion Simulation

This simulation example is for the case where matrix C is a "tall" matrix (columns less than rows). In this example, measurements from multiple sensors are available, and the goal is to recover the unknown states by fusing the data of sensors. The discrete-time linear time-invariant system model is given as follows:

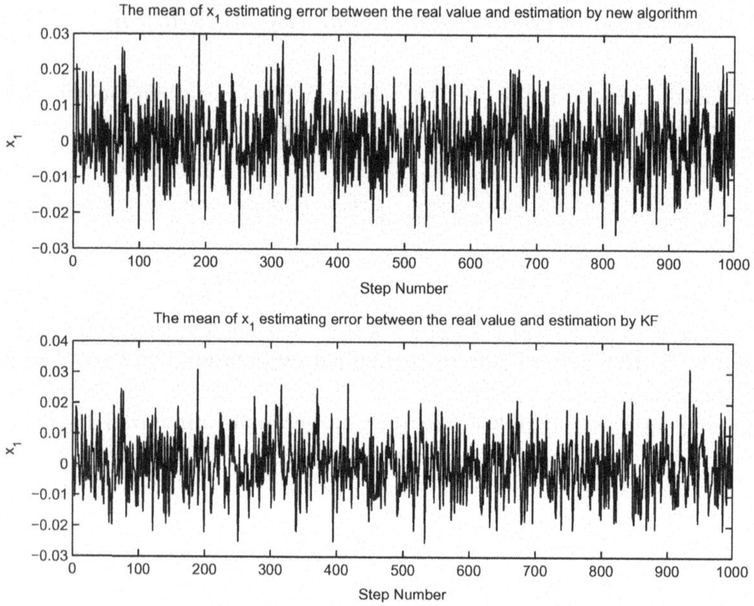

(a) x_1

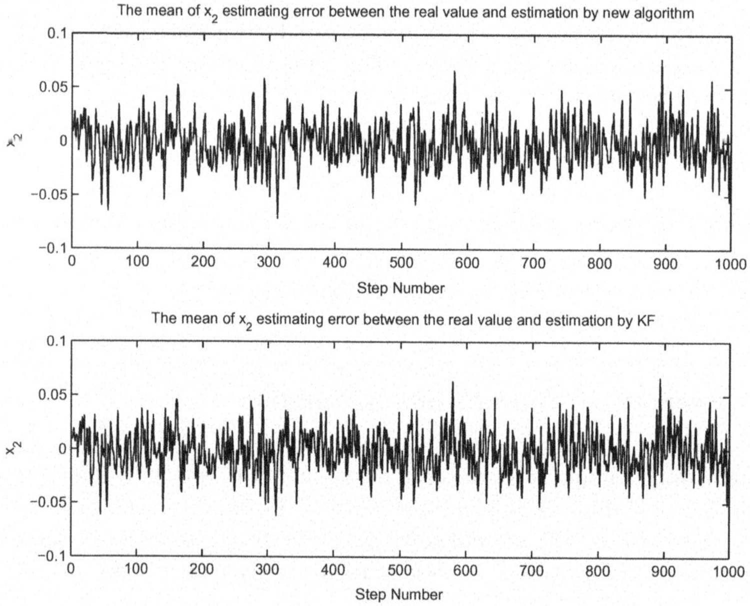

(b) x_2

Fig. 6.1 The mean of state estimation error at different times (Gaussian noise)

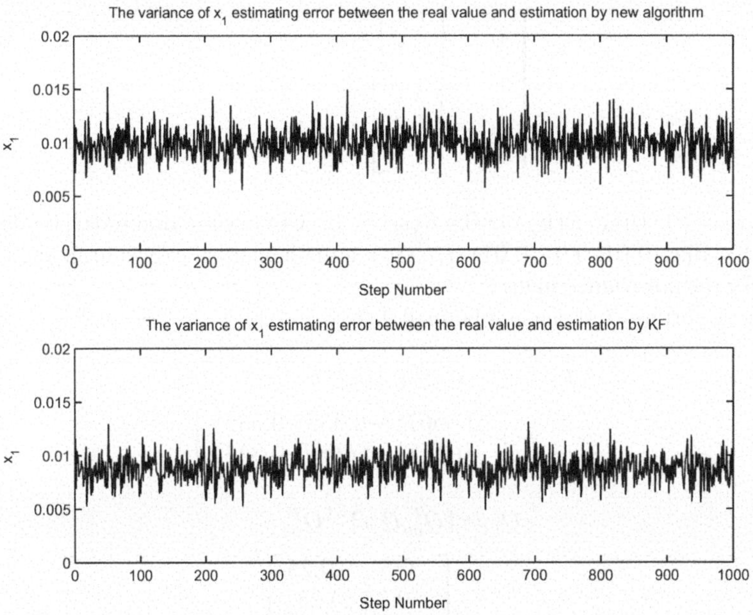

(a) x_1

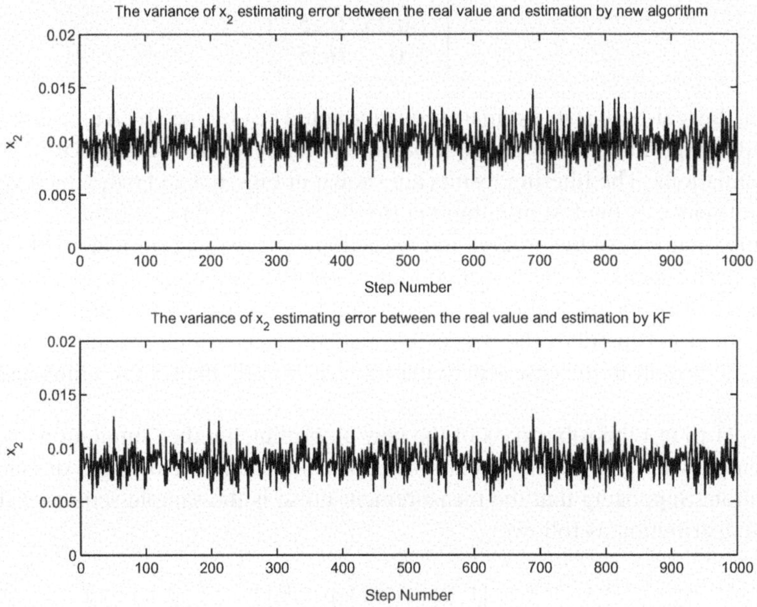

(b) x_2

Fig. 6.2 The variance of state estimation error at different times (Gaussian noise)

$$\begin{cases} x_k = \begin{bmatrix} 1 & 1 \\ 0 & 1 \end{bmatrix} x_{k-1} + w_{k-1} \\ y_k = \begin{bmatrix} 1 & 0 \\ 0 & 1 \\ 1 & 1 \end{bmatrix} x_k + v_k \end{cases} \tag{6.46}$$

where $x_0 = [0 \quad 0]^T$, $E(w_k) = 0$, $E(v_k) = 0$, $\mathrm{Cov}(w_k) = \mathrm{diag}(0.02, 0.02)$, and $\mathrm{Cov}(v_k) = \mathrm{diag}(0.01, 0.01, 0.01)$. $E(\cdot)$ is the mathematical expectation and $\mathrm{Cov}(\cdot)$ stands for the covariance matrix.

From Algorithm 7, it can be obtained that

$$\begin{aligned} K_k &= C^T(CC^T)^{-1} \\ &= \begin{bmatrix} 0.6667 & -0.3333 & 0.3333 \\ -0.3333 & 0.6667 & 0.3333 \end{bmatrix} \end{aligned} \tag{6.47}$$

$$\begin{aligned} Q_t &= [Q_{ob}^T Q_{ob}]^{-1} Q_{ob}^T \\ &= \begin{bmatrix} 0.5 & -0.25 \\ -0.25 & 0.25 \\ 0.25 & 0 \\ 0.25 & 0 \\ -0.25 & 0.25 \\ 0 & 0.25 \end{bmatrix}^T \end{aligned} \tag{6.48}$$

Then, the real-time state estimate sequence could be obtained by Algorithm 7. The simulation experiment is carried out based on Monte Carlo method with 100 random simulations. The filtering results are shown in Figs. 6.3 and 6.4, which verified the performance of the fast minimum norm filtering algorithm without knowing the covariance matrices of the process and measurement noises a priori. Roughly speaking, the performance of the fast minimum norm filtering algorithm is comparable with that of the standard Kalman filter (optimal filter) with exact system parameters. This simulation illustrates the acceptable performance of the fast minimum norm filtering algorithm in the case where matrix C is a "tall" matrix (columns less than rows).

In order to test the robustness of the new algorithm, another simulation example is presented when the measurement noise does not satisfy the demand of Gaussian distribution. Supposing that the measurement noise is the sample sequence from a uniform distribution as follows:

$$v_k^i \sim U[0, 1] \quad i \in \{1, 2, 3\} \quad \text{independent with each other} \tag{6.49}$$

The simulation experiment is carried out based on Monte Carlo method with 100 random simulations. The filtering results are shown in Figs. 6.5 and 6.6, which verified the performance of the fast minimum norm filtering algorithm when the measurement noise cannot satisfy the demand of Gaussian distribution.

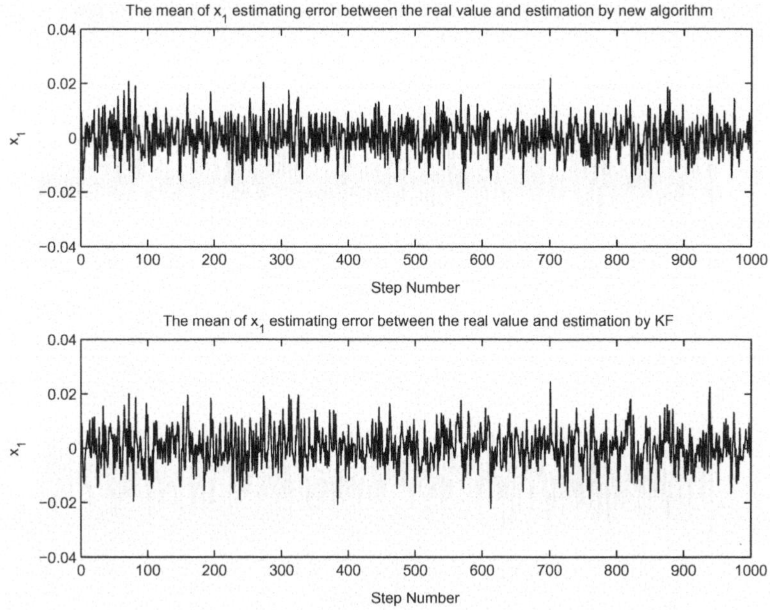

(a) x_1

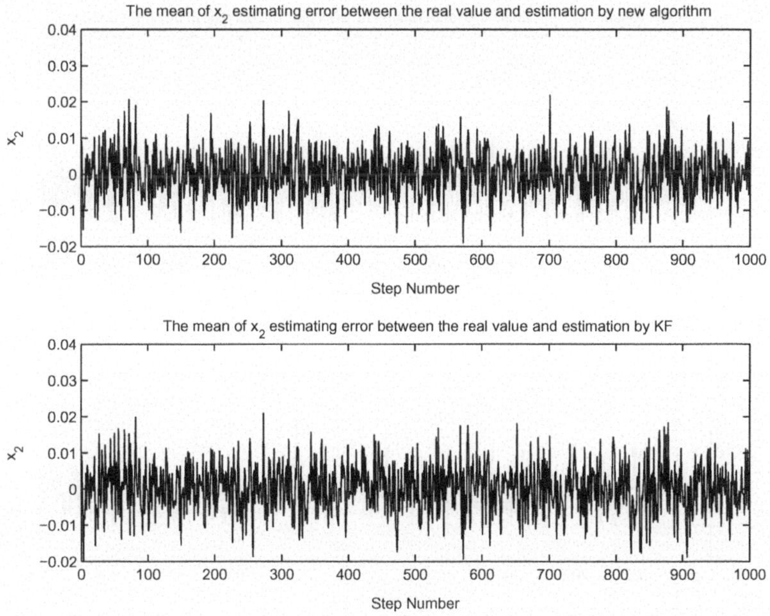

(b) x_2

Fig. 6.3 The mean of state estimation error at different times (Gaussian noise)

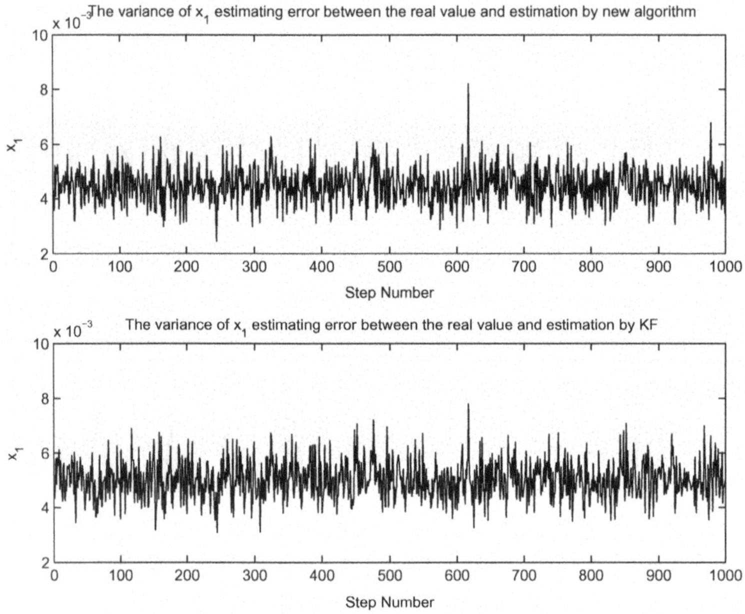

(a) x_1

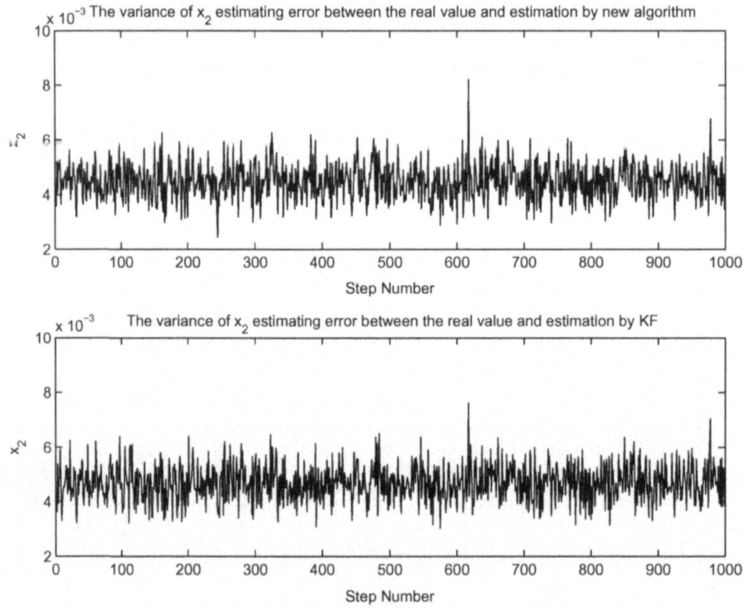

(b) x_2

Fig. 6.4 The variance of state estimation error at different times (Gaussian noise)

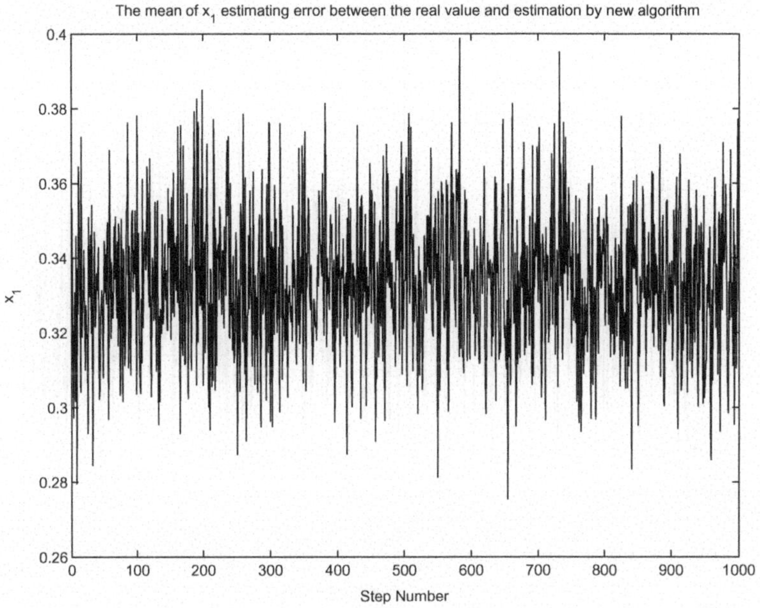

(a) x_1

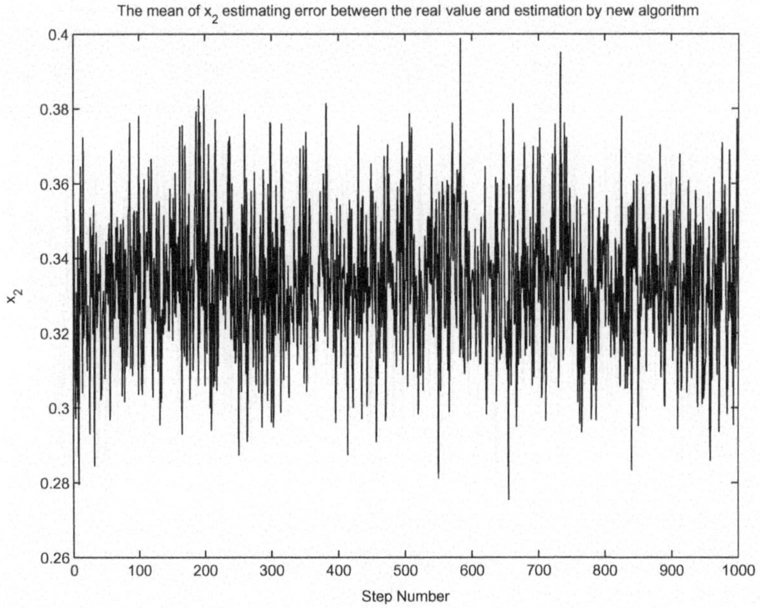

(b) x_2

Fig. 6.5 The mean of state estimation error at different times (Non-Gaussian noise)

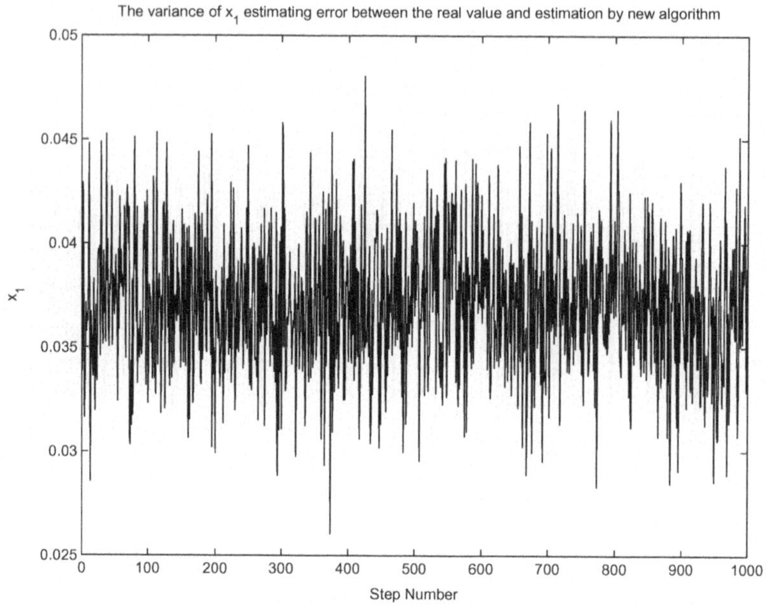

(a) x_1

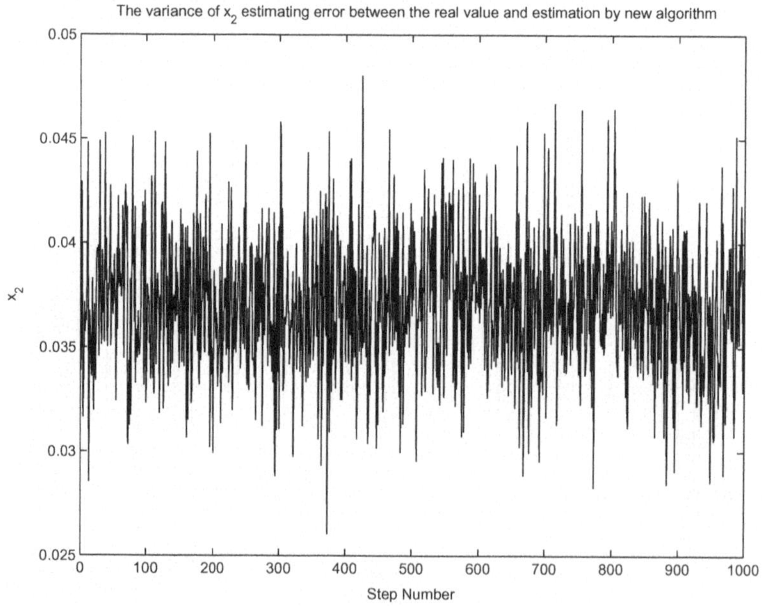

(b) x_2

Fig. 6.6 The variance of state estimation error at different times (Non-Gaussian noise)

6.4.3 Example III: Integrated Navigation Simulation

The data fusion algorithm plays an important role in the integrated navigation systems. For instance, some positioning systems are always used to be integrated with inertial navigation systems to suppress the navigation error of the inertial navigation systems divergence. The speed and acceleration information of the vehicle could be obtained by the inertial navigation systems, and the position information could be obtained by other aided positioning systems. Then the position information can be used to suppress the inertial navigation error divergence.

For brevity, we adopt a "lineland" (one-dimensional space) system model to demonstrate the application of the new algorithm in integrated navigation systems:

$$\begin{cases} \begin{bmatrix} x_k \\ \dot{x}_k \\ \ddot{x}_k \end{bmatrix} = \begin{bmatrix} 1 & T & 0.5T^2 \\ 0 & 1 & T \\ 0 & 0 & 1 \end{bmatrix} \begin{bmatrix} x_{k-1} \\ \dot{x}_{k-1} \\ \ddot{x}_{k-1} \end{bmatrix} + w_{k-1} \\ \begin{bmatrix} y_k \\ \dot{y}_k \\ \ddot{y}_k \end{bmatrix} = \begin{bmatrix} 1 & 0 & 0 \\ 0 & 1 & 0 \\ 0 & 0 & 1 \end{bmatrix} \begin{bmatrix} x_k \\ \dot{x}_k \\ \ddot{x}_k \end{bmatrix} + v_k \end{cases} \tag{6.50}$$

where T is the sample period, x_k is the vehicle position, $\dot{x}_k$ is the vehicle speed, $\ddot{x}_k$ is the acceleration, and $[y_k \quad \dot{y}_k \quad \ddot{y}_k]^T$ is the measurement signal.

Supposing the $E(w_k) = E(v_k) = 0$, $T = 0.05s$ and

$$\mathrm{Cov}(w_k) = Q = \begin{bmatrix} 0.25 & 0.04 & 0.04 \\ 0.04 & 0.25 & 0.04 \\ 0.04 & 0.04 & 0.25 \end{bmatrix} \tag{6.51}$$

$$\mathrm{Cov}(v_k) = R = \begin{bmatrix} 0.25 & 0.04 & 0.04 \\ 0.04 & 0.25 & 0.04 \\ 0.04 & 0.04 & 0.25 \end{bmatrix} \tag{6.52}$$

However, Q and R are completely unknown in the data fusion of the integrated navigation systems.

The simulation experiment is carried out based on Monte Carlo method with 100 random simulations. The simulation results are presented in Figs. 6.7 and 6.8. This simulation illustrates the acceptable performance of the fast minimum norm filtering algorithm in the case where matrix C is a square matrix.

6.5 Conclusion

To solve the state estimating problem when the covariance matrices of process and measurement noises were completely unknown for the discrete-time linear time-invariant systems with state-space model, a new algorithm, named as fast minimum

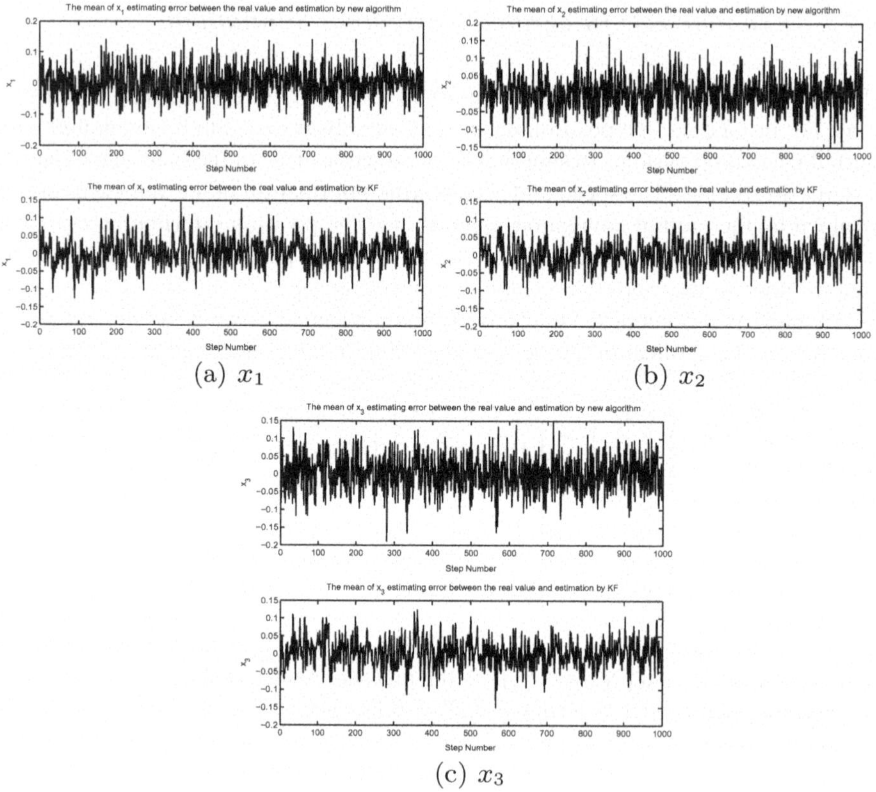

Fig. 6.7 The mean of state estimation error at different times (Gaussian noise)

norm filtering algorithm, has been presented in this chapter. For different four cases, various measurement updating methods have been proposed for the discrete-time linear time-invariant systems with state-space model.

Compared with the standard Kalman filter with accurate model, state estimation by the fast minimum norm filtering algorithm has an acceptable performance with less computing burden. Two mathematical theorems have been presented to illustrate that the covariance matrix of estimation error by the fast minimum norm filtering algorithm is determined by the covariance matrices of measurement noise and process noise.

The method proposed in this chapter does not require any a priori information on the knowledge of the covariance matrices of measurement noise and process noise; hence, this new filtering algorithm would be very useful in case where such knowledge are not available. And the extensive simulations have shown that this proposed filtering algorithm indeed has a comparable good performance in comparison with a standard optimal Kalman filter using the accurate covariance matrices of measure-

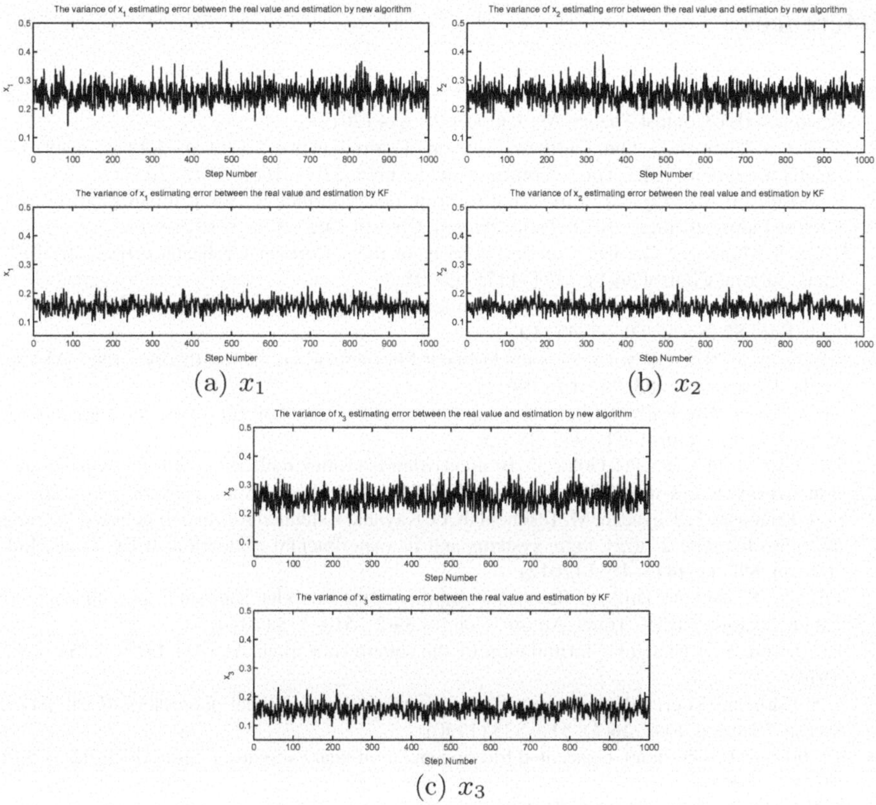

Fig. 6.8 The variance of state estimation error at different times (Gaussian noise)

ment noise and process noise. Therefore, the fast minimum norm filtering algorithm may potentially result in promising wide applications in practice like Kalman filter.

From several typical simulation examples, it can be seen that the new algorithm could obtain acceptable state estimate when the noise covariance matrices are either diagonal or almost diagonal with equal diagonal values. In practice, the constant noise covariance matrices are always used to replace the real covariance matrices of process and measurement noise. If the chosen covariance matrices are not suitable, the Kalman filter will degrade much or even not work. However, the new algorithm proposed in this chapter could be used to obtain the more acceptable state estimates than Kalman filter with unsuitable covariance matrices. Besides, the new algorithm has a Kalman-like style and is easy to be used in engineering applications and hence, it could obtain the real-time state estimate in certain conditions.

References

1. R. Cui, B. Ren, S.S. Ge, Synchronised tracking control of multi-agent system with high-order dynamics. IET Control Theory Appl. **6**(5), 603–614 (2012)
2. W. Wu, S. Chen, A.Q. Qin, Online estimation of ship dynamic flexure model parameters for transfer alignment. IEEE Trans. Control Syst. Technol. **21**(5), 1666–1678 (2013)
3. Y. Mostofi, R. Murray, To drop or not to drop: design principles for Kalman filtering over wireless fading channels. IEEE Trans. Autom. Control **54**(2), 376–381 (2009)
4. H. Xu, S. Mannor, A Kalman filter design based on the performance/robustness tradeoff. IEEE Trans. Autom. Control **54**(5), 1171–1175 (2009)
5. R.E. Kalman, A new approach to linear filtering and prediction problems. Trans. ASME-J. Basic Eng. **82** (Series D), 35–45 (1960)
6. R.E. Kalman, R.S. Bucy, New results in linear filtering and prediction theory. Trans. ASME. Ser. D, J. Basic Eng. **83**, 95–107 (1961)
7. S.F. Schmidt, The Kalman filter: its recognition and development for aerospace applications. AIAA J. Guid. Control **4**(1), 4–7 (1981)
8. S. Bogosyan, M. Barut, M. Gokasan, Braided extended Kalman filters for sensorless estimation in induction motors at high-low/zero speed. IET Control Theory Appl. **1**(4), 987–998 (2007)
9. M.A. Khanesar, E. Kayacan, M. Teshnehlab, O. Kaynak, Extended Kalman filter based learning algorithm for type-2 fuzzy logic systems and its experimental evaluation. IEEE Trans. Ind. Electron. **59**(11), 4443–4455 (2012)
10. S. Kluge, K. Reif, M. Brokate, Stochastic stability of the extended Kalman filter with intermittent observations. IEEE Trans. Autom. Control **55**(2), 514–518 (2010)
11. N. Carlson, Fast triangular formulation of the square root filter. AIAA J **11**(9), 1259–1265 (1973)
12. N.A. Carlson, Federated square root filter for decentralized parallel processors. IEEE Trans. Aerosp. Electron. Syst. **26**(3), 517–525 (1990)
13. S.J. Julier, J.K. Uhlmann, Unscented filtering and nonlinear estimation. Proc. IEEE **92**(3), 644 401–422 (2004)
14. S. Jafarzadeh, C. Lascu, M.S. Fadali, State estimation of induction motor drives using the unscented Kalman filter. IEEE Trans. Ind. Electron. **59**(11), 4207–4216 (2012)
15. H. Marina, F.J. Pereda, J.M. Giron-Sierra, F. Espinosa, UAV attitude estimation using unscented Kalman filter and TRIAD. IEEE Trans. Ind. Electron. **59**(11), 4465–4474 (2012)
16. S. Bhaumik, S. Sadhu, T.K. Ghoshal, Risk-sensitive formulation of unscented Kalman filter. IET Control Theory Appl. **3**(4), 375–382 (2009)
17. L. Chang, B. Hu, G. Chang, A. Li, Marginalised iterated unscented Kalman filter. IET Control Theory Appl. **6**(6), 847–854 (2012)
18. I.R. Petersen, A.V. Savkin, *Robust Kalman Filtering for Signals and Systems with Large Uncertainties* (Birkhause Boston, 1999)
19. B. Hassibi, A.H. Sayed, T. Kailath, H_∞ optimality of the LMS algorithm. IEEE Trans. Signal Process. **44**(2), 267–280 (1996)
20. W.S. Ra, I.H. Whang, J.B. Park, Non-conservative robust Kalman filtering using a noise corrupted measurement matrix. IET Control Theory Appl. **3**(9), 1226–1236 (2009)
21. B. Chen, L. Yu, W.A. Zhang, Robust Kalman filtering for uncertain state delay systems with random observation delays and missing measurements. IET Control Theory Appl. **5**(17), 1945–1954 (2011)
22. R.K. Mehra, On the identification of variances and adaptive Kalman filtering. IEEE Trans. Autom. Control **15**(2), 175–184 (1970)
23. K. Xiong, H. Zhang, L. Liu, Adaptive robust extended Kalman filter for nonlinear stochastic systems. IET Control Theory Appl. **2**(3), 239–250 (2008)

Chapter 7
A Framework of Finite-Model Kalman Filter with Case Study: MVDP-FMKF Algorithm

Kalman filtering techniques have been widely used in many applications; however, standard Kalman filters for linear Gaussian systems usually cannot work well or even diverge in the presence of large model uncertainty. To resolve this problem occurred frequently in practical applications, where it is expensive to have large number of high-cost experiments or even impossible to obtain the exact system model, motivated by our previous pioneering work on finite-model adaptive control, a framework of finite-model Kalman filtering is introduced in this contribution, which supposes that the large model uncertainty may be restricted by a finite set of known models, and the known models can be very different from each other, and the number of known models can be flexibly chosen so that the uncertain model may always be approximated by one of the known models. Within the presented framework of finite-model Kalman filter, according to the idea of adaptive switching via minimizing vector distance principle, a simple finite-model Kalman filter, termed as MVDP-FMKF, has been mathematically formulated and illustrated by extensive simulations. In addition, an experiment of MEMS gyroscope drift has verified the effectiveness of the proposed algorithm, which indicates that the mechanism of finite-model Kalman filter is useful and efficient in practical applications of Kalman filters, especially in inertial navigation systems.

7.1 Introduction

Inertial navigation systems (INS's) [1] are widely used in many applications, especially equipped on modern vehicles, ships, and aircraft. One high-performance INS is usually expected to have the characteristics of absolute independence, all-weather, and anti-interference. Gyroscope as a kernel device in the INS is of fundamental importance. The precision and cost of INS mainly depend on the precision and cost of inertial devices, such as gyroscopes and accelerometers, especially the drift of

H. Ma et al., *Kalman Filtering and Information Fusion*,
https://doi.org/10.1007/978-981-15-0806-6_7

gyroscope has cubic function of time affecting the growth of the position error of inertial navigation system.

In order to improve the precision of inertial navigation systems, an important class of theoretical and practical problems is to filter and estimate the errors of kernel devices. Estimating the errors to make a compensation for navigation errors is a main method of improving the precision of inertial navigation systems [2].

In his pioneering work [3], Kalman formulated one general state estimating problem and then he investigated the linear filtering and prediction problems; he also gave a method for the solution of these integral problems when system model is linear and precisely known and noise statistical natures are obtained.

Many extensions and generalizations followed Kalman's basic work [4, 5]. Stanley F. Schmidt adopted the idea of linearization around the trajectory to solve filtering and prediction problems of nonlinear systems in the 1960s, which was named as extend Kalman filter (EKF) later [6, 7]. According to Taylor expansion, EKF is trying to make nonlinear systems as linear model and then using traditional Kalman filter to estimate different states. Then Carlson presented federal Kalman filter (FKF) [8] algorithm to deal with multisystems' optimal data fusion.

When system model is uncertain, S. J. Julier and J. K. Uhlmann introduced the so-called unscented Kalman filter (UKF) [9, 10] for nonlinear estimation in 1997. In addition, robust Kalman filter (RKF) [11] was presented to solve the problems that system model cannot be obtained accurately. Then adaptive Kalman filter (AKF) [12, 13] was used to deal with the problem that process and observation noise statistical natures are uncertain.

In all these existing works, the objective is to obtain good denoising and estimating effects when system model has small uncertainty or the statistical properties of the process and observation noises are uncertain. The filtering and prediction problems when system models have large uncertainty are seldom addressed. In the presence of large model uncertainty, generally speaking, standard Kalman filter cannot work well and there is no way to ensure its stability and convergence. Due to this challenging problem, in the applications of Kalman filters, usually we need to make great efforts in the modeling of the errors by large number of high-cost experiments so as to gain a model accurate enough for use with standard Kalman filtering technique. This traditional approach also has the disadvantages of sensitivity to model drifting or model changes due to unexpected or unobserved uncertain issues like system faults.

Motivated by the above challenging problems, based on our previous pioneering work on finite-model adaptive control, a framework of finite-model Kalman filter is presented in this chapter. In this framework, a finite-model Kalman filter may adaptively adjust its active model or parameters so as to learn the model uncertainty and then make filtering and prediction based on a bank of parallel Kalman filters, which interchange information among one another and share some common information for the purpose of prediction. Within this framework, the finite-model Kalman filter algorithms are trying to solve filtering and prediction problems when the system has large model uncertainty.

To illustrate the idea of finite-model Kalman filter, one simple algorithm of such ideas is presented based on the idea of adaptive switching via minimizing vector distance principle. Mathematical form of the proposed algorithm is clearly given in this contribution, which consists of a group of iterations regarding the switching sequence, bank filter prediction, and the measurement update. The simulation and experiments show that the proposed method can effectively deal with MEMS gyroscope noise signal and improve the precision of the measurement data.

The main contributions of this chapter are highlighted as follows:

1. A framework of finite-model Kalman filter is presented, which aims to resolve the problem of performance degradation of standard Kalman filter, with the aid of measurement information rather than only innovation information, when the system model is uncertain in a large range.
2. A simple finite-model Kalman filter algorithm based on adaptive switching via minimizing vector distance principle, termed as FMKF-MVDP, is proposed and mathematically formulated to illustrate the presented framework of finite-model Kalman filter.
3. To verify the effectiveness of the proposed simple finite-model Kalman filter, extensive simulations and experiments are conducted and the results show that this algorithm has acceptable filtering effects when system model is inaccurate in large range.

The rest of this chapter is organized as follows: In Sect. 7.2, standard Kalman filter is reviewed with mathematical equations for convenience of later discussions; then, to resolve the challenging filtering problem for state estimation in the presence of large model uncertainties, based on the key idea borrowed from our previous studies on finite-model adaptive control, Sect. 7.3 introduces a general framework of finite-model Kalman filter, within which a simple finite-model Kalman filter, termed as MVDP-FMKF, according to the idea of adaptive switching via minimizing vector distance principle, is mathematically formulated in Sect. 7.4 and verified through simulations conducted in Sect. 7.5, which illustrated the usefulness of the proposed algorithm; and to further check whether the proposed algorithm can work well in practical navigation systems, some experimental tests on an MEMS gyroscope, MEMS130, are discussed in Sect. 7.6, which validates the effectiveness and usefulness of MVDP-FMKF algorithm; finally, Sect. 7.7 summarizes this chapter with some concluding remarks.

7.2 Kalman Filter

In 1960, R. E. Kalman published his famous paper describing a recursive solution to the discrete-time linear filtering problem. Since that time, due in large part to advances in digital computing, the Kalman filter has been the subject of extensive research and application, particularly in the area of autonomous or assisted navigation [14].

The Kalman filter is a set of mathematical equations that provides an efficient computational (recursive) means to estimate the state of a process, in a way that minimizes the mean of the squared error. The filter is very powerful since it supports estimations of past, present, and even future states.

It is the core of the program to design Kalman filter's observation update and state update. Suppose that estimated state x_k is controlled by system noise sequence w_k on time t_k, the system state equation is as follows:

$$\begin{cases} x_k = \phi_{k,k-1}x_{k-1} + \Gamma_{k-1}w_{k-1} \\ z_k = H_k x_k + v_k \end{cases} \tag{7.1}$$

where $\phi_{k,k-1}$ is the transfer matrix, H_k is the observation matrix, v_k is the measurement noise sequence, and z_k represents the observable measurements at time t_k.

If the system noise and observation noise are zero mean, Gaussian, and uncorrelated, then $\hat{x}_k$, the estimate of x_k, can be recursively obtained by the following equations [3]:

$$\begin{cases} \hat{x}_{k,k-1} = \phi_{k,k-1}\hat{x}_{k-1} \\ \hat{x}_k = \hat{x}_{k,k-1} + K_k(z_k - H_k\hat{x}_{k,k-1}) \\ P_{k,k-1} = \phi_{k,k-1}P_{k-1}\phi_{k,k-1}^{\mathrm{T}} + \Gamma_{k-1}Q_{k-1}\Gamma_{k-1}^{\mathrm{T}} \\ K_k = P_{k,k-1}H_k^{\mathrm{T}}(H_k P_{k,k-1}H_k^{\mathrm{T}} + R_k)^{-1} \\ P_k = (I - K_k H_k)P_{k,k-1} \end{cases} \tag{7.2}$$

7.3 Framework of Finite-Model Kalman Filter

In this chapter, our general framework of finite-model Kalman filter is mainly motivated by the thoughts to the difficult issue (b), which is relatively seldom discussed in the literature. It is well known that inaccurate model will degrade the performance of Kalman filter or even make it divergent. To address such a problem, besides the approach of system identification by expensive experiments, one approach called robust Kalman filter (RKF) is discussed in the literature [15]. The basic idea of RKF is rooted from the research on robust control, whose key concepts are nominal model and stability margin as well as certain performance index like H_∞ norm. Roughly speaking, robust Kalman filter can only deal with small model uncertainty due to its nature of worst-case analysis.

When the system model is unknown in a large range, the approach of RKF will fail. Noticing this challenging yet critical problem, in the sequel we present a kind of framework of finite-model Kalman filter (FMKF), which is motivated by our previous study on the finite-model adaptive control (FMAC), whose general framework was first established [16] and the essential idea is that large model uncertainty can be approximated by a finite bank of known models, and hence adaptive control law can be constructed from the control signals designed for the known finite models, no matter whether the known models are similar or not. To this end, several typical algorithms of FMAC are proposed, including LS-like algorithm [17], WLS-like algorithm [18], and other algorithms [19], whose rigorous closed-loop stability analysis can be found. Among these algorithms, the LS-like algorithm generates a switching signal among a bank of models at each step by minimizing the equally weighted sums of pth power of errors, while the WLS-like algorithm generates a switching signal at each step by minimizing a similar weighted performance with forgetting factor, and other types of FMAC algorithms adopt other ideas like adaptive combination rather than the idea of switching among models.

Following the idea of FMAC, we present the overall background of FMKF. Suppose that we have a finite bank of known models, and each model can be described by the following state-space equations:

$$\mathbb{M}_i : \begin{cases} x_{k+1} = f_i(x_k, u_k, w_k) \\ z_k \quad\; = h_i(x_k, u_{k-1}, v_k) \end{cases} \tag{7.3}$$

where x_k is the state at time k, u_k is the control at time k, z_k is the output at time k, w_k is the process noise at time k, and v_k is the measurement noise at time k. Here, functions f_i and h_i are known mappings to describe the state evolution and output measurement for model $\mathbb{M}_i$ $(i = 1, 2, \ldots, N)$. Hence, model $\mathbb{M}_i$ can be formally described by a triple $(f_i, h_i, \mathbb{N}_i)$, where $\mathbb{N}_i$ denotes the a priori knowledge on w_k and v_k, which may be presented by statistical properties or probability distributions of the process noise and the measurement noise.

For simplicity and brevity, we consider only linear Gaussian models without control like

$$\mathbb{M}_i : \begin{cases} x_{k+1} = \phi^{(i)}_{k+1,k} x_k + \Gamma^{(i)}_k w_k \\ z_k \quad\; = H^{(i)}_k x_k + v_k \end{cases} \tag{7.4}$$

where w_k and v_k are zero-mean jointly Gaussian noises with covariances $Q^{(i)}_k$ and $R^{(i)}_k$, respectively. Hence, model $\mathbb{M}_i$ can be presented by matrices $\phi^{(i)}_{k+1,k}$, $\Gamma^{(i)}_k$, $H^{(i)}_k$, and $Q^{(i)}_k$, $R^{(i)}_k$. Obviously, for each fixed model $\mathbb{M}_i$, since its model is accurate with exact parameters, we can design a standard Kalman filter to estimate its state x_k via its output sequence $\{z_k\}$.

Now we consider a practical uncertain system, whose true model is not known a priori. In this case, it is impossible to design a Kalman filter for this system since the implementation of Kalman filter requires to have all matrices $\phi_{k+1,k}$, Γ_k, H_k, Q_k, R_k available at each time step. For this challenging problem, from the perspective of finite models, supposing that we have enough known different models $\mathcal{M} \triangleq \{\mathbb{M}_1, \mathbb{M}_2, \ldots, \mathbb{M}_N\}$, we may find a model $\mathbb{M}_{\mathbb{I}} \in \mathcal{M}$, noting that the index $\mathbb{I}$ is not known a priori, such that the input–output behavior of model $\mathbb{M}_{\mathbb{I}}$ can approximate that of the true system. The key idea of FMKF is to use Kalman filters for known models to construct state estimates for the true unknown system, by making full use of posterior data at each step, that indicate which model may work well in the sense of better data matching.

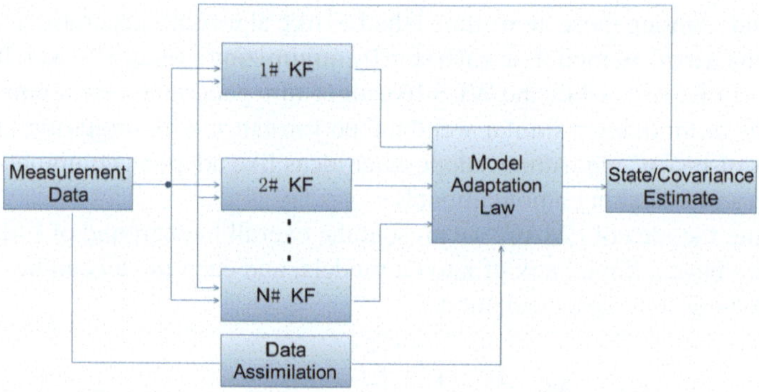

Fig. 7.1 Architecture of finite-model Kalman filter system

In the framework of FMKF, there may exist different methods for the purpose of constructing state estimates from the parallel Kalman filters for known models. At each time step k, the Kalman filter for model $\mathbb{M}_i$ involves two steps: the *prediction* step is to propagate $\hat{x}^{(i)}_{k-1,k-1}$ to $\hat{x}^{(i)}_{k,k-1}$ according to the state equation of model $\mathtt{i}$, while the *correction* step is to correct $\hat{x}^{(i)}_{k,k-1}$ to $\hat{x}^{(i)}_k$ with the available new measurement z_k using the optimal Kalman gain $K^{(i)}_k$, computed from know system matrices. Now, one natural problem will arise: *How to determine the best estimate $\hat{x}_k$ of true system state x_k from the N different estimates $\hat{x}^{(i)}_k$* $(i = 1, 2, \ldots, N)$. In other words, can we have some reasonable or useful rules, termed as *model switching* for convenience, to decide which $\hat{x}^{(i)}_k$ can serve as estimate $\hat{x}_k$ or can we have some reasonable or useful rules, termed as *model combination/fusion* for convenience, to combine or fuse all $\hat{x}^{(i)}_k$ $(i = 1, 2, \ldots, N)$ so as to yield $\hat{x}_k$? Once $\hat{x}_k$ can be determined at time k, then in next step, the Kalman filters for all models may reset their initial conditions $\hat{x}^{(i)}_k$ to the same $\hat{x}_k$ and use it to give new estimates $\hat{x}^{(i)}_{k+1,k}$ and $\hat{x}^{(i)}_{k+1,k+1}$, which continues the overall loop of filtering. The research of FMKF is to figure out such possible rules, to verify its usefulness or effectiveness, to analyze its closed-loop properties, and to explore its wide applications. Obviously, the FMKF architecture, which is briefly illustrated in Fig. 7.1, involves a kind of online learning/adaptation mechanism, and such mechanism will make it possible to cope with challenging problems of filtering, prediction, and control for systems with large model uncertainty or time-varying model drift.

Remark 7.1 In the general framework of FMKF, the true model can be time-invariant or slow time-varying yet unknown. Note that in practice usually we can't find out a governing rule or models transition probability matrix to describe the switching. And furthermore, in the framework of FMKF, the unknown true model need not be one member of the bank of known models. In our framework, the known models need not to have similar structures or same dimensions. The essential idea of FMKF borrows from the so-called "finite cover" concept, which make it possible to approximate the true model with the known finite models.

Remark 7.2 As depicted in Fig. 7.1, there may exist many ways to approximate the true model with the known finite models, that is to say, the so-called *model adaptation law* is not unique. In certain situations, several known existing algorithms like *autonomous multiple model* (AMM) [20] and *Interacting Multiple Model* (IMM) [21] may be also regarded as special cases of FMKF. For example, the AMM algorithm requires that the true operating mode, which does not change in time, matches a model among the ones in the set. For another example, the key idea behind the IMM algorithm is to assume that the mode jump process is a Markov process, with known mode transition probabilities. AMM algorithm and IMM algorithm are two ways of combining the estimates from known finite models using the concepts of model likelihood and mode probability.

Remark 7.3 In the framework of FMKF algorithm, we try to make full use of the measurement sequences to judge system model. Traditional multiple model (MM) algorithm, such as AMM and IMM, always tried to judge system model according to the innovation sequence. In this section, we give a new framework that judges the system model according to measurement sequence rather than the innovation sequence. Note that in Fig. 7.1, there is a certain block *Data Assimilation*, i.e., collection and processing of the measurement data for the purpose of model adaptation, which distinguishes FMKF from traditional MM framework since MM only uses innovation sequence. Such an idea will be demonstrated by an example of FMKF-MVDP algorithm in next example. This design is because in practice we believe that system model information was embedded in the input/output sequence and correct system model should match the measurement data, and generally speaking measurement sequence may contain more information than the innovation sequence.

7.4 MVDP Finite-Model Kalman Filter Algorithm

In this section, within the general framework of FMKF given in Sect. 7.3, a kind of FMKF algorithm, called MVDP finite-model Kalman filter (finite-model Kalman filter based on the minimum vector distance principle), will be presented.

Suppose that the true system model $\mathbb{S} \in \mathcal{M}$, where $\mathcal{M} \triangleq \{\mathbb{M}_1, \mathbb{M}_2, \ldots, \mathbb{M}_N\}$, here the finite known models can be very different in terms of their related matrices. Since we do not know which model is the true system, we try to guess which model may match the output data best at each step. As to the Kalman filter for model $\mathbb{M}_i$, at time step k it will generate a state estimate $\hat{x}_k^{(i)}$; consequently, we can define an output prediction $\hat{z}_k^{(i)}$ as

$$\hat{z}_k^{(i)} = H_k^{(i)} \hat{x}_k^{(i)} \tag{7.5}$$

Suppose that there exists such a function $d_i(\cdot)$ defined by

$$d_{i,k} = \left\| f_{z_k} - f_{\hat{z}_k^{(i)}} \right\|^2 \ (i = 1, 2, \ldots, N) \tag{7.6}$$

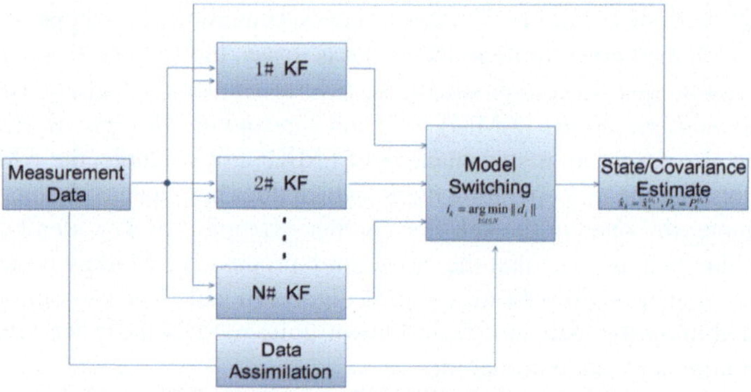

Fig. 7.2 Structure of MVDP finite-model Kalman filter system

where f_{z_k} and $f_{\hat{z}_k^{(i)}}$ are to be discussed later by one example. Intuitively speaking, we construct f_{z_k} and $f_{\hat{z}_k^{(i)}}$ at each time step k, which can be used to measure the discrepancy of the actual output data and the estimated output data using model $\mathbb{M}_i$. If we introduce

$$i_k = \arg\min_{1 \le i \le N} \left\| d_{i,k} \right\| \tag{7.7}$$

then the i_kth system model $\mathbb{M}_i$ can be treated as the active model, i.e., the most matching model in certain sense, at time step k. That is to say, we could try to find a function of the measurement z_k and the filtering output $\hat{z}_k^{(i)}$ of the ith system model, which was used to measure the extent of matching between the true system and the chosen ith models.

The system block diagram is shown in Fig. 7.2.

The rest of this section is to give an example of such a function $d_{i,k}$ and present the process of new algorithm. Note that there exist alternative meaningful approaches to construct $d_{i,k}$, f_{z_k}, and $f_{\hat{z}_k^{(i)}}$; hence, the principle introduced in this chapter can be flexibly used in various applications and further investigation in the future is necessary to explore more possible suitable schemes.

7.4.1 Derivation of d_i

First, expressions for f_{z_k} and $f_{\hat{z}_k^{(i)}}$ are given below. Let $j = 1, \ldots, 2\Delta + 1$ and let $z_{k,j} \stackrel{\Delta}{=} z_{k-j+1}$ be measurement data in the time window of length $2\Delta + 1$ before time step k. A second-order polynomial has the form

$$f_{z_k}(j) = \alpha_0 + \alpha_1 j + \alpha_2 j^2 \tag{7.8}$$

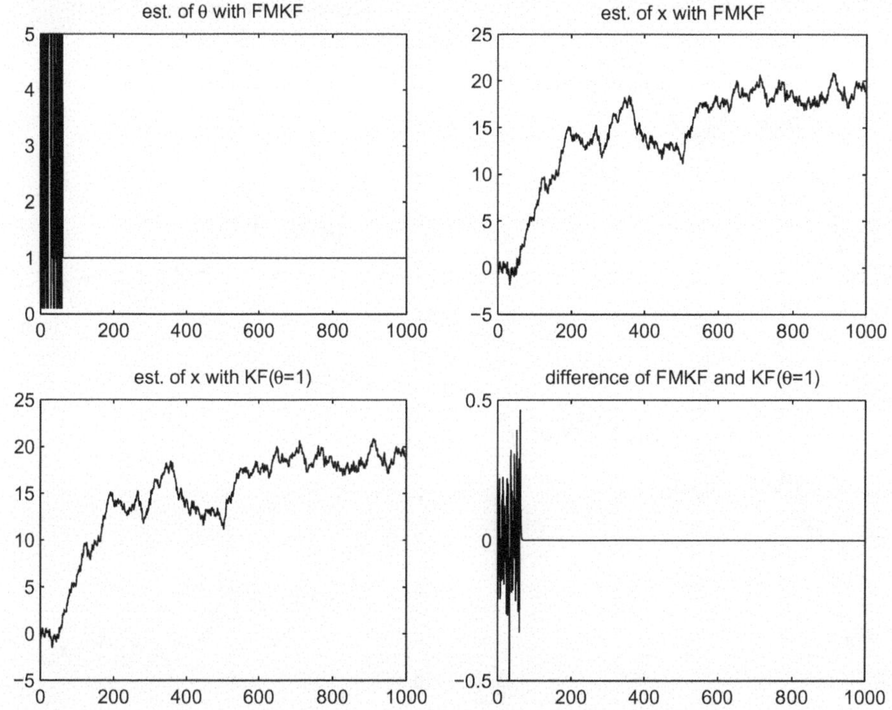

Fig. 7.3 Simulation results of one-dimensional MVDP-FMKF algorithm

and the coefficients α_0,α_1, and α_2 are determined in a least squares sense by minimizing

$$\sum_{j=1}^{2\Delta+1} (z_{k,j} - (\alpha_0 + \alpha_1 j + \alpha_2 j^2))^2 \tag{7.9}$$

Remark 7.4 The choices of f_{z_k} are not unique, and other forms except for second-order polynomial may be also okay. The intuitive meaning of f_{z_k} is to approximate local data of $z_{k,j}$ in a small time window. Note that in practice the sampling interval of measured data is very small; Hence, locally speaking, the data of $z_{k,j}$ will be smooth in certain sense. Considering the trade-off between the computation cost and approximation effect, here we choose a second-order polynomial to capture the basic nature of the data, and in this way only three parameters are needed to determine.

Using the notation (7.9) takes the form

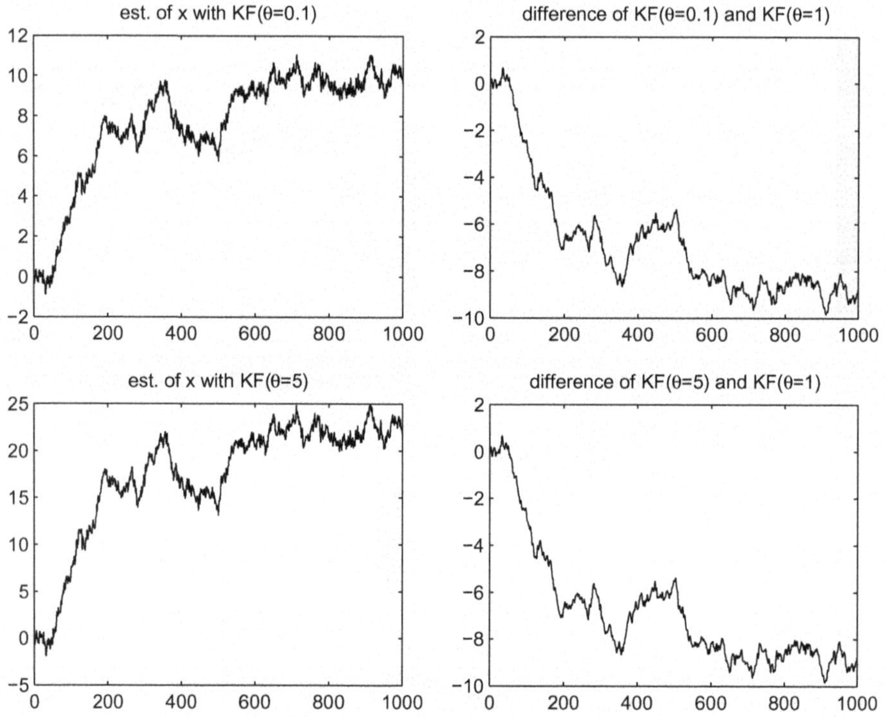

Fig. 7.4 Traditional Kalman filter using wrong model parameter θ can lead to divergent estimation errors

$$\begin{bmatrix} 1 & 1 & 1 \\ 1 & 2 & 4 \\ \vdots & \vdots & \vdots \\ 1 & 2\Delta+1 & (2\Delta+1)^2 \end{bmatrix} \begin{bmatrix} \alpha_0 \\ \alpha_1 \\ \alpha_2 \end{bmatrix} = \begin{bmatrix} z_k(1) \\ z_k(2) \\ \vdots \\ z_k(2\Delta+1) \end{bmatrix} \tag{7.10}$$

Introducing the notation

$$M \triangleq \begin{bmatrix} 1 & 1 & 1 \\ 1 & 2 & 4 \\ \vdots & \vdots & \vdots \\ 1 & 2\Delta+1 & (2\Delta+1)^2 \end{bmatrix} \tag{7.11}$$

$$\mathbb{Z}_k \triangleq \begin{bmatrix} z_k(1) \\ z_k(2) \\ \vdots \\ z_k(2\Delta+1) \end{bmatrix} \tag{7.12}$$

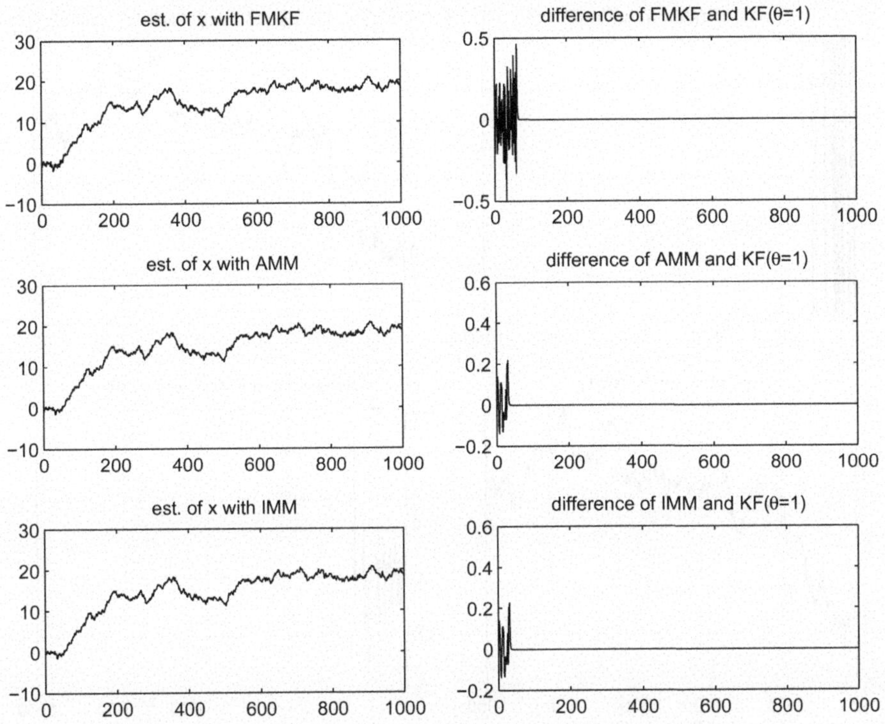

Fig. 7.5 Performance comparison of MVDP-FMKF, AMM, and IMM algorithms

and noting that $[\alpha_0, \alpha_1, \alpha_2]^{\mathrm{T}} = (M^{\mathrm{T}}M)^{-1}M^{\mathrm{T}}\mathbb{Z}_k$, we can get

$$f_{z_k} = M(M^{\mathrm{T}}M)^{-1}M^{\mathrm{T}}\mathbb{Z}_k \tag{7.13}$$

Similarly, we construct a second-order polynomial of form

$$f_{\hat{z}_{k,j}} = \hat{\alpha}_0 + \hat{\alpha}_1 j + \hat{\alpha}_2 j^2 \tag{7.14}$$

where the coefficients $\hat{\alpha}_0$, $\hat{\alpha}_1$, and $\hat{\alpha}_2$ are determined in a least squares sense by minimizing

$$\sum_{j=1}^{2\Delta+1} (\hat{z}_{k,j} - (\hat{\alpha}_0 + \hat{\alpha}_1 j + \hat{\alpha}_2 j^2))^2 \tag{7.15}$$

Consequently, we have similar results

$$[\hat{\alpha}_0, \hat{\alpha}_1, \hat{\alpha}_2]^{\mathrm{T}} = (M^{\mathrm{T}}M)^{-1}M^{\mathrm{T}}\hat{\mathbb{Z}}_k \tag{7.16}$$

and

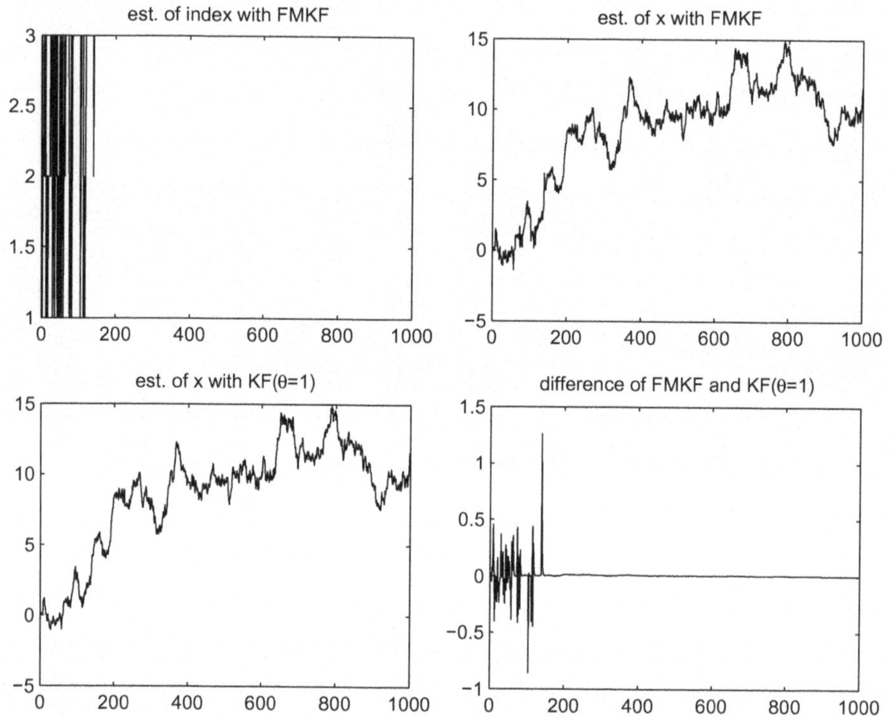

Fig. 7.6 Simulation results of mixing-dimensional MVDP-FMKF algorithm

$$f_{\hat{z}_k} = M(M^{\mathrm{T}}M)^{-1}M^{\mathrm{T}}\hat{\mathbb{Z}}_k \tag{7.17}$$

where

$$\hat{\mathbb{Z}}_k \triangleq \begin{bmatrix} \hat{z}_k(1) \\ \hat{z}_k(2) \\ \vdots \\ \hat{z}_k(2\Delta + 1) \end{bmatrix} \tag{7.18}$$

After that, we can write

$$d_{i,k} = \left\| M(M^{\mathrm{T}}M)^{-1}M^{\mathrm{T}}(\mathbb{Z}_k - \hat{\mathbb{Z}}_k) \right\|^2 \tag{7.19}$$

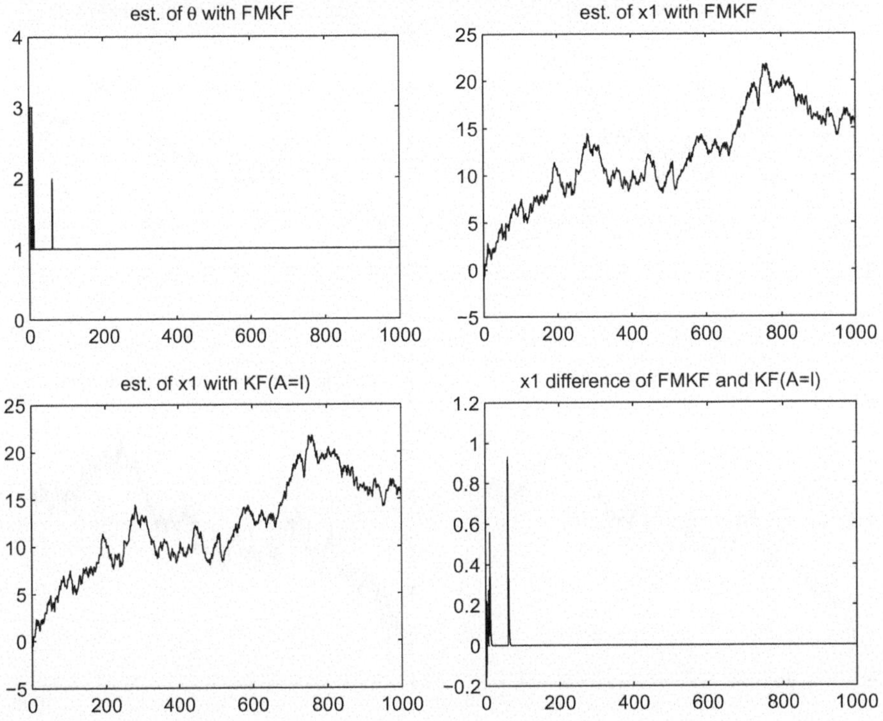

Fig. 7.7 Simulation results of multidimensional FMKF algorithm for state $x_1(k)$

7.4.2 Two-Model MVDP-FMKF Algorithm

In this subsection, the process of MVDP-FMKF algorithm was proposed when we have two alternative models. The process could be divided into two parts.

Initialization
The first step is initializing the filter parameters,

$$\begin{aligned} P_0^{(1)} &= P_0 \\ P_0^{(2)} &= P_0 \\ \hat{x}_0 &= \bar{x}_0 \end{aligned} \tag{7.20}$$

where P_0 is a positive definite matrix and $\bar{x}_0$ can be taken as $E[x_0]$ (if $E[x_0]$ is available a priori) or any vector of proper dimension (e.g., zero vector) (if we have no any knowledge on $E[x_0]$).

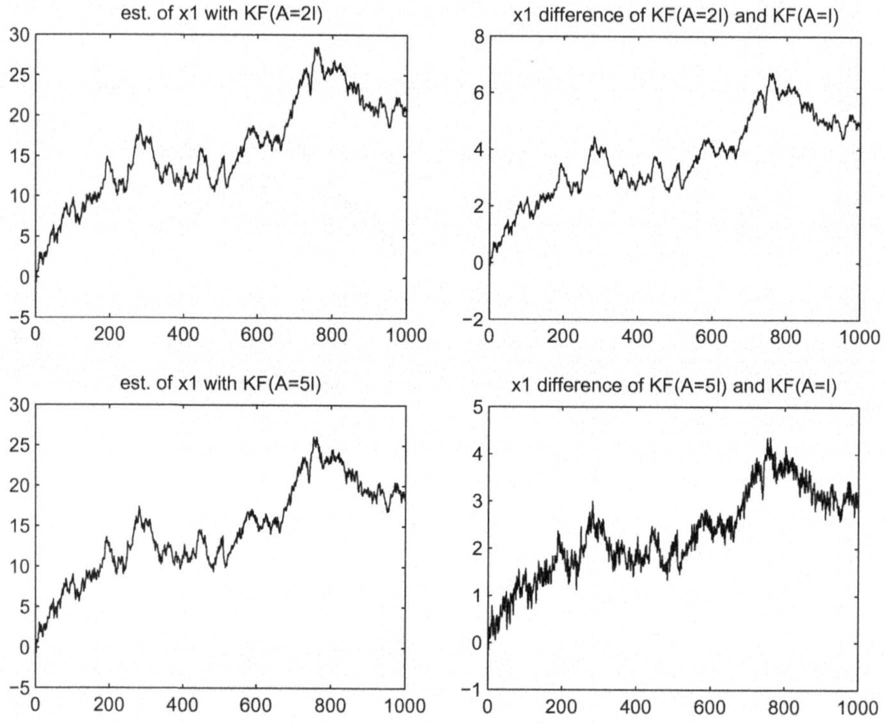

Fig. 7.8 Traditional Kalman filter using wrong model matrix A can lead to divergent estimation errors for state $x_1(k)$

Filtering Loop

The second step is filtering the original signals using the known linear system models for $k = 1, 2, 3, \ldots$:

$$\begin{cases} \hat{x}^{(1)}_{k,k-1} = \phi^{(1)}_{k,k-1}\hat{x}_{k-1} \\ P^{(1)}_{k,k-1} = \phi^{(1)}_{k,k-1}P_{k-1}\phi^{(1)T}_{k,k-1} + \Gamma^{(1)}_{k-1}Q^{(1)}_{k-1}\Gamma^{(1)T}_{k-1} \\ K^{(1)}_k = P^{(1)}_{k,k-1}H^{(1)T}_k(H^{(1)}_k P^{(1)}_{k,k-1}H^{(1)T}_k + R^{(1)}_k)^{-1} \\ P^{(1)}_k = (I - K^{(1)}_k H^{(1)}_k)P^{(1)}_{k,k-1} \\ \hat{x}^{(1)}_k = \hat{x}^{(1)}_{k,k-1} + K^{(1)}_k(z_k - H^{(1)}_k\hat{x}^{(1)}_{k,k-1}) \\ \hat{z}^{(1)}_k = H^{(1)}_k\hat{x}^{(1)}_k \end{cases} \tag{7.21}$$

$$\begin{cases} \hat{x}^{(2)}_{k,k-1} = \phi^{(2)}_{k,k-1}\hat{x}_{k-1} \\ P^{(2)}_{k,k-1} = \phi^{(2)}_{k,k-1}P_{k-1}\phi^{(2)T}_{k,k-1} + \Gamma^{(2)}_{k-1}Q^{(2)}_{k-1}\Gamma^{(2)T}_{k-1} \\ K^{(2)}_k = P^{(2)}_{k,k-1}H^{(2)T}_k(H^{(2)}_k P^{(2)}_{k,k-1}H^{(2)T}_k + R^{(2)}_k)^{-1} \\ P^{(2)}_k = (I - K^{(2)}_k H^{(2)}_k)P^{(2)}_{k,k-1} \\ \hat{x}^{(2)}_k = \hat{x}^{(2)}_{k,k-1} + K^{(2)}_k(z_k - H^{(2)}_k\hat{x}^{(2)}_{k,k-1}) \\ \hat{z}^{(2)}_k = H^{(2)}_k\hat{x}^{(2)}_k \end{cases} \tag{7.22}$$

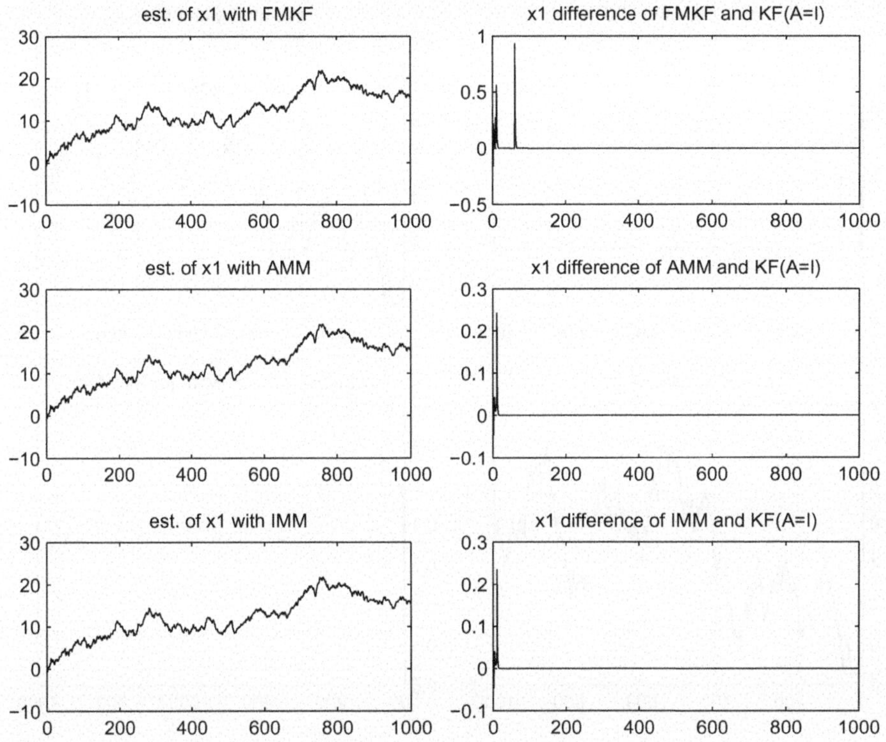

Fig. 7.9 Performance comparison of MVDP-FMKF, AMM, and IMM algorithms for state $x_1(k)$

Then we can calculate $d_{i,k}$ with the output of the different filters $\hat{z}_k^i$ as follows:

$$d_{i,k} = \left\| f_{z_k} - f_{\hat{z}_k^i} \right\|^2 \quad (i = 1, 2) \tag{7.23}$$

Then, we choose the active model by the minimizing vector distance principle and update $\hat{x}_k$ and P_k as follows:

- If

$$d_{1,k} \leq d_{2,k} \tag{7.24}$$

then $i_k = 1$ and we set

$$\begin{aligned} P_k &= P_k^{(1)} \\ \hat{x}_k &= \hat{x}_k^{(1)} \end{aligned} \tag{7.25}$$

- Otherwise, we have $i_k = 2$ since

$$d_{1,k} > d_{2,k} \tag{7.26}$$

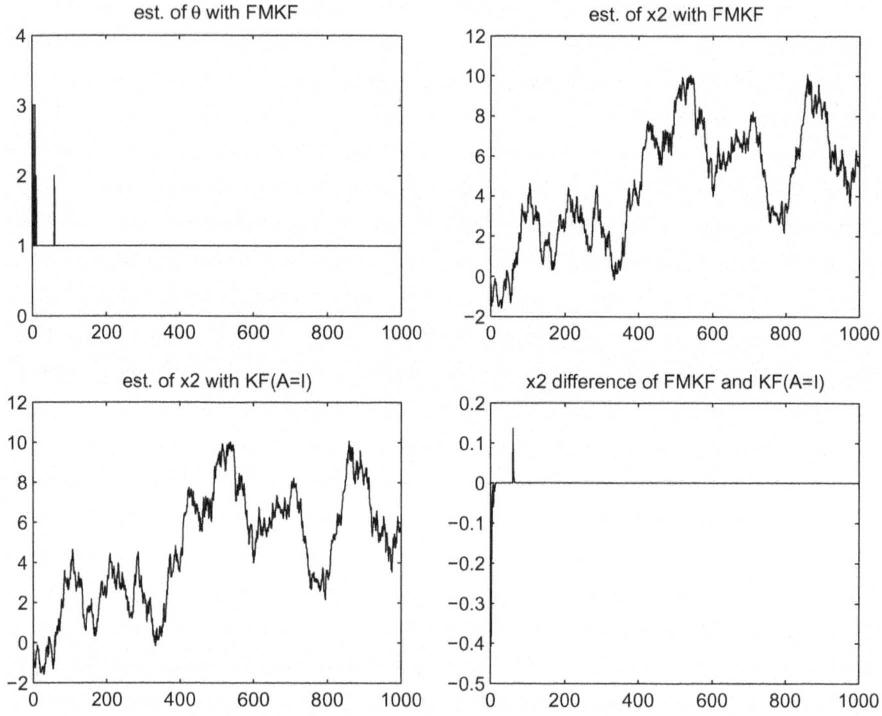

Fig. 7.10 Simulation results of multidimensional FMKF algorithm for state $x_2(k)$

Then, we set

$$\begin{aligned} P_k &= P_k^{(2)} \\ \hat{x}_k &= \hat{x}_k^{(2)} \end{aligned} \tag{7.27}$$

7.4.3 General MVDP-FMKF Algorithm

Similarly, we can introduce the general MVDP-FMKF algorithm as follows:

Initialization

$$\begin{aligned} P_0^{(i)} &= P_0 \quad (i = 1, 2, \ldots, N) \\ \hat{x}_0 &= \bar{x}_0 \end{aligned} \tag{7.28}$$

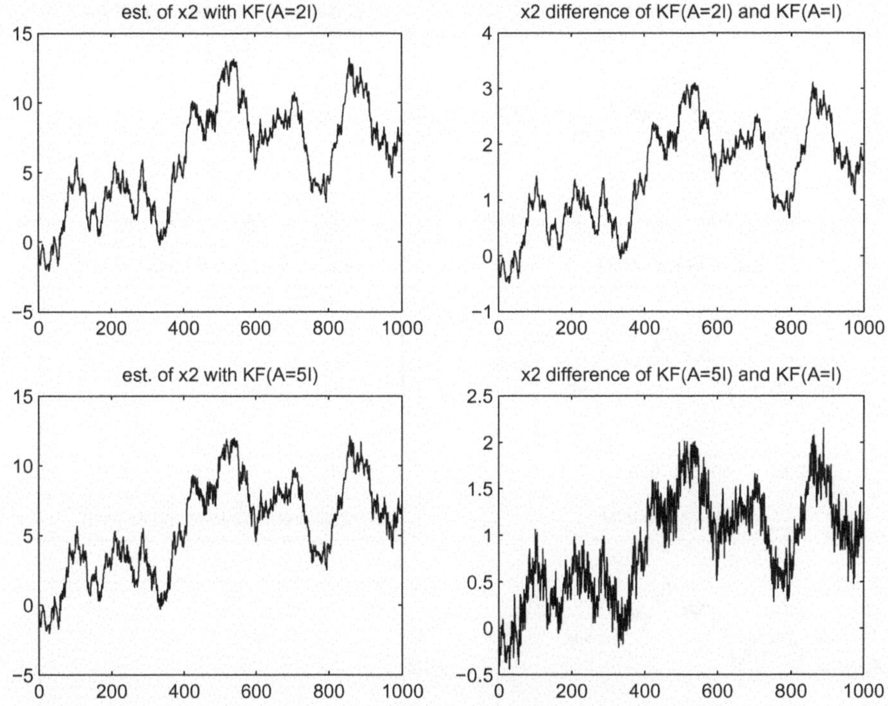

Fig. 7.11 Traditional Kalman filter using wrong model matrix A can lead to divergent estimation errors for state $x_2(k)$

Filtering Loop

At time step k, for each model $\mathbb{M}_i$ $(i = 1, 2, \ldots, N)$, the following Kalman iterations are made first:

$$
\begin{aligned}
\hat{x}_{k,k-1}^{(i)} &= \phi_{k,k-1}^{(i)} \hat{x}_{k-1} \\
P_{k,k-1}^{(i)} &= \phi_{k,k-1}^{(i)} P_{k-1} \phi_{k,k-1}^{(i)\mathrm{T}} + \Gamma_{k-1} Q_{k-1} \Gamma_{k-1}^{\mathrm{T}} \\
K_k^{(i)} &= P_{k,k-1}^{(i)} H_k^{\mathrm{T}} (H_k P_{k,k-1}^{(i)} H_k^{\mathrm{T}} + R_k)^{-1} \\
P_k^{(i)} &= (I - K_k^{(i)} H_k) P_{k,k-1}^{(i)} \\
\hat{x}_k^{(i)} &= \hat{x}_{k,k-1}^{(i)} + K_k^{(i)} (z_k - H_k \hat{x}_{k,k-1}^{(i)}) \\
\hat{z}_k^{(i)} &= H_k \hat{x}_k^{(i)}
\end{aligned}
\tag{7.29}
$$

Then we can calculate $d_{i,k}$

$$
d_{i,k} = \left\| f_{z_k} - f_{\hat{z}_k^i} \right\|^2 \quad (i = 1, 2) \tag{7.30}
$$

where f_{z_k} and $f_{\hat{z}_k^i}$ are introduced before. Then, according to the minimizing vector distance principle, $\hat{x}_k$ and P_k can be updated by

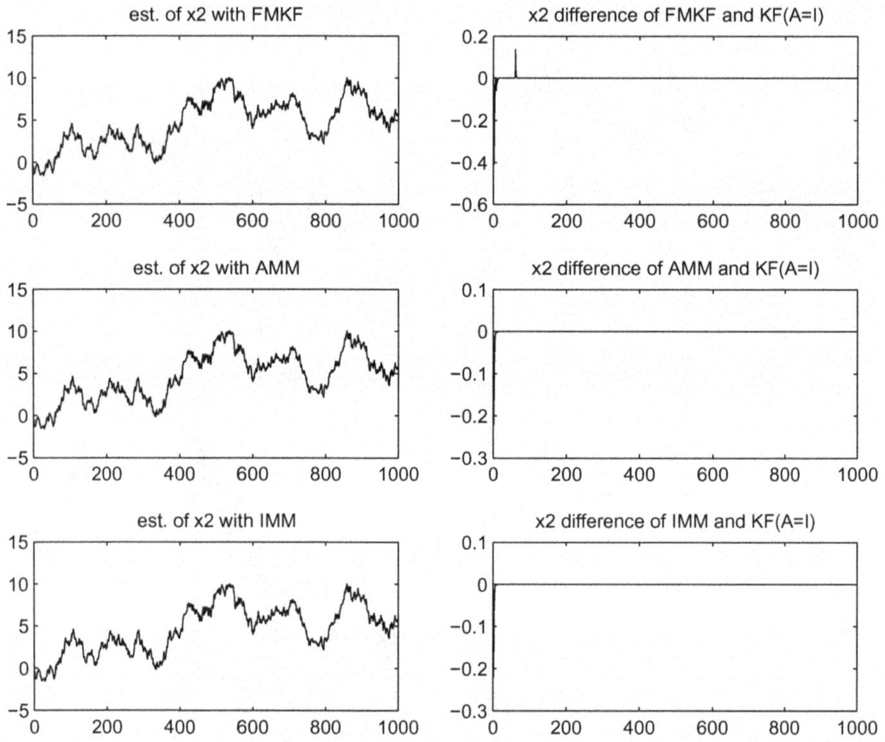

Fig. 7.12 Performance comparison of MVDP-FMKF, AMM, and IMM algorithms for state $x_2(k)$

$$
\begin{aligned}
P_k &= P_k^{(i_k)} \\
\hat{x}_k &= \hat{x}_k^{(i_k)}
\end{aligned}
\tag{7.31}
$$

where

$$
i_k = \underset{i \in \{1,2,\ldots,N\}}{\arg\min}\; d_{i,k} \tag{7.32}
$$

Repeat the above steps for each time step k.

7.5 Simulation of the MVDP-FMKF Algorithm

In this section, some simulation results are reported to illustrate the effects of MVDP-FMKF algorithm.

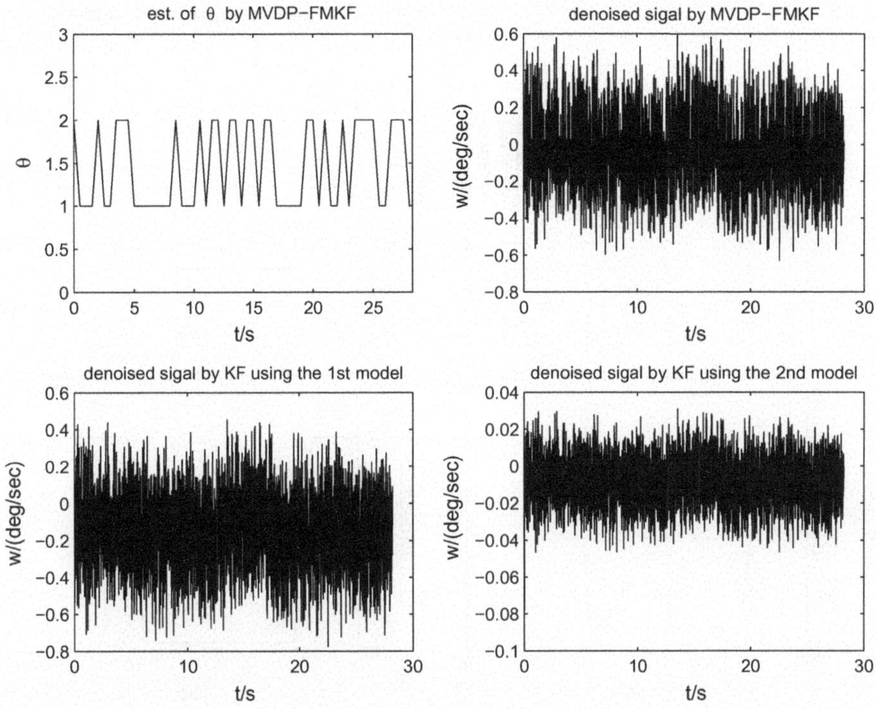

Fig. 7.13 For gyroscope constant angle rate output data, the online filtering results depict the consistence of MVDP-FMKF algorithm and the Kalman filter based on the first model

7.5.1 One-Dimensional Simulation

We consider the following simple scalar system:

$$\begin{cases} x_k = \theta x_{k-1} + w_{k-1} \\ z_k = x_k + v_k \end{cases} \tag{7.33}$$

where $E(w_k) = E(v_k) = 0$, $Var(w_k) = Var(v_k) = 0.1$, and $\theta = 1$.

Supposing that the precise value of θ is unknown a priori, we only know that $\theta \in \{0.1, 1, 5\}$. Obviously, a single Kalman filter cannot make sure to generate reliable state estimates. In this simulation, we have three models, and Kalman filters are designed for $\theta = 0.1, 1, 5$, respectively. The true value of θ is randomly taken from the set $\{0.1, 1, 5\}$ and then fixed for simulation. The simulation results of MVDP finite-model Kalman filter algorithm compared with KF, AMM, and IMM are shown in Figs. 7.3, 7.4, and 7.5, where the true value of θ is 1. For comparison, besides the results of MVDP finite-model Kalman filter algorithm, the ideal results of the optimal Kalman filter using the true system parameter ($\theta = 1$), the non-ideal results of the Kalman filter using the wrong system parameter ($\theta = 0.1$), the non-ideal results of

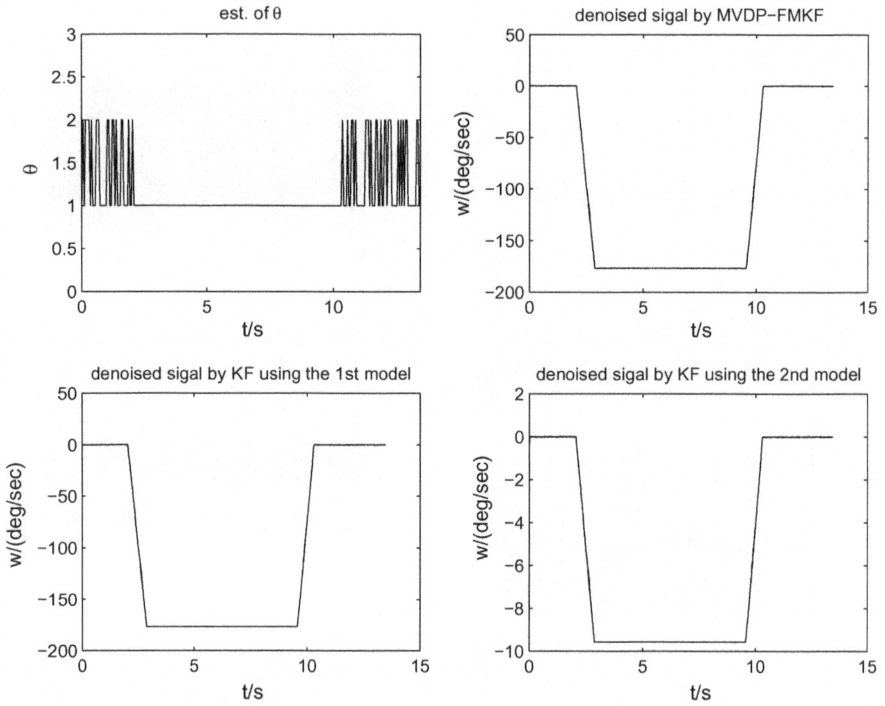

Fig. 7.14 For gyroscope constant angle rate output data, the online filtering results depict the better performance of MVDP-FMKF algorithm comparing with the Kalman filtering based on the second model

the Kalman filter using the wrong system parameter ($\theta = 5$), AMM algorithm, and IMM algorithm are also illustrated, which show that the Kalman filter using the wrong model performs worse than the Kalman filter using the precise model and the MVDP-FMKF algorithm.

From Figs. 7.3 and 7.4, it can be seen that the denoising effect of MVDP finite-model Kalman filter is almost equal to ideal Kalman filter with precise system parameters and state estimate errors will diverge with wrong system models ($\theta = 0.1$ and $\theta = 5$). That is to say, MVDP finite-model Kalman filter can choose the right system model in a large probability and get acceptable denoising effect when system parameters cannot be obtained precisely.

From Fig. 7.5, it can be seen that the denoising effect of MVDP-FMKF is comparable with that of AMM and IMM algorithms. However, MVDP finite-model Kalman filter algorithm does not need to calculate the model likelihood and mode probability, especially exponential functions. That is to say, MVDP finite-model Kalman filter algorithm has smaller computation cost compared with AMM and IMM algorithm. And MVDP-FMKF does not need to know transition probability matrix (TPM).

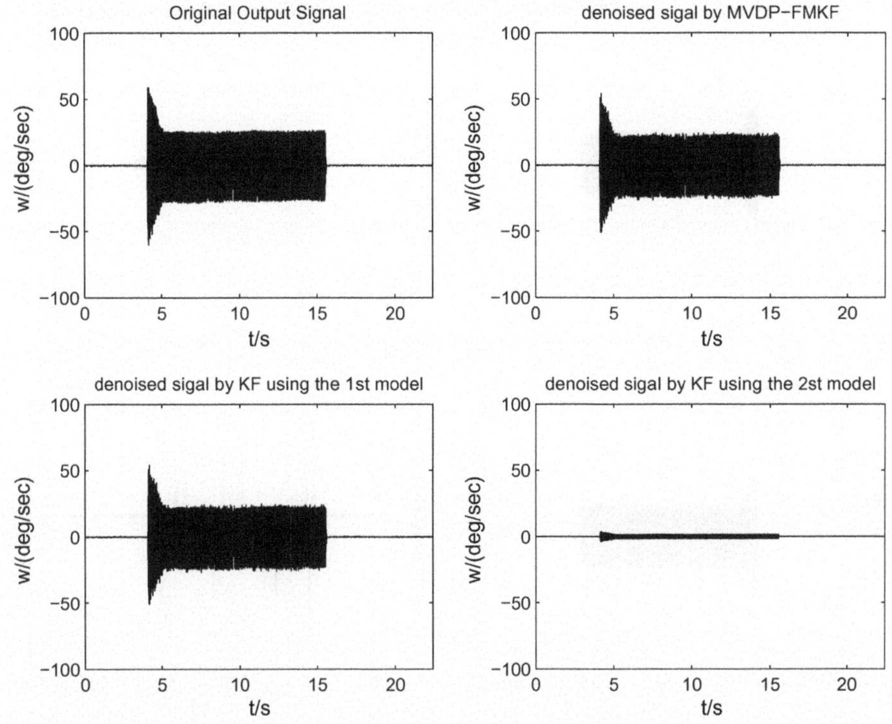

Fig. 7.15 For gyroscope angle swing output data (by 20 Hz frequency of 0.3° amplitude), online filtering results depict the success tracking of MVDP-FMKF algorithm for the original output data

Remark 7.5 Computation cost is very important for a practical system to deal with the signal real time. Especially, in an inertial integrated navigation system, there are many filtering blocks, and computation effort is more important than precision in some aspects. In the inertial integrated navigation system, MVDP-FMKF can be used to deal with the gyroscope output signal. With the low computation cost, the filter has been proved in practice that it can get acceptable results and satisfy the real-time performance.

Besides, AMM and IMM algorithms cannot deal with the estimating problem when the system models $\mathbb{M}_i$ have different dimensions.

Remark 7.6 Nowadays, the multiple models used by the existing multiple model estimators need not to have the same dimension or structure. A very typical example in the target tracking area is the use of the constant velocity motion model and constant acceleration model. For sure, these two models have different dimensions. But they just use the common state variable information to get estimating state and abandon private state variable information. That is to say, multiple model (MM) algorithms were designed to deal with system model set with same dimensions although they

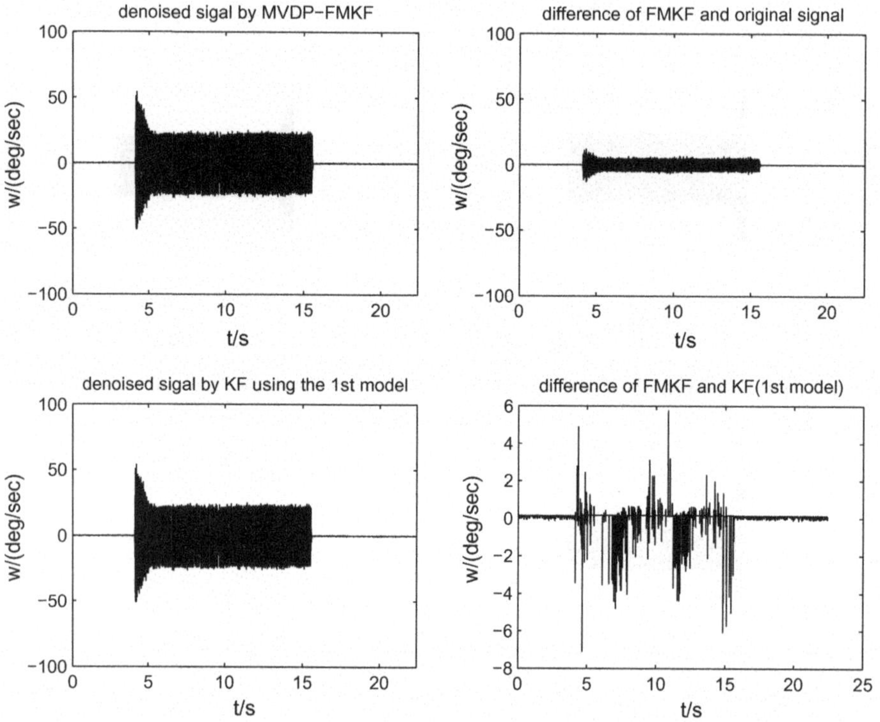

Fig. 7.16 For gyroscope angle swing output data (by 20 Hz frequency of 0.3° amplitude), the tracking performance of MVDP-FMKF algorithm is similar to that of Kalman filter based on the first model

can obtain good results in the model set with different dimensions. In the framework of FMKF, generally it is not necessary to require that all models share some common states.

For the system shown in Eq. (7.33), supposing that the precise value of θ is unknown a priori, we only know that $\theta \in \{0.1, 5\}$ or system model is as follows:

$$\begin{aligned} \begin{bmatrix} x_k \\ \delta_k \end{bmatrix} &= \begin{bmatrix} 1 & 1 \\ 0 & 1 \end{bmatrix} \begin{bmatrix} x_{k-1} \\ \delta_{k-1} \end{bmatrix} + \begin{bmatrix} 1 \\ 0 \end{bmatrix} w_{k-1} \\ z_k &= \begin{bmatrix} 1 & 0 \end{bmatrix} \begin{bmatrix} x_k \\ \delta_k \end{bmatrix} + v_k \end{aligned} \tag{7.34}$$

Remark 7.7 The model in (7.34) is a different system model. MVDP-FMKF demands that the model set has to have at least one or more model that can describe the system motion exactly or roughly at every moment. But in this example, the first and second cannot describe the system motion. So we build the model (7.34) to describe the system motion to satisfy MVDP-FMKF demand. We use this example to demonstrate

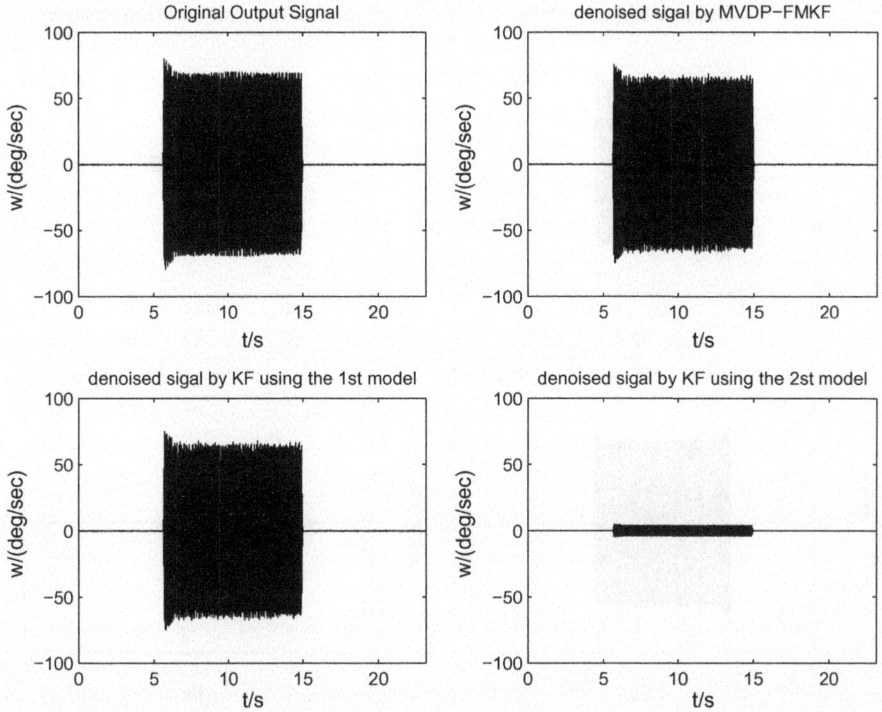

Fig. 7.17 For gyroscope angle swing output data (by 10 Hz frequency of 1° amplitude), the online filtering results depict the success tracking of MVDP-FMKF algorithm for the original output data

that MVDP-FMKF method can be applied in the cases of multiple different types of models.

Under this situation, AMM algorithm and IMM algorithm cannot generate system state estimating because the system model set contains models with different dimensions, which obviously makes it difficult to introduce and calculate the likelihood function for each model candidate and the transition probability matrix (TPM) of the system model set is also unknown, which is another factor to limit using AMM and IMM algorithms. While MVDP finite-model Kalman filter can deal with these estimating problems, the principle of MVDP optimization can be applied to models of any dimension and transition probability matrix (TPM) is not asked to be obtained. The simulation results are shown in Fig. 7.6.

In Fig. 7.6, after finite time steps $index = 1$ when the first system model ($\theta = 0.1$) is active, $index = 2$ when the second system model ($\theta = 5$) is active, and $index = 3$ when the third system model is active.

From Fig. 7.6, it can be seen that MVDP finite-model Kalman filter can deal with mixing-dimensional system state estimating problems. And it can also yield an

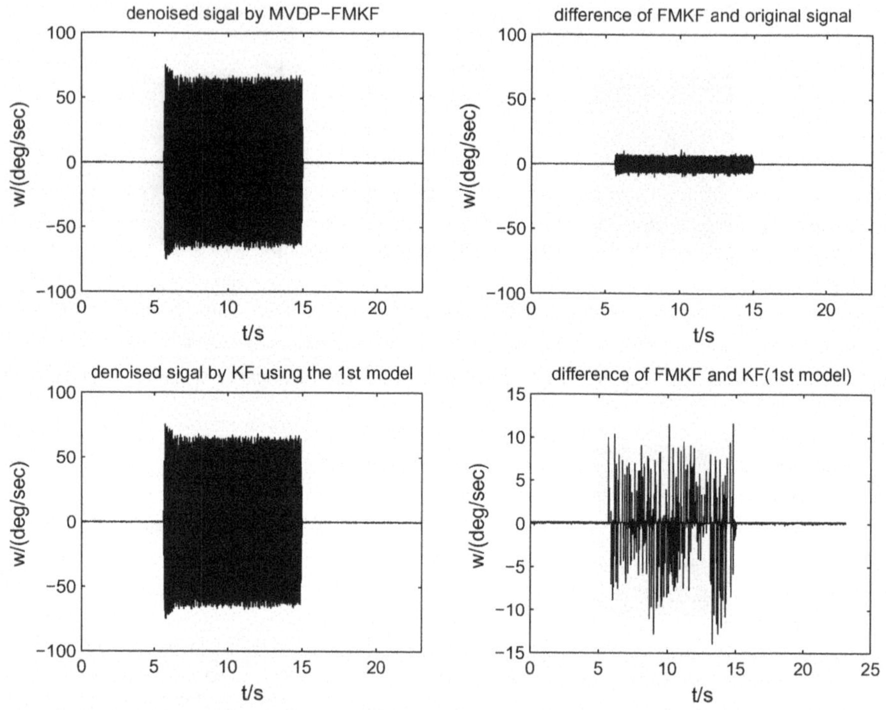

Fig. 7.18 For gyroscope angle swing output data (by 10 Hz frequency of 1° amplitude), the tracking performance of MVDP-FMKF algorithm is similar to that of Kalman filter based on the first model

acceptable filtering performance even when the system model does not contain the precise system model.

7.5.2 *Multidimensional Simulation*

In this part, multidimensional MVDP finite-model Kalman filter simulation is presented. The true system model is given as follows:

$$\begin{aligned}
\begin{bmatrix} x_1(k) \\ x_2(k) \end{bmatrix} &= \begin{bmatrix} 1 & 0 \\ 0 & 1 \end{bmatrix} \begin{bmatrix} x_1(k-1) \\ x_2(k-1) \end{bmatrix} + \begin{bmatrix} w_1(k) \\ w_2(k) \end{bmatrix} \\
\begin{bmatrix} z_1(k) \\ z_2(k) \end{bmatrix} &= \begin{bmatrix} 1 & 0 \\ 0 & 1 \end{bmatrix} \begin{bmatrix} x_1(k) \\ x_2(k) \end{bmatrix} + \begin{bmatrix} v_1(k) \\ v_2(k) \end{bmatrix}
\end{aligned} \tag{7.35}$$

where $E(w_i) = E(v_i) = 0$, and $Var(w_i) = Var(v_i) = 0.1$ $(i = 1, 2)$. Then finite known models include three models, among which one is the true system model, and the other ones are similar to the true system model except that its system transition matrix $\phi_{k,k-1}$ is $\begin{bmatrix} 5 & 0 \\ 0 & 5 \end{bmatrix}$ or $\begin{bmatrix} 2 & 0 \\ 0 & 2 \end{bmatrix}$ rather than $\begin{bmatrix} 1 & 0 \\ 0 & 1 \end{bmatrix}$. The filtering problem is to estimate the unknown states $x_{1,k}$ and $x_{2,k}$ in the presence of large model uncertainty since the exact system model is completely not known a priori. The simulation results are shown in Figs. 7.7, 7.8, 7.9, 7.10, 7.11, and 7.12 where the results of ideal Kalman filter using the precise system model are also plotted for the purpose of comparison.

From the simulation results, it can be seen that in this case of multidimensional simulation, the new algorithm could choose the right system model ($A = I$) as the active one with big probability and obtain state estimates almost as precisely as Kalman filter with the accurate system model ($A = I$). Besides, state estimate errors have a diverging trend with wrong system models ($A = 2I$ and $A = 5I$). In other words, the new algorithm can also get acceptable denoising effect when the true system model is not precisely available.

7.6 Experimental Test

In this section, we choose MEMS gyroscope to test the proposed MVDP-FMKF algorithm. For MEMS gyroscope, the existence of random error is the most important factor to influence the precision. Because of the nature of uncertain random error, it cannot be simply compensated as unknown yet constant drift error; hence, it is the main error to limit measurement precision. Therefore, it is particularly important to reduce random error to improve measurement accuracy of MEMS gyroscope.

We use MEMS130 gyroscope in the condition of 25 °C to test the practical performance of MVDP-FMKF algorithm. Shaft angle rate information of the output data of MEMS gyroscope is used as original data. The sampling period is $T = 5$ ms.

In this experiment, two different models are used to test MVDP finite-model Kalman filter algorithm. The first model is taken as an ARMA(2,2) (Auto-Regressive and Moving Average) model based on the zero drift data. And the second model is an ARMA(2,2) model based on the constant output and swing data.

The real-time filtering results of gyroscope zero drift are depicted in Fig. 7.13. From Fig. 7.13, it can be seen that compared with the first model and the second model, MVDP finite-model Kalman filter has good denoising effect for the gyroscope zero drift output. From the value of θ, it can be seen that MVDP-FMKF algorithm switched between the first and second system models. For the gyroscope signal generated with zero drift (i.e., the first model), the MVDP-FMKF algorithm has similar performance with the corresponding optimal Kalman filter using the first system model, while the denoised signal by the Kalman filter using the wrong model (second model) is almost incorrect from its magnitude.

Testing gyroscope constant output, the real-time filtering results are shown in Fig. 7.14. From Fig. 7.14, it can be seen that when the gyroscope has a high constant output, the second model has amplitude attenuation. But MVDP finite-model Kalman filter does not have amplitude attenuation. The better filtering results are automatically chosen by MVDP finite-model Kalman filter. From the value of θ, we can see that MVDP-FMKF algorithm adaptively choose the first system model when the second model has big amplitude attenuation. For zero drift output signal, the algorithm can choose the second model to improve denoising effect at some time.

Testing the gyroscope swing by 20 Hz frequency of 0.3° amplitude, the real-time filtering results are shown in Figs. 7.15 and 7.16. From Figs. 7.15 and 7.16, it can be seen that when the gyroscope wings by 20 Hz frequency of 0.3° amplitude, the second model has amplitude attenuation. But MVDP finite-model Kalman filter does not have amplitude attenuation and it can follow the gyroscope real-time output. That is to say, MVDP finite-model Kalman filter has good real-time performance and the bandwidth of the algorithm is more than or equal to 20 Hz at least.

Testing the gyroscope swing by 10 Hz frequency of 1° amplitude, the real-time filtering results are shown in Figs. 7.17 and 7.18. From Figs. 7.17 and 7.18, it can be seen that when the gyroscope swings by 10 Hz frequency of 1° amplitude, the second model has amplitude attenuation. MVDP finite-model Kalman filter does not have amplitude attenuation and it can follow large-amplitude high-frequency real-time output.

7.7 Conclusion

Kalman filtering techniques have been widely used in many applications, especially the navigation problems. However, successful applications of standard Kalman filters often require that the state-space model in use is accurate enough with exact system matrices or parameters as well as proper a priori information of the process noise and the measurement noise, which in fact request people to make great efforts in the modeling procedure before the real application of Kalman filter. Generally speaking, exact or perfect modeling is impossible or expensive in most practical applications, and inaccurate model often results that the performance of Kalman filter is very poor or completely not acceptable.

To address the crucial filtering degradation problem due to the large model uncertainty, motivated by our previous pioneering work on finite-model adaptive control, we introduce the idea of finite-model Kalman filtering, which supposes that the large model uncertainty may be restricted by a finite set of known models. This assumption is not as restrictive as it looks, since the known models can be very different from each other, and the number of known models can be flexibly chosen so that the uncertain model may always be approximated by one of the known models, although the matched model is not a priori known.

Within the presented framework of finite-model Kalman filter, we have mathematically formulated and illustrated a simple finite-model Kalman filter, termed as MVDP-FMKF, which adopts the idea of adaptive switching via minimizing vector distance principle. Despite the intuitive and easy-to-implement algorithm, its closed-loop system is rather complex, which needs further investigation in the future.

In this contribution, extensive simulations and an experiment of gyroscope drift have verified the effectiveness of the proposed algorithm, which indicates that the mechanism of finite-model Kalman filter is useful and efficient in dealing with large model uncertainties which may be frequently seen in practical applications of Kalman filters and may degrade the performance or even destroy the convergence or stability of standard Kalman filter. In summary, this introduced method of finite-model Kalman filter has practical significance in the inertial navigation systems and may be extended to other types of practical systems where Kalman filtering techniques are required, yet the system model is not accurately known a priori.

In the future, to gain deeper understanding to finite-model Kalman filters, it is necessary to investigate more FMKF algorithms, to mathematically establish their stability properties, and to explore their wide applications.

References

1. L. Zhao, Z.F. He, An in-coordinate interval adaptive Kalman filtering algorithm for INS/GPS/SMNS, in *Proceedings of 2012 10th IEEE International Conference on Industrial Informatics*, pp. 41–44
2. B. Feng, Q. Li, P.P. Yang, G.D. Fu, On-line realization of multi-scale soft-threshold de-noising of MEMS gyro by wavelet, in *Proceedings of the 2009 9th International Conference on Electronic Measurement & Instruments*, pp. 2–442–445
3. R.E. Kalman, A new approach to linear filtering and prediction problems. Trans. ASME-J. Basic Eng. **82**(Series D), 35–45 (1960)
4. Z. Deng, S. Sun, Wiener state estimatiors based on Kalman filtering. Acta Autom. Sin. **30**(1), 126–130 (2004)
5. B.S. Koh, J. Junkins, Kalman filter for linear fractional order systems. J. Guid., Control Dyn. **35**(6), 1816–1827 (2012)
6. S. Kluge, K. Reif, M. Brokate, Stochastic stability of the extended Kalman filter with intermittent observations. IEEE Trans. Autom. Control **55**(2), 514–518 (2010)
7. Q. Ge, W. Li, R. Sun, Z. Xu, Research on centralized fusion algorithms based on EKF for multisensor non-linear systems. Acta Autom. Sin. **38**(12), 1–10 (2012)
8. N.A. Carlson, Federated square root filter for decentralized parallel processors. IEEE Trans. Aerosp. Electron. Syst. **26**(3), 517–525 (1990)
9. L. Zhao, X. Wang, M. Sun, J. Ding, C. Yan, Adaptive UKF filtering algorithm based on maximum a posterior estimation and exponential weighting. Acta Autom. Sin. **36**(7), 1007–1019 (2010)
10. S.J. Julier, J.K. Uhlmann, Unscented filtering and nonlinear estimation. Proc. IEEE **92**(3), 401–422 (2004)
11. X. Kai, C. Wei, L. Liu, Robust extended Kalman filtering for nonlinear systems with stochastic uncertainties. IEEE Trans. Syst., Man, Cybern. Part A: Syst. Hum. **40**(2), 399–405 (2010)
12. M. Karasalo, X. Hu, An optimization approach to adaptive Kalman filtering. Automatica **47**(8), 1785–1793 (2011)
13. O. Yeste Ojeda, J. Grajal, Adaptive-fresh filters for compensation of cycle-frequency errors. IEEE Trans. Signal Process. **58**(1), 1–10 (2010)
14. X. Xiao, B. Feng, B. Wang, On-line realization of SVM Kalman filter for MEMS gyro, in *Proceedings of the 3rd International Conference on Measuring Technology and Mechatronics Automation*, pp. 768–770
15. L. Xie, Y.C. Soh, Robust Kalman filtering for uncertain systems. Syst. Control Lett. **22**(2), 123–129 (1994)

16. H.B. Ma, Capability and limitation of feedback mechanism in dealing with uncertainties of some discrete-time control systems. Ph.D. thesis, Graduate School of Chinese Academy of Sciences, Beijing, China, June 2006
17. H.B. Ma, Finite-model adaptive control using LS-like algorithm. Int. J. Adapt. Control Signal Process. **21**(5), 391–414 (2006)
18. H.B. Ma, Finite-model adaptive control using WLS-like algorithm. Automatica **43**, 677–684 (2007)
19. H.B. Ma, Several algorithms for finite-model adaptive control problem, Math. Control, Signals, Syst. **20**, 271–303 (2008). Available online from 28 May 2008
20. X.R. Li, V.P. Jilkov, A survey of maneuvering target tracking-part v: multiple-model methods. IEEE Trans. Aerosp. Electron. Syst. **41**(4), 1255–1321 (2005)
21. Y. Bar-Shalom, X. Li, T. Kirubarajan, *Estimation with Applications to Tracking and Navigation: Theory Algorithms and Software* (Wiley, New York, 2001)

Chapter 8
Kalman Filters for Continuous Parametric Uncertain Systems

Kalman filter algorithm is an optimal state estimator in the sense of minimum mean squared errors and maximum likelihood estimation. However, standard Kalman filter is highly restricted by strict a priori requirements on the information of dynamic model and statistical information of process noises. In practice, the covariance matrix of process noise is almost impossible to be directly determined a priori due to the intrinsic coupling of process noise and system state dynamics. Considering such background of Kalman filter algorithm, one algorithm, named as recursive covariance estimation (RCE), is introduced to estimate the covariance matrix when there is no information of process noise covariance matrix in the linear time-invariant (LTI) systems. Based on the framework of standard Kalman filter, a new algorithm named as Kalman filter with recursive covariance estimation (KF-RCE) is introduced to resolve this challenging problem of state estimation without the covariance matrix of process noise. Then the rigorous stability analysis is presented to show that the algorithm is optimal and the covariance matrix and state estimations are asymptotically consistent with the ideal Kalman filter using the exact covariance matrix of process noise. Finally, some simulation results of navigation system are given to examine theory analysis theorem and algorithm optimality.

8.1 Introduction

The problem of signal estimation deals with recovering some desired variables of a dynamic system from available measurements. A renown estimation approach is the so-called Kalman filtering. The idea of this algorithm is to design a filter to minimize the estimation error covariance, see [1]. The Kalman filtering has demonstrated its power in state estimation, system identification, adaptive control, and signal processing [2] and has been widely used in many engineering problems [3] since it was originally developed in the 1960s [4]. One of the key assumptions in standard Kalman filtering is that the dynamic model and the noises of the system under consideration are exactly known and the noises are zero-mean Gaussian white noises.

H. Ma et al., *Kalman Filtering and Information Fusion*,
https://doi.org/10.1007/978-981-15-0806-6_8

However, in many practical cases, the assumption abovementioned is difficult to be guaranteed. This mainly because any real system usually has various degrees of parametric uncertainty and the statistics of the noise usually can't be known. More recently, the assumption above, which is needed for the standard Kalman filtering approach, has been relaxed so that the designer will be able to cope with parameter uncertainty and unknown noise statistics. A kind of filters named H_∞ filter was proposed to deal with the above situation. The H_∞ filter does not make any assumption about the noise, and it minimizes the worst-case estimation error [5]. Hence, it is also called minimax filter. Many other results on the robust filtering problem are available in the literature to date (see [6–9]). These papers share the same model chosen to describe the parameter uncertainty, namely, norm-bounded uncertainty acting in the state and/or in the output matrices. After 40 years of development, the Kalman filtering theory has developed into a more comprehensive theoretical system. There are some nonlinear extensions of the Kalman filter, such as the extended Kalman filter(EKF)[10] and unscented Kalman filter (UKF) [11].

On the other hand, although various kinds of Kalman filtering algorithms have been proposed, most of them are based on equal-spaced observations, i.e., the sensors pick a sequence of observations $f(nT), n = 1, 2, \ldots$ every other a fixed sampled period T. This is an ideal situation. In fact, limited by the sampling equipments and sampled signals, exactly equal sampling period can't be achieved. Hence, in a real system, except for the uncertain parameters, there exist nonuniformly sampled observations.

Recently, a great deal of attention has been paid to multirate uncertain systems. A lift technique which can change a time-varying periodic system to a time-invariant periodic system was given in [12]. The data-driven Kalman filters for nonuniformly sampled multirate (NUSM) systems were proposed in [13–15]. The polynomial transformation technique studied the problem of identification and output estimation for dual-rate systems and analyzed the rate of convergence of the parameter estimation in the stochastic framework was put forward in literature [16–18]. However, for nonuniformly sampled system, if there exist uncertain parameters in system models, how shall we handle the above situation. This is just the main research objective of this chapter.

In the present note, we consider designing a filter for linear continuous systems with uncertain parameters of the system model based on Kalman filtering theory under nonuniformly sampled data. We assume that the measurement noises are stationary white signals with known statistics. The problem we address is the design of the filter such that the filtering errors are guaranteed to be minimized. We are interested in an adaptive Kalman filter which has the essential structure of a standard Kalman filter but with the necessary modifications to account for the parameter uncertainty and nonuniformly sampled observations. The main idea of this algorithm proposed has two key points:

- Discretize the continuous system to a discrete-time (DT) system with nonuniformly sampled data by definition.

Table 8.1 The notation of this chapter

$E\langle * \rangle$	Expected value of the vector $*$
I	The identity matrix of appropriate size
$*^T$	The transpose of the matrix $*$
$\mathrm{tr}(*)$	The trace of the matrix $*$
$\hat{*}$	The estimated value of the state vector $*$

- During iterations of the algorithm, we utilize the states of last iteration to estimate the current parameter and use estimated parameter to estimate the states of next iteration, which make the parameter estimation and state filtering in a closed loop.

At last, a numerical example is presented to demonstrate the efficiency of the algorithm.

Briefly, this chapter is organized as follows: Sect. 8.2 utilizes the discretization method to get the discrete-time state-space model under nonuniformly sampled data. Section 8.3 derives the parameter estimating and state filtering algorithms of model based on the Kalman filtering principle. Section 8.4 gives the convergence analysis of the proposed algorithms. Section 8.6 provides an illustrative example to show effectiveness of the algorithms proposed. Finally, we offer some concluding remarks in Sect. 8.6.

We use the notation as in Table 8.1.

8.2 Problem Formulation

For conventional discrete-time systems, the periods of zero-order holders and samplers are the same with period T; for nonuniformly sampled multirate systems, the updating and sampling intervals are $T_1, T_2, \ldots, T_n$. We set $T_i = T + \tau_i, i = 1, 2, 3, \ldots$. And τ_i is a random variable which subjects to normal distribution $N(0, \sigma^2)$ where σ is a limited constant.

Consider the following class of uncertain continuous-time systems:

$$\begin{cases} \dot{x}(t) = (A + \delta A_\delta)x(t) + Bu(t) \\ y(t) = Cx(t) + v(t) \end{cases} \tag{8.1}$$

where $x(t) \in \boldsymbol{R}^n$ is the state vector, $\delta \in \boldsymbol{R}^1$ represents the unknown parameter which keeps unchanging, $u(t) \in \boldsymbol{R}^1$ is the controlled input, $y(t) \in \boldsymbol{R}^1$ is the output, $v(t) \in \boldsymbol{R}^1$ is a stochastic noise with zero mean, and A, A_δ, B, C are constant matrices of appropriate sizes. From (8.1) according to the definition of the discretization, we have

$$x(t_{k+1}) = e^{(A+\delta A_\delta)T_k} x(t_k) + \int_{t_k}^{t_{k+1}} e^{(A+\delta A_\delta)(t_{k+1}-\tau)} Bu(\tau)d\tau \tag{8.2}$$

Assume that $u(t)$ keep unchanging during the period $[t_k, t_{k+1}]$. Substituting $t = t_{k+1} - \tau$ into (8.2), we can yield

$$\begin{aligned} & x(t_{k+1}) \\ &= e^{(A+\delta A_\delta)T_k} x(t_k) + \int_0^{T_k} e^{(A+\delta A_\delta)t} dt Bu(t_k) \\ &= A_{k,\delta} x(t_k) + B_{k,\delta} u(t_k) \end{aligned} \tag{8.3}$$

where

$$A_{k,\delta} = e^{(A+\delta A_\delta)T_k} \tag{8.4}$$

$$B_{k,\delta} = \int_0^{T_k} e^{(A+\delta A_\delta)t} dt B \tag{8.5}$$

Denoting that $x(t_{k+1}) := x_{k+1}$, $u(t_k) := u_k$, we obtain the discrete-time space model of the abovementioned continuous-time parametric uncertain system

$$\begin{cases} x_{k+1} = A_{k,\delta} x_k + B_{k,\delta} u_k \\ y_k = C x_k + v_k \end{cases} \tag{8.6}$$

8.3 The Estimation Algorithm

In this section, we derive a state estimation algorithm for the system in consideration based on Kalman filtering theory. Therefore, the first step we need to do is to estimate the unknown parameter δ.

8.3.1 The Kalman Filtering-Based Parameter Estimation

In this subsection, we derive an estimation algorithm for the unknown parameter of system in consideration. From the assumptions about the system in consideration, we can assume that the unknown parameter keep invariant during the whole iterations. That is to say, we can set the state equation of the unknown parameter as follows:

$$\delta_{k+1} = \delta_k$$

In order to apply the Kalman filtering theory, the next we need to do is to construct a measurement equation of the unknown parameter δ.

From the discrete-time space model (8.6), we have

$$y_{k+1} = CA_{k,\delta}x_k + CB_{k,\delta}u_k + v_k \tag{8.7}$$

Substituting (8.4) and (8.5) into (8.7) yields

$$\begin{aligned} y_{k+1} &= Ce^{(A+\delta A_\delta)T_k}x_k \\ &\quad + C\int_0^{T_k} e^{(A+\delta A_\delta)t}dt\,Bu_k \end{aligned} \tag{8.8}$$

Obviously, Eq. (8.8) is a nonlinear function about δ. Taking the Taylor first-order approximation on δ of (8.8), we can get

$$\begin{aligned} y_{k+1} &= CA_k(I + A_\delta T_k \delta_k)x_k \\ &\quad + CB_k u_k + CB_\delta u_k \delta_k + v_k \end{aligned}$$

where

$$\begin{aligned} A_k &= e^{AT_k} \\ B_k &= \int_0^{T_k} e^{At}dt\,B \\ B_\delta &= \int_0^{T_k} e^{A_\delta t}A_\delta t dt\,B \end{aligned}$$

Thus, we can get the state-space equation of δ as follows:

$$\begin{cases} \delta_{k+1} &= \delta_k \\ y_{\delta,k+1} &= C_{\delta,k+1}\delta_{k+1} + v_{k+1} \end{cases} \tag{8.9}$$

where

$$y_{\delta,k} = y_{k+1} - CA_k x_k - CB_k u_k \tag{8.10}$$

$$C_{\delta,k} = CA_k A_\delta T_k x_k + CB_\delta u_k \tag{8.11}$$

Remark 8.1 In Eq. (8.11), the true states x_k can't be obtained. Therefore, we usually use the estimated values of x_k to represent the true values of x_k.

The basic idea of Kalman filtering theory is to minimize the estimation error covariance matrix P to get an optimal gain vector L such that the estimation errors arrive at least. The algorithm we will propose is based on the derived state-space model in (8.9). According to the Kalman filtering principle, suppose that the state variable δ_{k+1} is estimated by the recursive equation,

$$\hat{\delta}_{k+1} = \hat{\delta}_k + L_{\delta,k+1}(y_{\delta,k+1} - C_{\delta,k+1}\hat{\delta}_{k+1|k}) \tag{8.12}$$

where

$$\hat{\delta}_{k+1|k} = \hat{\delta}_k$$

Define the parameter estimation error

$$\tilde{\delta}_{k+1} := \delta_{k+1} - \hat{\delta}_{k+1} \tag{8.13}$$

Replacing (8.13) with (8.12), we have

$$\tilde{\delta}_{k+1} = (I - L_{\delta,k+1}C_{\delta,k+1})\tilde{\delta}_k - L_{\delta,k+1}v_k \tag{8.14}$$

The purpose of the algorithm is to choose an optimal gain vector $L_{\delta,k+1}$ to minimize the estimation mean square error

$$\min E[|\tilde{\delta}_{k+1}|^2]$$

It is easy to note that

$$\begin{aligned} E[|\tilde{\delta}_{k+1}|^2] &= \mathrm{tr}(E\langle\tilde{\delta}_{k+1}\tilde{\delta}_{k+1}^T\rangle) \\ &= \mathrm{tr}(P_{\delta,k+1}) \end{aligned} \tag{8.15}$$

Since the state estimation error is uncorrelated with $L_{\delta,k+1}$, expanding out the term $P_{\delta,k+1}$ in the equation above and using Eqs. (8.14), (8.15) yields

$$\begin{aligned} &P_{\delta,k+1} \\ &= (I - L_{\delta,k+1}C_{\delta,k+1})P_{\delta,k}(I - L_{\delta,k+1}C_{\delta,k+1})^T \\ &\quad + L_{\delta,k+1}R_v L_{\delta,k+1}^T \\ &= P_{\delta,k} - L_{\delta,k+1}C_{\delta,k+1}P_{\delta,k} \\ &\quad - P_{\delta,k}C_{\delta,k+1}^T L_{\delta,k+1}^T + L_{\delta,k+1}S_{\delta,k+1}L_{\delta,k+1}^T \end{aligned}$$

where

$$\begin{aligned} S_{\delta,k+1} &= C_{\delta,k+1}P_{\delta,k}C_{\delta,k+1}^T + R_v, \\ R_v &= E\langle v(t)v^T(t)\rangle \end{aligned}$$

The trace is minimized when the matrix derivation is zero

$$\frac{\partial \mathrm{tr}(P_{\delta,k+1})}{\partial L_{\delta,k+1}} = -2(C_{\delta,k+1}P_{\delta,k})^T + 2L_{\delta,k+1}S_{\delta,k+1} = 0$$

Solving above equation for $L_{\delta,k+1}$ yields the Kalman gain

$$L_{\delta,k+1} = P_{\delta,k}C_{\delta,k+1}^T S_{\delta,k+1}^{-1} \tag{8.16}$$

Hence, the parameter estimation algorithm for the system in (8.9) may be expressed as

$$\begin{aligned}
\hat{\delta}_{k+1} &= \hat{\delta}_k + L_{\delta,k+1}(y_{\delta,k+1} - C_{\delta,k+1}\hat{\delta}_k) \\
S_{\delta,k+1} &= C_{\delta,k+1}P_{\delta,k}C_{\delta,k+1}^T + R_v \\
L_{\delta,k+1} &= P_{\delta,k}C_{\delta,k+1}^T S_{\delta,k+1}^{-1} \\
P_{\delta,k+1} &= (I - L_{\delta,k+1}C_{\delta,k+1})P_{\delta,k}(I - L_{\delta,k+1}C_{\delta,k+1})^T \\
&\quad + L_{\delta,k+1}R_v L_{\delta,k+1}^T
\end{aligned} \tag{8.17}$$

8.3.2 The Kalman Filtering-Based State Estimation

The key objective of the algorithm proposed is to get the precise estimated states of the system, which make it can be used for designing controllers more conveniently. Analyzing the system in (8.6), we have found that the Kalman filtering algorithm is suitable for the construction of the system except for the unknown parameter, while we have derived the estimation algorithm for the unknown parameter in the above subsection. Hence, the next we need to do is to formulate a state estimation algorithm. Here, the state estimation algorithm we will propose is also based on Kalman filtering theory. Rewrite the state-space model of the system (8.6)

$$\begin{cases} x_{k+1} = A_{k,\delta}x_k + B_{k,\delta}u_k \\ y_k \quad\; = Cx_k + v_k \end{cases} \tag{8.18}$$

Similarly, applying the Kalman filtering theory to the system (8.18), we can get the recursive equation of estimated state vector $\hat{x}_{k+1}$ as follows:

$$\hat{x}_{k+1} = \hat{x}_{k+1|k} + L_{x,k+1}(y_{k+1} - C\hat{x}_{k+1|k}) \tag{8.19}$$

where

$$\hat{x}_{k+1|k} = A_{k,\delta}\hat{x}_k + B_{k,\delta}u_k$$

Define the pre- and prior- estimation errors as follows:

$$\begin{cases} \tilde{x}_{k+1} \quad = x_{k+1} - \hat{x}_{k+1} \\ \tilde{x}_{k+1|k} = x_{k+1} - \hat{x}_{k+1|k} \end{cases} \tag{8.20}$$

Using (8.19) and (8.20), the state estimation errors satisfy

$$\begin{aligned}\tilde{x}_{k+1} &= \tilde{x}_{k+1|k} - L_{x,k+1}(C\tilde{x}_{k+1|k} + v_{k+1}) \\ &= (I - L_{x,k+1}C)\tilde{x}_{k+1|k} - L_{x,k+1}v_{k+1}\end{aligned}$$

According to the definition of covariance matrix, we can yield the covariance of the state estimation errors,

$$P_{x,k+1} := E\langle \tilde{x}_{k+1}\tilde{x}_{k+1}^T \rangle \tag{8.21}$$

Since the state estimation errors are uncorrelated with $L_{x,k+1}$, replacing k in (8.32) with $k+1$ gives

$$\begin{aligned}P_{x,k+1} &= E\langle \tilde{x}_{k+1}\tilde{x}_{k+1}^T \rangle \\ &= (I - L_{x,k+1}C)P_{x,k+1|k}(I - L_{x,k+1}C)^T \\ &\quad + L_{x,k+1}R_v L_{x,k+1}^T\end{aligned}$$

where

$$\begin{aligned}P_{x,k+1|k} &= E\langle \tilde{x}_{k+1|k}\tilde{x}_{k+1|k}^T \rangle \\ P_{x,k+1|k} &= A_{k,\delta}P_{x,k}A_{k,\delta}^T\end{aligned}$$

According to the Kalman theory, the estimation mean square errors $E[|\tilde{x}_{k+1}|^2]$ are minimized when the following equation

$$\frac{\partial \mathrm{tr} P_{x,k+1}}{\partial L_{x,k+1}} = 0$$

is satisfied. That is to say,

$$-2P_{x,k}C^T + 2L_{x,k+1}S_{x,k+1} = 0 \tag{8.22}$$

where

$$S_{x,k+1} = R_v + CP_{x,k+1|k}C^T$$

Therefore, the Kalman filter for the system (8.18) is as follows:

$$\begin{aligned}\hat{x}_{k+1} &= \hat{x}_{k+1|k} + L_{x,k+1}(y_{k+1} - C\hat{x}_{k+1|k}) \\ P_{x,k+1|k} &= A_{k,\delta}P_{x,k}A_{k,\delta}^T \\ L_{x,k+1} &= P_{x,k+1|k}C^T S_{x,k+1}^{-1} \\ S_{x,k+1} &= R_v + CP_{x,k+1|k}C^T \\ P_{x,k+1} &= (I - L_{x,k+1}C)P_{x,k+1|k}(I - L_{x,k+1}C)^T \\ &\quad + L_{x,k+1}R_v L_{x,k+1}^T\end{aligned} \tag{8.23}$$

Remark 8.2 In Eq. (8.23), the parameter δ can't be obtained. Usually, we use the estimated value $\hat{\delta}$ to represent the true value during iterations.

Now we have derived both the parameter estimation and state estimation algorithms. From (8.17) and (8.23), it is easy to note that when the unknown parameter is estimated, the state estimation is needed and when the states are estimated, the parameter estimation is needed. Therefore, the next we need to do is to make the two estimation algorithms in a closed loop. That is to say, at iteration k, we use the estimated states $\hat{x}_{k-1}$ to represent the true states x_{k-1} and use the estimated parameter $\hat{\delta}_k$ as the true value δ_k to estimate the current states $\hat{x}_k$. Therefore, combining (8.17) and (8.23), we can get the whole filtering algorithms for the system (8.1) as the following equations:

$$\begin{aligned}
\hat{\delta}_k &= \hat{\delta}_{k-1} + L_{\delta,k}(y_{\delta,k} - C_{\delta,k}\hat{\delta}_{k-1}) \\
P_{\delta,k} &= (I - L_{\delta,k}C_{\delta,k})P_{\delta,k-1}(I - L_{\delta,k}C_{\delta,k})^T \\
&\quad + L_{\delta,k}R_v L_{\delta,k}^T \\
L_{\delta,k} &= P_{\delta,k-1}C_{\delta,k}^T S_{\delta,k}^{-1} \\
S_{\delta,k} &= (C_{\delta,k}P_{\delta,k-1}C_{\delta,k}^T + R_v) \\
\hat{x}_{k+1} &= \hat{x}_{k+1|k} + L_{x,k+1}(y_{k+1} - C\hat{x}_{k+1|k}) \\
P_{x,k+1|k} &= A_{k,\hat{\delta}_k} P_{x,k} A_{k,\hat{\delta}_k}^T \\
L_{x,k+1} &= P_{x,k+1|k}C^T(R_v + CP_{x,k+1|k}C^T)^{-1} \\
P_{x,k+1} &= (I - L_{x,k+1}C)P_{x,k+1|k}(I - L_{x,k+1}C)^T \\
&\quad + L_{x,k+1}R_v L_{x,k+1}^T
\end{aligned} \tag{8.24}$$

where

$$\begin{aligned}
A_{k-1} &= e^{AT_{k-1}} \\
B_{k-1} &= \int_0^{T_{k-1}} e^{At}\,dt\,B \\
B_{\delta} &= \int_0^{T_{k-1}} e^{A_\delta t} A_\delta t\,dt\,B \\
y_{\delta,k} &= y_{k+1} - CA_k\hat{x}_k - CB_k u_k \\
C_{\delta,k} &= CA_kA_\delta T_k\hat{x}_k + CB_\delta u_k \\
A_{k,\hat{\delta}_k} &= e^{(A+\hat{\delta}_k A_\delta)T_k} \\
B_{k,\hat{\delta}_k} &= \int_0^{T_k} e^{(A+\hat{\delta}_k A_\delta)t}\,dt\,B
\end{aligned} \tag{8.25}$$

8.4 Convergence Analysis

In this section, the convergence analysis of the proposed algorithm (8.24) will be performed by the Lyapunov approach. In the following, we show that a strictly decreasing Lyapunov function means resolving an LMI problem whose solution depends closely on designing the matrices R_v and $A_{x,k}$. First of all, we rewrite the recursive equation of the states:

$$\hat{x}_{k+1} = \hat{x}_{k+1|k} + L_{x,k+1}(y_{k+1} - C\hat{x}_{k+1|k})$$

Setting $e_{k+1} = y_{k+1} - C\hat{x}_{k+1|k}$, we can get

$$\hat{x}_{k+1} = \hat{x}_{k+1|k} + L_{x,k+1}e_{k+1} \tag{8.26}$$

Now, let's rewrite the error vectors $\tilde{x}_{k+1}$ and $\tilde{x}_{k+1|k}$ as (8.20),

$$\begin{aligned}\tilde{x}_{k+1} &= x_{k+1} - \hat{x}_{k+1}\\ \tilde{x}_{k+1|k} &= x_{k+1} - \hat{x}_{k+1|k}\end{aligned}$$

Then define a candidate Lyapunov function V_{k+1} as follows:

$$V_{k+1} = \tilde{x}_{k+1}^T P_{x,k+1}^{-1} \tilde{x}_{k+1}. \tag{8.27}$$

By subtracting both sides of (8.26) from x_{k+1}, we obtain

$$\tilde{x}_{k+1} = \tilde{x}_{k+1|k} - L_{x,k+1}e_{k+1} \tag{8.28}$$

where

$$\tilde{x}_{k+1|k} = A_{k,\delta}\tilde{x}_k \tag{8.29}$$

$$e_{k+1} = C\tilde{x}_{k+1|k} \tag{8.30}$$

In addition, according to the Kalman filtering theory [5] and (8.24), we have

$$\begin{aligned}L_{x,k+1} &= P_{x,k+1}C^T R_v^{-1}\\ &= P_{k+1|k}C^T(CP_{k+1|k}C^T + R_v)^{-1}\end{aligned} \tag{8.31}$$

and

$$P_{x,k+1}^{-1} = P_{x,k+1|k}^{-1} + C^T R_v^{-1} C \tag{8.32}$$

where

$$P_{x,k+1|k} = E\langle \tilde{x}_{k+1|k}\tilde{x}_{k+1|k}^T\rangle.$$

Substituting (8.31) into (8.28) and (8.32) into (8.27), the Lyapunov function V_{k+1} becomes

$$\begin{aligned} V_{k+1} = &\tilde{x}_{k+1|k}^T P_{x,k+1}^{-1} \tilde{x}_{k+1|k} - \tilde{x}_{k+1|k}^T C^T R_v^{-1} e_{k+1} \\ &- e_{k+1}^T R_v^{-1} C \tilde{x}_{k+1|k} \\ &+ e_{k+1}^T R_v^{-1} C P_{x,k+1} C^T R_v^{-1} e_{k+1} \end{aligned} \tag{8.33}$$

From (8.32), we can get

$$\begin{aligned} V_{k+1} = &V_{k+1|k} + \tilde{x}_{k+1|k}^T C R_v^{-1} C \tilde{x}_{k+1|k} \\ &- \tilde{x}_{k+1|k}^T C^T R_v^{-1} e_{k+1} - e_{k+1}^T R_v^{-1} C \tilde{x}_{k+1|k} \\ &+ e_{k+1}^T R_v^{-1} C P_{x,k+1} C^T R_v^{-1} e_{k+1} \end{aligned}$$

with

$$V_{k+1|k} = \tilde{x}_{k+1|k}^T P_{x,k+1|k}^{-1} \tilde{x}_{k+1|k}$$

According to Eqs. (8.29) and (8.30), (8.33) becomes

$$\begin{aligned} V_{k+1} = &V_{k+1|k} + e_{k+1}^T (R_v^{-1} \\ &- 2R_v^{-1} + R_v^{-1} C P_{x,k+1} C^T R_v^{-1}) e_{k+1} \end{aligned}$$

On the other hand, $V_{k+1|k}$ can be written as

$$\begin{aligned} V_{k+1|k} &= \tilde{x}_k^T A_{x,k}^T (P_{x,k+1|k})^{-1} A_{x,k} \tilde{x}_k \\ &= \tilde{x}_k P_{x,k}^{-1} \tilde{x}_k \end{aligned}$$

where

$$P_{k+1|k} = A_{k,\delta} P_{x,k} A_{k,\delta}^T$$

A decreasing sequence $\{V_k\}_{k=1,...}$ means that

$$V_{k+1} - V_k \leq 0$$

or equivalently

$$\begin{aligned} &V_{k+1} - V_k \\ &= e_{k+1}^T (R_v^{-1} - 2R_v^{-1} + R_v^{-1} C P_{x,k+1} C^T R_v^{-1}) e_{k+1} \\ &\quad + \tilde{x}_k^T [P_{x,k}^{-1} - P_{x,k}^{-1}] \tilde{x}_k \leq 0 \\ &= e_{k+1}^T (-R_v^{-1} + R_v^{-1} C P_{x,k+1} C^T R_v^{-1}) e_{k+1} \\ &\leq 0 \end{aligned} \tag{8.34}$$

A sufficient condition to ensure (8.34) leads to the following LMI:

$$R_v^{-1} C P_{x,k+1} C^T R_v^{-1} - R_v^{-1} \leq 0 \tag{8.35}$$

That is to say, if the inequality (8.35) is satisfied, the convergence of the proposed algorithm can be guaranteed.

8.5 Numerical Example

In this section, we will give an example to demonstrate the feasibility of the proposed algorithms. Give an uncertain continuous-time process model to be

$$\begin{aligned} \dot{x}(t) &= (A + \delta A_\delta)x(t) + Bu(t) \\ y(t) &= Cx(t) + v(t) \end{aligned} \tag{8.36}$$

where

$$A = \begin{bmatrix} 0 & -0.5 \\ 1 & -1 \end{bmatrix}$$

$$B = \begin{bmatrix} 1 \\ 1 \end{bmatrix}$$

$$C = [1, 1]$$

$$A_\delta = \begin{bmatrix} 1 & 0 \\ 0 & 1 \end{bmatrix}$$

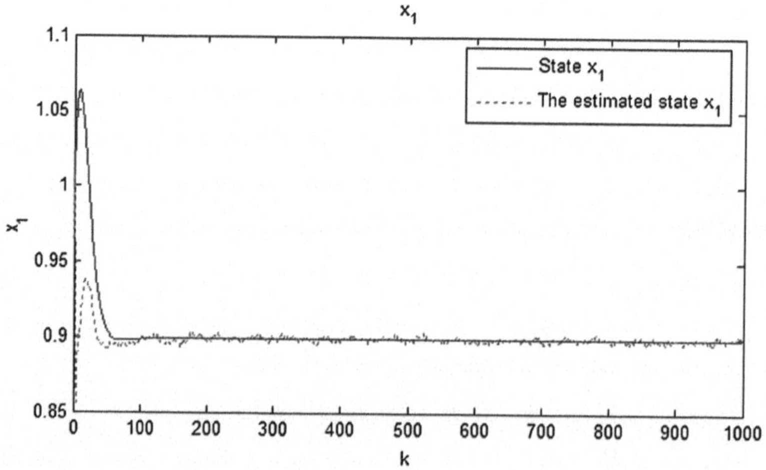

Fig. 8.1 The state $x_1(k)$ and estimated $\hat{x}_1(k)$

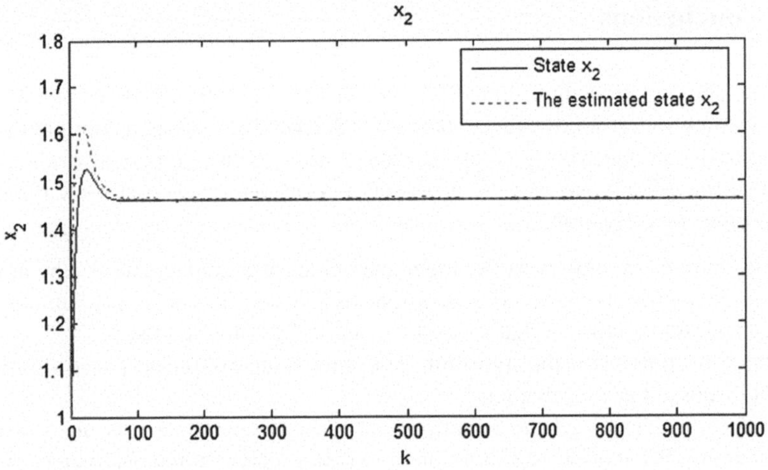

Fig. 8.2 The state $x_2(k)$ and estimated $\hat{x}_2(k)$

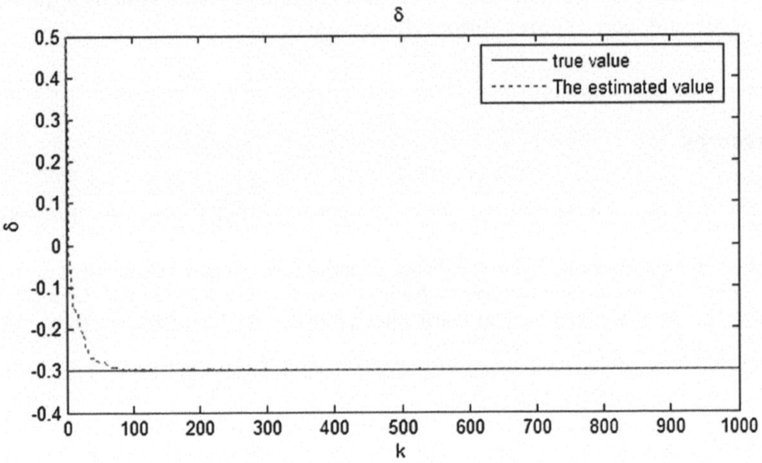

Fig. 8.3 The true value δ and estimated value $\hat{\delta}$

Set that $T_k = 0.1 + 0.01\tau$, $\tau \sim \mathcal{N}(0, 1)$ and $v(t)$ is a white noise with zero mean and variance $R_v = 0.02$. And assume that the true value $\delta_{\text{true}} = -0.3$ and the control input is $u_k = 1$. We apply the state estimation algorithms in (8.24) and (8.25) to estimate the parameter δ and the state vector $x = [x_1, x_2]^T$ of the system (8.36). The true values and the estimated values of the parameter and the states are shown in Figs. 8.1, 8.2, and 8.3.

From Figs. 8.1, 8.2, and 8.3, it is clear that the state and parameter estimation is gradually close to the true states and true parameter as k increases. This validates the effectiveness of the proposed algorithms.

8.6 Conclusions

A parameter estimation algorithm and a state estimation algorithm based on minimizing the estimation error covariance matrix are derived for uncertain continuous-time systems under nonuniformly sampled data. Although the numerical example has illustrated the effectiveness of the proposed algorithms, there exist some questions worth further investigation:

- In this chapter, we only consider the noise in measurement equation. The noise in the state equation is not in consideration. In addition, the noise is assumed to be zero-mean white noise. What's the situation of the colored noise?
- We have just only given the algorithm. The convergence of the proposed algorithms is not discussed in this chapter.
- We have just discussed the situation that only one parameter is unknown and unchanging. What about the situation of two parameters, three parameters or even more? if the parameter is changing, how do we handle this problem?

In a word, the state estimation of uncertain continuous-time system is quite interesting, nontrivial, and requires further discussion.

References

1. B.D.O. Anderson, J.B. Moore, *Optimal filtering* (Prentice-Hall, Englewood Cliffs, 2005), pp. 71–76
2. S. Haykin, *Adaptive Filter Theory*, 3rd edn. (Prentice-Hall, New Jersey, 1996)
3. H. Sorenson, Least-squares estimation: from gauss to Kalman. IEEE Spectr. **7**, 63–68 (1970)
4. R.E. Kalman, A new approach to linear filtering and prediction problems. Trans. ASME-J. Basic Eng. **82**(Series D), 35–45 (1960)
5. D. Simon, *Optimal State Estimation: Kalman, H Infinity, and Nonlinear Approaches*, 1st edn. (Wiley-Interscience, New York, 2006)
6. B.N. Jain, Guaranteed error estimation in uncertain systems. IEEE Trans. Autom. Control **AC-20**, 230–232 (1975)
7. I.R. Petersen, D.C. MacFarlane, *Robust Estimation for Uncertain Systems* (Brighton, U.K., 1991)
8. U. Shaked, C.E. de Souza, Robust minimum variance filtering. IEEE Trans. Signal Process. **43**, 2474–2483 (1995)
9. L. Xie, C.E. de Souza, Y. Wang, *Worst-Case Filter Design for a Class of Uncertain Dynamical Systems* (Tucson, AZ, 1992)
10. R.E. Kalman, R.S. Bucy, New results in linear filtering and prediction theory. Trans. ASME Ser. D, J. Basic Eng. **83**, 95–107 (1961)
11. S.J. Julier, The scaled unscented transformation. Automatica (2000)
12. G. Kranc, Input-output analysis of multirate feedback systems. IRE Trans. Autom. Control **3**(1), 21–28 (1957)
13. W. Li, S. Shah, *Kalman Filters for Non-uniformly Sampled Systems* (Czec Republic, Prague, 2005)
14. W. Li, S. Shah, *Data-Driven Kalman Filters for Non-uniformly Sampled Multirate Systems with Application to Fault Diagnosis* (Portland, OR, USA, 2005)

15. J.H. Wang, H.X. Jiang, *Filtering of Non-uniformly Multirate Sampled-data Systems Using the Lifting Technique* (Jinan, China, 2007)
16. W.P. Lu, D.G. Fisher, Least-squares output estimation with multirate sampling. IEEE Trans. Autom. Control **34**(6), 669–672 (1989)
17. F. Ding, T. Chen, *Parameter Identification and Intersample Output Estimation of a Class of Dual-Rate Systems* (Hawaii, USA, 2003), pp. 5555–5560
18. F. Ding, T. Chen, Parameter estimation for dual-rate systems with finite measurement data. Dyn. Contin., Discret. Impuls. Syst., Ser. B: Appl. Algorithms **11**(1), 101–121 (2004)

Part III
Kalman Filtering for Multi-sensor Systems

Chapter 9
Optimal Centralized, Recursive, and Distributed Fusion for Stochastic Systems with Coupled Noises

In this chapter, optimal state estimation for linear systems is introduced, where the noises of different sensors are cross-correlated and also coupled with the system noise of the previous step. The main result is that the optimal linear estimation is generated in recursive form (Optimal Recursive Fusion: ORF) and distributed fusion (Optimal Distributed Fusion: ODF). They are both compared with the optimal batch fusion (OBF) algorithm by the use of many measures; besides the traditional statistical estimation error and the trace of the estimation error covariances, some measures related to the normalized estimation error square or the Mahalanobis distance are introduced and analyzed. Simulation on a target tracking example is done to show the effectiveness of the presented algorithms and to illustrate the usefulness of the novel measure indices.

9.1 Introduction

Estimation fusion is the problem of how to best utilize useful information contained in multiple sets of data for the purpose of estimating a quantity, e.g., a parameter or process [1]. It originated in the military field and is now widely used in military and civilian fields, e.g., target tracking and localization, guidance and navigation, surveillance and monitoring, due to its improved estimation accuracy, enhanced reliability, and survivability.

Most of the earlier works and many works recently were based on the cross-independent sensor noise [2–4]. In practical applications, however, most multisensor systems often have correlated noises when the dynamic process is observed in a common noisy environment [5]. Moreover, most of the real systems are described in continuous time. To get the state estimation online, discretization is needed, and in this process, the system noise and the measurement noise are shown to be coupled [6].

When the sensor noises are cross-correlated, optimal batch fusion and distributed fusion are generated in a unified form in [1]. By using the Cholesky factorization, the coupled noises are decoupled and the optimal state estimation is generated based

H. Ma et al., *Kalman Filtering and Information Fusion*,
https://doi.org/10.1007/978-981-15-0806-6_9

on the Kalman filter with batch fusion [7] and distributed fusion [8], respectively. By use of the inverse of the filtering error covariance and the information form of Kalman filter, Song et al. deal with the state estimation problem with cross-correlated sensor noises and proved that it is optimal under a mild condition [5]. In [9–11], distributed fusion with unavailable cross-correlation of local estimates is studied without considering the coupling between sensor noise and the system noise.

In the abovementioned papers, the correlation between the system noise and the measurement noise was not considered. Reference [12] studied the distributed fusion when the measurement noises are coupled and correlated with the system noise at the same time step. In [6, 13], it is shown that if the discrete-time linear system is obtained from discretizing the continuous-time system, then the measurement noise is shown to be correlated with the system noise of *the previous time step*. In this case, when a single sensor is concerned, the optimal recursive state estimation is generated by modifying the classical Kalman filter [13]. When multisensor dynamic system is concerned, [6] presents some general idea of estimation, including the idea of decoupling and recursive fusion. However, the concrete form of optimal state estimation is not given, and the computing of the correlation between the local state estimation for the optimal distributed fusion and the computation of the correlation between the state prediction and the measurement for the recursive fusion is important when the measurement noise are correlated and coupled with the system noise.

In this chapter, optimal state estimation for a class of discrete time-varying linear systems is studied when the measurement noises are correlated and also coupled with the system noise of the previous time step. The optimal state estimation is generated in recursive form and distributed fusion, respectively. Their performances are analyzed in terms of estimation accuracy, computational complexity, and communication, and are compared with the optimal batch fusion by the use of statistical estimation errors, the trace of the estimation covariances, and some Mahalanobis distance-related measures. An example on target tracking shows the effectiveness of the presented algorithms.

This chapter is organized as follows. In Sect. 9.2, the problem is formulated. Section 9.3 generates the optimal estimation. In Sect. 9.4, the performance of the presented algorithms is analyzed, and the effectiveness of the algorithms is shown through an example simulation. Section 9.5 draws conclusions.

9.2 Problem Formulation

Consider the following linear dynamic system:

$$x_{k+1} = A_k x_k + w_k, \quad k = 0, 1, \ldots \tag{9.1}$$

$$z_{i,k} = C_{i,k} x_k + v_{i,k}, \quad i = 1, 2, \ldots, N \tag{9.2}$$

where $x_k \in R^n$ is the system state, $A_k \in R^{n\times n}$ is the state transition matrix. w_k is the system noise and is assumed to be Gaussian distributed with

$$\begin{cases} E\{w_k\} = 0 \\ E\{w_k(w_j)^T\} = Q_k\delta_{ij} \end{cases} \tag{9.3}$$

where $Q_k \geq 0$ and

$$\delta_{kj} = \begin{cases} 1, \ k = j \\ 0, \ k \neq j \end{cases} \tag{9.4}$$

$z_{i,k} \in R^{m_i}$ is the measurement of sensor i at time k. $C_{i,k} \in R^{m_i\times n}$ is the measurement matrix. The measurement noise $v_{i,k}$ is assumed to be white Gaussian with

$$\begin{cases} E\{v_{i,k}\} = 0 \\ E\{v_{i,k}v_{j,l}^T\} = R_{ij,k}\delta_{kl} \\ E\{w_{k-1}v_{i,k}^T\} = S_{i,k} \end{cases} \tag{9.5}$$

Note that the measurement noises are correlated with the system noise of the previous step, that is, $v_{i,k}$ are dependent on w_{k-1} for $k = 1, 2, \ldots$ and $i = 1, 2, \ldots, N$. The measurement noises of different sensors are correlated at each time step, that is, $v_{i,k}$ and $v_{j,k}$ are cross-correlated at time k with $E[v_{i,k}v_{j,k}^T] = R_{ij,k} \neq 0$ for $i, j = 1, 2, \ldots, N$. For simplicity, we denote $R_{i,k} := R_{ii,k} > 0$ in the sequel for $i = 1, 2, \ldots, N$.

The initial state x_0 is independent of w_k and $v_{i,k}$ for $k = 1, 2, \ldots$ and $i = 1, 2, \ldots, N$, and is assumed to be Gaussian distributed with

$$\begin{cases} E\{x_0\} = \bar{x} \\ \text{cov}\{x_0\} = E\{[x_0 - \bar{x}][x_0 - \bar{x}]^T\} = P_0 \end{cases} \tag{9.6}$$

where $\text{cov}\{x_0\}$ denotes the covariance of x_0.

The objective of this chapter is to generate the optimal estimation of state x_k by recursive use of the observations based on the above description.

9.3 Optimal Fusion Algorithms

To get the optimal estimation, the most straightforward method is the centralized fusion. Let

$$z_k = \left[z_{1,k}^T \ z_{2,k}^T \ \cdots \ z_{N,k}^T \right]^T \tag{9.7}$$

$$C_k = \left[C_{1,k}^T \ C_{2,k}^T \ \cdots \ C_{N,k}^T \right]^T \tag{9.8}$$

$$v_k = \left[v_{1,k}^T \ v_{2,k}^T \ \cdots \ v_{N,k}^T \right]^T \tag{9.9}$$

From (9.2), we have

$$z_k = C_k x_k + v_k \tag{9.10}$$

For (9.1) and (9.17), from [13, 14], we have the optimal state estimation.

Theorem 9.1 (The optimal batch fusion (OBF)) *By centralized batch fusion, the optimal state estimation of x_k is given by the following formulas:*

$$\begin{cases} \hat{x}_{b,k|k} & = \hat{x}_{b,k|k-1} + K_{b,k}[z_k - C_k \hat{x}_{b,k|k-1}] \\ P_{b,k|k} & = P_{b,k|k-1} - K_{b,k}[C_k P_{b,k|k-1} + S_k^T] \\ \hat{x}_{b,k|k-1} & = A_{k-1}\hat{x}_{b,k-1|k-1} \\ P_{b,k|k-1} & = A_{k-1} P_{b,k-1|k-1} A_{k-1}^T + Q_{k-1} \\ K_{b,k} & = [P_{b,k|k-1} C_k^T + S_k][C_k P_{b,k|k-1}) C_k^T \\ & \quad + R_k + C_k S_k + S_k^T C_k^T]^{-1} \end{cases} \tag{9.11}$$

where subscript "b" denotes batch fusion, and

$$R_k = \begin{bmatrix} R_{11,k} & R_{12,k} & \cdots & R_{1N,k} \\ R_{21,k} & R_{21,k} & \cdots & R_{2N,k} \\ \vdots & \vdots & & \vdots \\ R_{N1,k} & R_{N1,k} & \cdots & R_{NN,k} \end{bmatrix} \tag{9.12}$$

$$S_k = \begin{bmatrix} S_{1,k} & S_{2,k} & \cdots & S_{N,k} \end{bmatrix} \tag{9.13}$$

Proof Denote

$$z_k = \begin{bmatrix} z_{1,k}^T & z_{2,k}^T & \cdots & z_{N,k}^T \end{bmatrix}^T \tag{9.14}$$

$$C_k = \begin{bmatrix} C_{1,k}^T & C_{2,k}^T & \cdots & C_{N,k}^T \end{bmatrix}^T \tag{9.15}$$

$$v_k = \begin{bmatrix} v_{1,k}^T & v_{2,k}^T & \cdots & v_{N,k}^T \end{bmatrix}^T \tag{9.16}$$

We can rewrite (9.2) as

$$z_k = C_k x_k + v_k \tag{9.17}$$

From (9.1) to (9.17), by projection theorem, the optimal state prediction is given by

$$\begin{aligned} \hat{x}_{b,k|k-1} &= E\{x_k | z_1, z_2, \ldots, z_{k-1}\} \\ &= E\{A_{k-1} x_{k-1} + w_{k-1} | z_1, z_2, z_{k-1}\} \\ &= A_{k-1} \hat{x}_{b,k-1|k-1} \end{aligned} \tag{9.18}$$

and the optimal state estimation is

$$\begin{aligned}\hat{x}_{b,k|k} &= E\{x_k|z_1, z_2, \ldots, z_{k-1}, z_k\} \\ &= \hat{x}_{b,k|k-1} + \mathrm{cov}\{\tilde{x}_{b,k|k-1}, \tilde{z}_{k|k-1}\} \\ &\quad \cdot \mathrm{var}\{\tilde{z}_{k|k-1}\}^{-1}\tilde{z}_{k|k-1} \\ &= \hat{x}_{b,k|k-1} + K_{b,k}[z_k - C_k\hat{x}_{b,k|k-1}]\end{aligned} \tag{9.19}$$

where

$$\begin{aligned}\tilde{z}_{k|k-1} &= z_k - \hat{z}_{k|k-1} \\ &= z_k - E\{z_k|z_1, z_2, \ldots, z_{k-1}\} \\ &= z_k - E\{C_k x_k + v_k|z_1, z_2, \ldots, z_{k-1}\} \\ &= z_k - C_k\hat{x}_{b,k|k-1} \\ &= C_k\tilde{x}_{b,k|k-1} + v_k\end{aligned} \tag{9.20}$$

and

$$\tilde{x}_{b,k|k-1} = x_k - \hat{x}_{b,k|k-1} \tag{9.21}$$

$$\begin{aligned}K_{b,k} &= \mathrm{cov}\{\tilde{x}_{b,k|k-1}, \tilde{z}_{k|k-1}\}\mathrm{var}\{\tilde{z}_{k|k-1}\}^{-1} \\ &= E\{\tilde{x}_{b,k|k-1}\tilde{z}^T_{k|k-1}\} \\ &\quad \cdot E\{\tilde{z}_{k|k-1}\tilde{z}^T_{k|k-1}\}^{-1} \\ &= E\{\tilde{x}_{b,k|k-1}[C_k\tilde{x}_{b,k|k-1} + v_k]^T\} \\ &\quad \cdot E\{[C_k\tilde{x}_{b,k|k-1} + v_k] \\ &\quad \cdot [C_k\tilde{x}_{b,k|k-1} + v_k]^T\}^{-1} \\ &= [P_{b,k|k-1}C_k^T + \Lambda_k][C_k P_{b,k|k-1}C_k^T \\ &\quad + R_k + C_k\Lambda_k + \Lambda_k^T C_k^T]^{-1}\end{aligned} \tag{9.22}$$

where

$$\begin{aligned}\Lambda_k &= E\{\tilde{x}_{b,k|k-1}v_k^T\} \\ &= E\{[x_k - \hat{x}_{b,k|k-1}]v_k^T\} \\ &= E\{[A_{k-1}x_{k-1} + w_{k-1} - A_{k-1}\hat{x}_{b,k-1|k-1}]v_k^T\} \\ &= E\{[A_{k-1}\tilde{x}_{b,k-1|k-1}]v_k^T\} + E\{w_{k-1}v_k^T\} \\ &= S_k\end{aligned} \tag{9.23}$$

and

$$\begin{aligned}S_k &= E\{w_{k-1}v_k^T\} \\ &= E\{w_{k-1}\begin{bmatrix} v_{1,k}^T & v_{2,k}^T & \cdots & v_{N,k}^T \end{bmatrix}\} \\ &= \begin{bmatrix} S_{1,k} & S_{2,k} & \cdots & S_{N,k} \end{bmatrix}\end{aligned} \tag{9.24}$$

Substituting (9.23) into (9.22) yields

$$K_{b,k} = [P_{b,k|k-1}C_k^T + S_k][C_k P_{b,k|k-1} C_k^T + R_k + C_k S_k + S_k^T C_k^T]^{-1} \tag{9.25}$$

The state prediction error covariance is

$$\begin{aligned} P_{b,k|k-1} &= E\{\tilde{x}_{b,k|k-1}\tilde{x}_{b,k|k-1}^T\} \\ &= E\{[x_k - \hat{x}_{b,k|k-1}][x_k - \hat{x}_{b,k|k-1}]^T\} \\ &= E\{[A_{k-1}\tilde{x}_{b,k-1|k-1} + w_{k-1}] \\ &\quad \cdot [A_{k-1}\tilde{x}_{b,k-1|k-1} + w_{k-1}]^T\} \\ &= A_{k-1} P_{b,k-1|k-1} A_{k-1}^T + Q_{k-1} \end{aligned} \tag{9.26}$$

By (9.19)–(9.26), the state estimation error covariance is

$$\begin{aligned} P_{b,k|k} &= E\{\tilde{x}_{b,k|k}\tilde{x}_{b,k|k}^T\} \\ &= E\{[x_k - \hat{x}_{b,k|k}][x_k - \hat{x}_{b,k|k}]^T\} \\ &= E\{[(I - K_{b,k}C_k)\tilde{x}_{b,k|k-1} - K_{b,k}v_k] \\ &\quad \cdot [(I - K_{b,k}C_k)\tilde{x}_{b,k|k-1} - K_{b,k}v_k]^T\} \\ &= [I - K_{b,k}C_k]P_{b,k|k-1}[I - K_{b,k}C_k]^T \\ &\quad + K_{b,k}R_k K_{b,k}^T - [I - K_{b,k}C_k]S_k K_{b,k}^T \\ &\quad - K_{b,k}S_k^T[I - K_{b,k}C_k]^T \\ &= P_{b,k|k-1} - K_{b,k}[C_k P_{b,k|k-1} + S_k^T] \end{aligned} \tag{9.27}$$

From (9.11), one can see that the formula in generating the prediction $\hat{x}_{b,k|k-1}$, $P_{b,k|k-1}$, and the estimation $\hat{x}_{b,k|k}$ have the same form as the classical Kalman filter, except that the innovation matrix $K_{b,k}$ is different in the updating step, which leads to a different estimation covariance $P_{b,k|k}$. The modification of $K_{b,k}$ depends on the correlation between the measurement noise and the system noise S_k. When $S_k = 0$, it is the classical Kalman filter.

The batch fusion has a simple form. However, recursive fusion is necessary for virtually all real-time applications due to computational limitations. In [15], recursibility and optimal linear estimation and filtering is discussed, and concluded that for many linear systems, the LMMSE filter does not have a recursive form, and the recursive form filter is not necessarily equivalent to the batch fusion in performance. In the sequel, we will generate the optimal recursive estimation based on Sect. 9.2 formulation and discuss the performance between the optimal recursive fusion and the **OBF**. We would like to give the algorithm here and leave the performance analysis for Sect. 9.4 for clearness consideration.

Theorem 9.2 (The optimal recursive fusion (ORF)) *Given the optimal state estimation* $\hat{x}_{r,k-1|k-1}$ *and its estimation error covariance* $P_{r,k-1|k-1}$*, the optimal state estimation of* x_k *by recursively use of the measurements is given by*

$$\begin{cases} \hat{x}_{i,k|k} & = \hat{x}_{i-1,k|k} + K_{i,k}[z_{i,k} - C_{i,k}\hat{x}_{i-1,k|k}] \\ P_{i,k|k} & = P_{i-1,k|k} - K_{i,k}[C_{i,k}P_{i-1,k|k} + \Delta_{i-1,k}^T] \\ K_{i,k} & = [P_{i-1,k|k}C_{i,k}^T + \Delta_{i-1,k}^T][C_{i,k}P_{i-1,k|k}C_{i,k}^T + R_{i,k} \\ & \quad +C_{i,k}\Delta_{i-1,k} + \Delta_{i-1,k}^T C_{i,k}^T]^{-1} \\ \Delta_{i-1,k} & = \prod_{p=i-1}^{1} [I - K_{p,k}C_{p,k}]S_{i,k} \\ & \quad -\sum_{q=2}^{i-1} \prod_{p=i-1}^{q} [I - K_{p,k}C_{p,k}]K_{q-1,k}R_{q-1,i,k} \\ & \quad -K_{i-1,k}R_{i-1,i,k} \end{cases} \tag{9.28}$$

where $\prod_{p=i-1}^{q} [I - K_{p,k}C_{p,k}] = [I - K_{i-1,k}C_{i-1,k}] \times [I - K_{i-2,k}C_{i-2}(k)] \times \cdots \times [I - K_{q,k}C_{q,k}]$ *is the product of* $(i-q)$ *terms from larger index* $(i-1)$ *to the smaller index p. Note that because the commutative law is not applicable for the product of matrices,* $\prod_{p=i-1}^{q} [I - K_{p,k}C_{p,k}]$ *shouldn't be mistaken as* $\prod_{p=q}^{i-1} [I - K_{p,k}C_{p,k}]$, $i = 1, 2, \ldots, N$.

For $i = 0$*, we have*

$$\begin{cases} \hat{x}_{0,k|k} & = \hat{x}_{k|k-1} = A_{k-1}\hat{x}_{k-1|k-1} \\ P_{0,k|k} & = P_{k|k-1} \\ & = A_{k-1}P_{k|k-1}A_{k-1}^T + Q_{k-1} \\ \Delta_{0,k} & = S_{1,k} \end{cases} \tag{9.29}$$

In (9.28), $\hat{x}_{i,k|k}$ *and* $P_{i,k|k}$ *denote the state estimation of* x_k *and the corresponding estimation error covariance based on the observations of sensors from* $1, 2, \ldots$ *to* i*. When* $i = N$*, we have* $\hat{x}_{r,k|k} = \hat{x}_{N,k|k}$ *and* $P_{r,k|k} = P_{N,k|k}$*, which are the optimal fused estimation of the state and the corresponding estimation error covariance, where subscript "r" denotes the recursive fusion, also known as sequential Kalman filter.*

Proof For $i = 1, 2, \ldots, N$, denote

$$\mathbb{Z}_{i,k} = \{z_{i,1}, z_{i,2}, \ldots, z_{i,k}\} \tag{9.30}$$

$$\mathbb{Z}_{1,k}^i = \{z_{1,k}, z_{2,k}, \ldots, z_{i,k}\} \tag{9.31}$$

$$\bar{\mathbb{Z}}_{1,k}^i = \{\mathbb{Z}_{1,l}^i\}_{l=1}^k \tag{9.32}$$

Then, $\mathbb{Z}_{i,k}$ stands for the measurements observed by sensor i up to and including time k. $\mathbb{Z}_{1,k}^i$ denotes the measurements observed by sensors $1, 2, \ldots, i$ at time k. $\bar{\mathbb{Z}}_{1,k}^i$ denotes the measurements observed by sensors $1, 2, \ldots, i$ up to and including time k. $\bar{\mathbb{Z}}_{1,k}^{N}$ then stands for the measurements observed by all the sensors up to and including time k.

We prove Theorem 9.2 deductively by applying the projection theorem.

By the projection theorem, the state prediction of x_k based on data $z_{i,l}, i = 1, 2, \ldots, N; l = 1, 2, \ldots, k-1$ is given by

$$\begin{aligned}\hat{x}_{k|k-1} &= E\{x_k|\bar{\mathbb{Z}}_{1,k-1}^{N}\}\\ &= E\{A_{k-1}x_{k-1} + w_{k-1}|\bar{\mathbb{Z}}_{1,k-1}^{N}\}\\ &= A_{k-1}\hat{x}_{k-1|k-1}\end{aligned} \tag{9.33}$$

The state prediction error covariance is

$$P_{k|k-1} = E\{\tilde{x}_{k|k-1}\tilde{x}_{k|k-1}^{T}\} \tag{9.34}$$

From Eqs. (9.1) to (9.33), the state prediction error is

$$\begin{aligned}\tilde{x}_{k|k-1} &= x_k - \hat{x}_{k|k-1}\\ &= A_{k-1}x_{k-1} + w_{k-1} - A_{k-1}\hat{x}_{k-1|k-1}\\ &= A_{k-1}\tilde{x}_{k-1|k-1} + w_{k-1}\end{aligned} \tag{9.35}$$

Substituting (9.35) into (9.34) yields

$$P_{k|k-1} = A_{k-1}P_{k-1|k-1}A_{k-1}^{T} + Q_{k-1} \tag{9.36}$$

Without loss of generality, we assume that the measurements are processed in the sequence of $i = 1, 2 \ldots, N$ at each time step.

Applying the projection theorem, the update by the measurement of Sensor 1 is

$$\begin{aligned}\hat{x}_{1,k|k} &= E\{x_k|\bar{\mathbb{Z}}_{1,k-1}^{N}, z_{1,k}\}\\ &= \hat{x}_{k|k-1} + \mathrm{cov}\{\tilde{x}_{k|k-1}, \tilde{z}_{1,k|k-1}\}\\ &\quad \cdot \mathrm{var}\{\tilde{z}_{1,k|k-1}\}^{-1}\tilde{z}_{1,k|k-1}\\ &= \hat{x}_{k|k-1} + K_{1,k}\tilde{z}_{1,k|k-1}\end{aligned} \tag{9.37}$$

where $\mathrm{cov}\{\tilde{x}_{k|k-1}, \tilde{z}_{1,k|k-1}\}$ denotes the covariance of $\tilde{x}_{k|k-1}$ and $\tilde{z}_{1,k|k-1}$, $\mathrm{var}\{\tilde{z}_{1,k|k-1}\}$ denotes the variance of $\tilde{z}_{1,k|k-1}$, $\tilde{z}_{1,k|k-1}$ is the measurement prediction error of $z_{1,k}$ based on $\bar{\mathbb{Z}}_{1,k-1}^{N}$, and $K_{1,k}$ is the innovation matrix given by

$$K_{1,k} = \mathrm{cov}\{\tilde{x}_{k|k-1}, \tilde{z}_{1,k|k-1}\}\mathrm{var}\{\tilde{z}_{1,k|k-1}\}^{-1} \tag{9.38}$$

$\tilde{z}_{1,k|k-1}$ is computed by

$$\begin{aligned}\tilde{z}_{1,k|k-1} &= z_{1,k} - \hat{z}_{1,k|k-1} \\ &= z_{1,k} - E\{z_{1,k}|\bar{\mathbb{Z}}_{1,k-1}^{N}\} \\ &= z_{1,k} - C_{1,k}\hat{x}_{k|k-1} \\ &= C_{1,k}\tilde{x}_{k|k-1} + v_{1,k}\end{aligned} \tag{9.39}$$

By (9.35) and (9.39), we obtain

$$\begin{aligned}\operatorname{cov}\{\tilde{x}_{k|k-1}, \tilde{z}_{1,k|k-1}\} &= E\{\tilde{x}_{k|k-1}\tilde{z}_{1,k|k-1}^{T}\} \\ &= E\{\tilde{x}_{k|k-1}[C_{1,k}\tilde{x}_{k|k-1} + v_{1,k}]^{T}\} \\ &= P_{k|k-1}C_{1,k}^{T} + \Delta_{0,k}\end{aligned} \tag{9.40}$$

where

$$\begin{aligned}\Delta_{0,k} &= E\{\tilde{x}_{k|k-1}v_{1,k}^{T}\} \\ &= E\{[A_{k-1}\tilde{x}_{k-1|k-1} + w_{k-1}]v_{1,k}^{T}\} \\ &= S_{1,k}\end{aligned} \tag{9.41}$$

and (9.35) is used.

Similarly, we have

$$\begin{aligned}\operatorname{var}\{\tilde{z}_{1,k|k-1}\} &= E\{\tilde{z}_{1,k|k-1}\tilde{z}_{1,k|k-1}^{T}\} \\ &= E\{[C_{1,k}\tilde{x}_{k|k-1} + v_{1,k}][C_{1,k}\tilde{x}_{k|k-1} + v_{1,k}]^{T}\} \\ &= C_{1,k}P_{0,k|k}C_{1,k}^{T} + R_{1,k} + C_{1,k}\Delta_{0,k} + \Delta_{0,k}^{T}C_{1,k}^{T}\end{aligned} \tag{9.42}$$

Substituting (9.40) and (9.42) into (9.38), we have

$$\begin{aligned}K_{1,k} = [P_{k|k-1}C_{1,k}^{T} + \Delta_{0,k}][C_{1,k}P_{k|k-1}C_{1,k}^{T} \\ + R_{1,k} + C_{1,k}\Delta_{0,k} + \Delta_{0,k}^{T}C_{1,k}^{T}]^{-1}\end{aligned} \tag{9.43}$$

Substituting (9.43) and the third equation of (9.39) into (9.37), we obtain

$$\hat{x}_{1,k|k} = \hat{x}_{k|k-1} + K_{1,k}[z_{1,k} - C_{1,k}\hat{x}_{k|k-1}] \tag{9.44}$$

By (9.2) and (9.44), we have

$$\begin{aligned}\tilde{x}_{1,k|k} &= x_{k} - \hat{x}_{1,k|k} \\ &= x_{k} - \hat{x}_{k|k-1} - K_{1,k}[z_{1,k} - C_{1,k}\hat{x}_{k|k-1}] \\ &= [I - K_{1,k}C_{1,k}]\tilde{x}_{k|k-1} - K_{1,k}v_{1,k}\end{aligned} \tag{9.45}$$

Therefore,

$$\begin{aligned}P_{1,k|k} &= E\{\tilde{x}_{1,k|k}\tilde{x}_{1,k|k}^T\}\\ &= [I - K_{1,k}C_{1,k}]P_{k|k-1}[I - K_{1,k}C_{1,k}]^T\\ &\quad + K_{1,k}R_{1,k}K_{1,k}^T - [I - K_{1,k}C_{1,k}]\Delta_{0,k}K_{1,k}^T\\ &\quad - K_{1,k}\Delta_{0,k}^T[I - K_{1,k}C_{1,k}]^T \end{aligned} \tag{9.46}$$

By (9.43), (9.46) can be rewritten as

$$P_{1,k|k} = P_{k|k-1} - K_{1,k}[C_{1,k}P_{k|k-1} + \Delta_{0,k}^T] \tag{9.47}$$

From the above deduction, if we denote $\hat{x}_{0,k|k} = \hat{x}_{k|k-1}$ and $P_{0,k|k} = P_{k|k-1}$, we generate

$$\begin{cases} \hat{x}_{1,k|k} &= \hat{x}_{0,k|k} + K_{1,k}[z_{1,k} - C_{1,k}\hat{x}_{0,k|k}]\\ P_{1,k|k} &= P_{0,k|k} - K_{1,k}[C_{1,k}P_{0,k|k} + \Delta_{0,k}^T]\\ K_{1,k} &= [P_{0,k|k}C_{1,k}^T + \Delta_{0,k}][C_{1,k}P_{0,k|k}C_{1,k}^T\\ &\qquad + R_{1,k} + C_{1,k}\Delta_{0,k} + \Delta_{0,k}^T C_{1,k}^T]^{-1}\\ \hat{x}_{0,k|k} &= \hat{x}_{k|k-1} = A_{k-1}\hat{x}_{k-1|k-1}\\ P_{0,k|k} &= P_{k|k-1}\\ &= A_{k-1}P_{k-1|k-1}A_{k-1}^T + Q_{k-1}\\ \Delta_{0,k} &= S_{1,k} \end{cases} \tag{9.48}$$

Next, we deduce the state estimation $\hat{x}_{2,k|k}$ and the corresponding estimation error covariance $P_{2,k|k}$ that fused sensors $i = 1, 2$.

Applying the projection theorem yields

$$\begin{aligned}\hat{x}_{2,k|k} &= E\{x_k|\bar{\mathbb{Z}}_{1,k-1}^N, \mathbb{Z}_{1,k}^2\}\\ &= E\{x_k|\bar{\mathbb{Z}}_{1,k-1}^N, z_{1,k}, z_{2,k}\}\\ &= \hat{x}_{1,k|k} + K_{2,k}\tilde{z}_{2,k}\end{aligned} \tag{9.49}$$

where

$$\begin{aligned}\tilde{z}_{2,k} &= z_{2,k} - C_{2,k}\hat{x}_{1,k|k}\\ &= C_{2,k}\tilde{x}_{1,k|k} + v_{2,k}\end{aligned} \tag{9.50}$$

and

$$\begin{aligned}
K_{2,k} &= \mathrm{cov}\{\tilde{x}_{1,k|k}, \tilde{z}_{2,k}\}\mathrm{var}\{\tilde{z}_{2,k}\}^{-1} \\
&= E\{\tilde{x}_{1,k|k}\tilde{z}_{2,k}^T\}E\{\tilde{z}_{2,k}\tilde{z}_{2,k}^T\}^{-1} \\
&= E\{\tilde{x}_{1,k|k}[C_{2,k}\tilde{x}_{1,k|k} + v_{2,k}]^T\}E\{[C_{2,k}\tilde{x}_{1,k|k} + v_{2,k}] \\
&\quad \cdot [C_{2,k}\tilde{x}_{1,k|k} + v_{2,k}]^T\}^{-1} \\
&= [P_{1,k|k}C_{2,k}^T + \Delta_{1,k}][C_{2,k}P_{1,k|k}C_{2,k}^T + R_{2,k} \\
&\quad + C_{2,k}\Delta_{1,k} + \Delta_{1,k}^T C_{2,k}^T]^{-1}
\end{aligned} \tag{9.51}$$

By (9.45) and (9.35), we have

$$\begin{aligned}
\Delta_{1,k} &= E\{\tilde{x}_{1,k|k}v_{2,k}^T\} \\
&= E\{[(I - K_{1,k}C_{1,k})\tilde{x}_{k|k-1} - K_{1,k}v_{1,k}]v_{2,k}^T\} \\
&= [I - K_{1,k}C_{1,k}]E\{\tilde{x}_{k|k-1}v_{2,k}^T\} - K_{1,k}R_{12,k} \\
&= [I - K_{1,k}C_{1,k}]E\{[A_{k-1}\tilde{x}_{k-1|k-1} + w_{k-1}]v_{2,k}^T\} \\
&\quad - K_{1,k}R_{12,k} \\
&= [I - K_{1,k}C_{1,k}]S_{2,k} - K_{1,k}R_{12,k}
\end{aligned} \tag{9.52}$$

Substituting (9.51) and the first equation of (9.50) into (9.49) yields

$$\hat{x}_{2,k|k} = \hat{x}_{1,k|k} + K_{2,k}[z_{2,k} - C_{2,k}\hat{x}_{1,k|k}] \tag{9.53}$$

The estimation error is

$$\begin{aligned}
\tilde{x}_{2,k|k} &= x_k - \hat{x}_{2,k|k} \\
&= x_k - \hat{x}_{1,k|k} - K_{2,k}[z_{2,k} - C_{2,k}\hat{x}_{1,k|k}] \\
&= [I - K_{2,k}C_{2,k}]\tilde{x}_{1,k|k} - K_{2,k}v_{2,k}
\end{aligned} \tag{9.54}$$

So, the estimation error covariance is

$$\begin{aligned}
P_{2,k|k} &= E\{\tilde{x}_{2,k|k}\tilde{x}_{2,k|k}^T\} \\
&= [I - K_{2,k}C_{2,k}]P_{1,k|k}[I - K_{2,k}C_{2,k}]^T \\
&\quad + K_{2,k}R_{2,k}K_{2,k}^T - [I - K_{2,k}C_{2,k}]\Delta_{1,k}K_{2,k}^T \\
&\quad - K_{2,k}\Delta_{1,k}^T[I - K_{2,k}C_{2,k}]^T \\
&= P_{1,k|k} - K_{2,k}[C_{2,k}P_{1,k|k} + \Delta_{1,k}^T]
\end{aligned} \tag{9.55}$$

where (9.51) is used and $\Delta_{1,k}$ is given by (9.52).

Combining (9.53), (9.55), (9.51), and (9.52) yields

$$\begin{cases} \hat{x}_{2,k|k} &= \hat{x}_{1,k|k} + K_{2,k}[z_{2,k} - C_{2,k}\hat{x}_{1,k|k}] \\ P_{2,k|k} &= P_{1,k|k} - K_{2,k}[C_{2,k}P_{1,k|k} + \Delta_{1,k}^T] \\ K_{2,k} &= [P_{1,k|k}C_{2,k}^T + \Delta_{1,k}][C_{2,k}P_{1,k|k}C_{2,k}^T \\ & \quad +R_{2,k} + C_{2,k}\Delta_{1,k} + \Delta_{1,k}^T C_{2,k}^T]^{-1} \\ \Delta_{1,k} &= [I - K_{1,k}C_{1,k}]S_{2,k} - K_{1,k}R_{12,k} \end{cases} \tag{9.56}$$

Suppose that we have obtained $\hat{x}_{i-1,k|k}$ and the corresponding estimation error covariance $P_{i-1,k|k}$, next we show how to generate $\hat{x}_{i,k|k}$ and $P_{i,k|k}$.

Applying the projection theorem yields

$$\begin{aligned} \hat{x}_{i,k|k} &= E\{x_k|\bar{\mathbb{Z}}_{1,k-1}^N, \mathbb{Z}_{1,k}^i\} \\ &= E\{x_k|\bar{\mathbb{Z}}_{1,k-1}^N, \mathbb{Z}_{1,k}^{i-1}, z_{i,k}\} \\ &= \hat{x}_{i-1,k|k} + \text{cov}\{\tilde{x}_{i-1,k|k}, \tilde{z}_{i,k}\}\text{var}\{\tilde{z}_{i,k}\}^{-1}\tilde{z}_{i,k} \end{aligned} \tag{9.57}$$

where

$$\tilde{z}_{i,k} = z_{i,k} - \hat{z}_{i,k} \tag{9.58}$$

and

$$\begin{aligned} \hat{z}_{i,k} &= E\{z_{i,k}|\bar{\mathbb{Z}}_{1,k-1}^N, \mathbb{Z}_{1,k}^{i-1}\} \\ &= E\{C_{i,k}x_k + v_{i,k}|\bar{\mathbb{Z}}_{1,k-1}^N, \mathbb{Z}_{1,k}^{i-1}\} \\ &= C_{i,k}\hat{x}_{i-1,k|k} \end{aligned} \tag{9.59}$$

So,

$$\begin{aligned} \tilde{z}_{i,k} &= z_{i,k} - \hat{z}_{i,k} \\ &= z_{i,k} - C_{i,k}\hat{x}_{i-1,k|k} \\ &= C_{i,k}x_{i,k} + v_{i,k} - C_{i,k}\hat{x}_{i-1,k|k} \\ &= C_{i,k}\tilde{x}_{i-1,k|k} + v_{i,k} \end{aligned} \tag{9.60}$$

Therefore,

$$\begin{aligned} \text{cov}\{\tilde{x}_{i-1,k|k}, \tilde{z}_{i,k}\} &= E\{\tilde{x}_{i-1,k|k}\tilde{z}_{i,k}^T\} \\ &= E\{\tilde{x}_{i-1,k|k}[C_{i,k}\tilde{x}_{i-1,k|k} + v_{i,k}]^T\} \\ &= P_{i-1,k|k}C_{i,k}^T + \Delta_{i-1,k} \end{aligned} \tag{9.61}$$

and

$$\begin{aligned}\mathrm{var}\{\tilde{z}_{i,k}\} &= E\{\tilde{z}_{i,k}\tilde{z}_{i,k}^T\} \\ &= E\{[C_{i,k}\tilde{x}_{i-1,k|k} + v_{i,k}][C_{i,k}\tilde{x}_{i-1,k|k} + v_{i,k}]^T\} \\ &= C_{i,k}P_{i-1,k|k}C_{i,k}^T + R_{i,k} + C_{i,k}\Delta_{i-1,k} \\ &\quad + \Delta_{i-1,k}^T C_{i,k}^T \end{aligned} \tag{9.62}$$

where

$$\Delta_{i-1,k} = E\{\tilde{x}_{i-1,k|k}v_{i,k}^T\} \tag{9.63}$$

$$\begin{aligned}\tilde{x}_{i-1,k|k} &= x_k - \hat{x}_{i-1,k|k} \\ &= x_k - \hat{x}_{i-2,k|k} - K_{i-1,k}[z_{i-1,k} - C_{i-1,k}\hat{x}_{i-2,k|k}] \\ &= \tilde{x}_{i-2,k|k} - K_{i-1,k}[C_{i-1,k}x_k + v_{i-1,k} \\ &\quad - C_{i-1,k}\hat{x}_{i-2,k|k}] \\ &= [I - K_{i-1,k}C_{i-1,k}]\tilde{x}_{i-2,k|k} - K_{i-1,k}v_{i-1,k}\end{aligned} \tag{9.64}$$

Substituting (9.64) into (9.63) yields

$$\begin{aligned}\Delta_{i-1,k} &= E\{\tilde{x}_{i-1,k|k}v_{i,k}^T\} \\ &= [I - K_{i-1,k}C_{i-1,k}]E\{\tilde{x}_{i-2,k|k}v_{i,k}^T\} \\ &\quad - K_{i-1,k}E\{v_{i-1,k}v_{i,k}^T\} \\ &= \prod_{p=i-1}^{1}[I - K_{p,k}C_{p,k}]S_{i,k} \\ &\quad - \sum_{q=2}^{i-1}\prod_{p=i-1}^{q}[I - K_{p,k}C_{p,k}]K_{q-1,k}R_{q-1,i,k} \\ &\quad - K_{i-1,k}R_{i-1,i,k}\end{aligned} \tag{9.65}$$

Substituting (9.61), (9.62) and the second equation of (9.60) into (9.57) yields

$$\hat{x}_{i,k|k} = \hat{x}_{i-1,k|k} + K_{i,k}[z_{i,k} - C_{i,k}\hat{x}_{i-1,k|k}] \tag{9.66}$$

where

$$\begin{aligned}K_{i,k} &= \mathrm{cov}\{\tilde{x}_{i-1,k|k}, \tilde{z}_{i,k}\}\mathrm{var}\{\tilde{z}_{i,k}\}^{-1} \\ &= [P_{i-1,k|k}C_{i,k}^T + \Delta_{i-1,k}][C_{i,k}P_{i-1,k|k}C_{i,k}^T + R_{i,k} \\ &\quad + C_{i,k}\Delta_{i-1,k} + \Delta_{i-1,k}^T C_{i,k}^T]^{-1}\end{aligned} \tag{9.67}$$

The estimation error covariance is

$$
\begin{aligned}
P_{i,k|k} &= E\{\tilde{x}_{i,k|k}\tilde{x}_{ik|k}^T\} \\
&= E\{[x_k - \hat{x}_{i,k|k}][x_k - \hat{x}_{i,k|k}]^T\} \\
&= E\{[(I - K_{i,k}C_{i,k})\tilde{x}_{i-1,k|k} - K_{i,k}v_{i,k}] \\
&\quad \cdot [(I - K_{i,k}C_{i,k})\tilde{x}_{i-1,k|k} - K_{i,k}v_{i,k}]^T\} \\
&= [I - K_{i,k}C_{i,k}]P_{i-1,k|k}[I - K_{i,k}C_{i,k}^T] \\
&\quad + K_{i,k}R_{i,k}K_{i,k}^T - [I - K_{i,k}C_{i,k}]\Delta_{i-1,k}K_{i,k}^T \\
&\quad - K_{i,k}\Delta_{i-1,k}^T[I - K_{i,k}C_{i,k}^T] \\
&= P_{i-1,k|k} - K_{i,k}[C_{i,k}P_{i-1,k|k} + \Delta_{i-1,k}^T]
\end{aligned}
\tag{9.68}
$$

where Eq. (9.67) is used.

Combining (9.66)–(9.68) and (9.65), we obtain

$$
\begin{cases}
\hat{x}_{i,k|k} = \hat{x}_{i-1,k|k} + K_{i,k}[z_{i,k} - C_{i,k}\hat{x}_{i-1,k|k}] \\
P_{i,k|k} = P_{i-1,k|k} - K_{i,k}[C_{i,k}P_{i-1,k|k} + \Delta_{i-1,k}^T] \\
K_{i,k} = [P_{i-1,k|k}C_{i,k}^T + \Delta_{i-1,k}][C_{i,k}P_{i-1,k|k}C_{i,k}^T \\
\qquad +R_{i,k} + C_{i,k}\Delta_{i-1,k} + \Delta_{i-1,k}^T C_{i,k}^T]^{-1} \\
\Delta_{i-1,k} = \prod_{p=i-1}^{1}[I - K_{p,k}C_{p,k}]S_{i,k} \\
\qquad -\sum_{q=2}^{i-1}\prod_{p=i-1}^{q}[I - K_{p,k}C_{p,k}]K_{q-1,k}R_{q-1,i,k} \\
\qquad -K_{i-1,k}R_{i-1,i,k}
\end{cases}
\tag{9.69}
$$

If we denote $\hat{x}_{r,k|k} = \hat{x}_{N,k|k}$ and $P_{r,k|k} = P_{N,k|k}$, then by combining (9.48) and (9.69), we have the state estimation of recursive fusion $\hat{x}_{r,k|k}$ and $P_{r,k|k}$, and the proof is completed.

From Theorem 9.2, it can be seen that the ORF is distinguished from the classical sequential Kalman filter in that there is one more term: $\Delta_{i,k}$, which is related to the correlation between a different sensor noise $R_{ij,k}$, $i \neq j$ and the correlation between the sensor noise and the system noise $S_{i,k}$, $i = 1, 2, \ldots, N$. When the measurement noise of different sensors is not coupled and not correlated with the system noise, we have $R_{ij,k} = 0$ for $i \neq j$, and $S_{i,k} = 0$, $i = 1, 2, \ldots, N$. In this case, from Eqs. (9.28) to (9.29), we obtain $\Delta_{i,k} = 0$ for $i = 0, 1, 2, \ldots, N-1$, and the ORF is the classical sequential Kalman filter (CSKF) then. So, the ORF in this chapter is the generalization of the CSKF.

Briefly, we have introduced the optimal batch fusion (OBF) and the optimal recursive fusion (ORF) up to now. They both use the measurements directly to get the state estimation. Next, we will generate the optimal distributed fusion (ODF), which is superior for its flexibility, robustness, and survivability. Without noise correlation, the ODF can be generated from the batch fusion and is therefore optimal in the sense of linear minimum mean square error (LMMSE) [14]. When the measurement noise and the system noise are correlated, however, the distributed fusion is hard to generate by minimizing the MMSE directly. In this case, it can be derived by minimizing F_x, where

$$F_x = (y_k - I_k x_k)^T \Sigma_k (y_k - I_k x_k) \tag{9.70}$$

$$y_k = \begin{bmatrix} \hat{x}_{d,1,k|k} \\ \hat{x}_{d,2,k|k} \\ \vdots \\ \hat{x}_{d,N,k|k} \end{bmatrix} \tag{9.71}$$

where $\hat{x}_{d,i,k|k}$, $i = 1, 2, \ldots, N$ are the local estimates. Hence, ODF is optimal in the sense of weighted least squares (WLS). From [1], it is shown that the WLS estimation and the LMMSE estimation are generally not equivalent. However, under certain circumstances, without prior information, they are equivalent.

Theorem 9.3 (Optimal Distributed Fusion: ODF) *Based on system (9.1)–(9.2), the optimal state estimation of x_k by distributed fusion is given by*

$$\begin{cases} \hat{x}_{d,k|k} = \sum_{i=1}^{N} \alpha_{i,k} \hat{x}_{d,i,k|k} \\ P_{d,k|k} = (I_k{}^T \Sigma_k^{-1} I_k)^{-1} \\ \alpha_k = (I_k{}^T \Sigma_k^{-1} I_k)^{-1} I_k{}^T \Sigma_k^{-1} \end{cases} \tag{9.72}$$

where $\alpha_k = [\alpha_{1,k} \ \alpha_{2,k} \ \cdots \ \alpha_{N,k}]$ is the $n \times nN$ matrix. $I_k = [\, I_n \ I_n \ \cdots \ I_n \,]^T$ is the $nN \times n$ matrix and I_n denotes the identity matrix of dimension n. $\Sigma_k = (P_{d,ij,k|k})$ is an $nN \times nN$ matrix whose ijth block is $P_{d,ij,k|k}$, $i, j = 1, 2, \ldots, N$, and where

$$\begin{aligned} P_{d,ij,k|k} = {} & [I - K_{d,i,k} C_{i,k}][A_{k-1} P_{d,ij,k-1|k-1} A_{k-1}^T + Q_{k-1}] \\ & \cdot [I - K_{d,j,k} C_{j,k}]^T - [I - K_{d,i,k} C_{i,k}] S_{j,k} K_{d,j,k}^T \\ & - K_{d,i,k} S_{i,k}^T [I - K_{d,j,k} C_{j,k}]^T \\ & + K_{d,i,k} R_{ij,k} K_{d,j,k} \end{aligned} \tag{9.73}$$

The local state estimates are computed by

$$\begin{cases} \hat{x}_{d,i,k|k} & = \hat{x}_{d,i,k|k-1} + K_{d,i,k}[z_{i,k} - C_{i,k} \hat{x}_{d,i,k|k-1}] \\ P_{d,i,k|k} & = P_{d,i,k|k-1} - K_{d,i,k}[C_{i,k} P_{d,i,k|k-1} + S_{i,k}] \\ \hat{x}_{d,i,k|k-1} & = A_{k-1} \hat{x}_{d,i,k-1|k-1} \\ P_{d,i,k|k-1} & = A_{k-1} P_{d,i,k-1|k-1} A_{k-1}^T + Q_{k-1} \\ K_{d,i,k} & = [P_{d,i,k|k-1} C_{i,k}^T + S_{i,k}][C_{i,k} P_{d,i,k|k-1} C_{i,k}^T \\ & \quad + R_{i,k} + C_{i,k} S_{i,k} + S_{i,k}^T C_{i,k}^T]^{-1} \end{cases} \tag{9.74}$$

where the subscript "d" stands for the distributed fusion, and 'd, i' denotes the local estimates from the observations of sensor i.

It can be shown that

$$P_{d,k|k} \leq P_{d,i,k|k}, \quad i = 1, 2, \ldots, N \tag{9.75}$$

Proof Similar to Theorem 9.1, for each sensor $i = 1, 2, \ldots, N$ from system (9.1) and (9.2), we obtain the local state estimates $\hat{x}_{d,i,k|k}$ and $P_{d,i,k|k}$ by (9.74). The cross-correlation between the local estimation errors is given by

$$P_{d,ij,k|k} = E\{[x_k - \hat{x}_{d,i,k|k}][x_k - \hat{x}_{d,j,k|k}]^T\} \tag{9.76}$$

Substituting the first equation of (9.74) into the above equation and by use of (9.1) and (9.2), we have

$$\begin{aligned} P_{d,ij,k|k} &= E\{[x_k - \hat{x}_{d,i,k|k}][x_k - \hat{x}_{d,j,k|k}]^T\} \\ &= E\{[(I - K_{d,i,k}C_{i,k})\tilde{x}_{d,i,k|k-1} - K_{d,i,k}v_{i,k}] \\ &\quad \cdot [(I - K_{d,j,k}C_{j,k})\tilde{x}_{d,j,k|k-1} - K_{d,j,k}v_{j,k}]^T\} \\ &= [I - K_{d,i,k}C_{i,k}]E\{\tilde{x}_{d,i,k|k-1}\tilde{x}^T_{d,j,k|k-1}\} \\ &\quad \cdot [I - K_{d,j,k}C_{j,k}]^T \\ &\quad - [I - K_{d,i,k}C_{i,k}]E\{\tilde{x}_{d,i,k|k-1}v^T_{j,k}\}K^T_{d,j,k} \\ &\quad - K_{d,i,k}E\{v_{i,k}\tilde{x}^T_{d,j,k|k-1}\}[I - K_{d,j,k}C_{j,k}]^T \\ &\quad + K_{d,i,k}E\{v_{i,k}v^T_{j,k})\}K^T_{d,j,k} \end{aligned} \tag{9.77}$$

where

$$\begin{aligned} \tilde{x}_{d,i,k|k-1} &= x_k - \hat{x}_{d,i,k|k-1} \\ &= A_{k-1}\tilde{x}_{d,i,k-1|k-1} + w_{k-1} \end{aligned} \tag{9.78}$$

and $\tilde{x}_{d,i,k-1|k-1} = x_k - \hat{x}_{d,i,k-1|k-1}$. From (9.78) and by (9.5), we obtain

$$\begin{cases} E\{\tilde{x}_{d,i,k|k-1}\tilde{x}^T_{d,j,k|k-1}\} = A_{k-1}P_{d,ij,k-1|k-1}A^T_{k-1} + Q_{k-1} \\ E\{\tilde{x}_{d,i,k|k-1}v^T_j(k)\} = S_{j,k} \\ E\{v_{i,k}\tilde{x}^T_{d,j,k|k-1}\} = S^T_{i,k} \end{cases} \tag{9.79}$$

Substituting (9.79) into (9.77) and by use of (9.5), we have

$$
\begin{aligned}
P_{d,ij,k|k} = {} & [I - K_{d,i,k}C_{i,k}][A_{k-1}P_{d,ij,k-1|k-1}A_{k-1}^T \\
& + Q_{k-1}][I - K_{d,j,k}C_{j,k}]^T \\
& - [I - K_{d,i,k}C_{i,k}]S_{j,k}K_{d,j,k}^T \\
& - K_{d,i,k}S_{i,k}^T[I - K_{d,j,k}C_{j,k}]^T \\
& + K_{d,i,k}R_{ij,k}K_{d,j,k}^T
\end{aligned} \tag{9.80}
$$

Denote

$$
y_k = \begin{bmatrix} \hat{x}_{d,1,k|k} \\ \hat{x}_{d,2,k|k} \\ \vdots \\ \hat{x}_{d,N,k|k} \end{bmatrix}, \quad I_k = \begin{bmatrix} I_n \\ I_n \\ \vdots \\ I_n \end{bmatrix}, \quad \tilde{v}_k = \begin{bmatrix} -\tilde{x}_{d,1,k|k} \\ -\tilde{x}_{d,2,k|k} \\ \vdots \\ -\tilde{x}_{d,N,k|k} \end{bmatrix} \tag{9.81}
$$

we have

$$
y_k = I_k x_k + \tilde{v}_k \tag{9.82}
$$

The optimal fused state estimate in the sense of weighted least squares (WLS) can be generated by $\min\{F_x\}$, where [1]

$$
F_x = (y_k - I_k x_k)^T \Sigma_k^{-1} (y_k - I_k x_k) \tag{9.83}
$$

and $\Sigma_k = \left[P_{d,ij,k|k}\right]$ is an $nN \times nN$ symmetrical positive definite matrix.

For each $k = 1, 2, \ldots$, it is obvious that F_x is a quadratic form of x_k, and therefore a convax function of x_k. So the minimum value of F_x exists and can be found by $\frac{\mathrm{d}F_x}{\mathrm{d}x} = 0$, from which we obtain

$$
x_k = (I_k{}^T \Sigma_k^{-1} I_k)^{-1} I_k{}^T \Sigma_k^{-1} y_k \tag{9.84}
$$

So we have the optimal distributed estimation

$$
\hat{x}_{d,k|k} = \alpha_k y_k = \sum_{i=1}^{N} \alpha_{i,k} \hat{x}_{d,i,k|k} \tag{9.85}
$$

where

$$
\begin{aligned}
\alpha_k &= [\alpha_{1,k} \;\; \alpha_{2,k} \;\; \alpha_{N,k}] \\
&= (I_k{}^T \Sigma_k^{-1} I_k)^{-1} I_k{}^T \Sigma_k^{-1}
\end{aligned} \tag{9.86}
$$

is an $nN \times n$ matrix.

From (9.85) and the unbiasedness property, we have

$$\tilde{x}_{d,k|k} = \sum_{i=1}^{N} \alpha_{i,k}\tilde{x}_{d,i,k|k} \tag{9.87}$$

So the error covariance of the optimal distributed estimation is

$$P_{d,k|k} = \sum_{i=1}^{N} \alpha_{i,k} P_{d,ij,k|k}\alpha_{j,k}^T = \alpha_K^T \Sigma_k \alpha_k \tag{9.88}$$

Substituting (9.86) into the above equation yields

$$P_{d,k|k} = ({I_k}^T \Sigma_k^{-1} I_k)^{-1} \tag{9.89}$$

Notice that setting $\alpha_{i,k} = I_n$ and $\alpha_{j,k} = 0$ $(j = 1, 2 \ldots, N,\ j \neq i)$ in (9.88), we have $P_{d,i,k|k}$. Therefore, we obtain $P_{d,k|k} \leq P_{d,i,k|k}$.

From Theorem 9.3, when the cross-correlation between local state estimates is ignored, the suboptimal distributed estimation is obtained.

Corollary 9.4 (Suboptimal Distributed Fusion: SODF) *Based on system (9.1)–(9.2), by setting $P_{d,ij,k|k} = 0$ for $i \neq j$, the state estimation of x_k could be generated with suboptimal distributed fusion by*

$$\begin{cases} \hat{x}^s_{d,k|k} = \sum\limits_{i=1}^{N} \alpha_{i,k}\hat{x}_{d,i,k|k} \\ P^s_{d,k|k} = (\sum\limits_{i=1}^{N} P^{-1}_{d,i,k|k})^{-1} \\ \alpha_{i,k} = P^s_{d,k|k} P^{-1}_{d,i,k|k} \end{cases} \tag{9.90}$$

where we use superscript "s" and subscript "d" jointly to denote the suboptimal distributed fusion, and the local state estimation is computed by (9.74). From the second equation in (9.90), it is obvious that $P^s_{d,k|k} \leq P_{d,i,k|k}$ for each $i = 1, 2, \ldots, N$.

Remark 9.5 In this chapter, the inverses of many matrices are used, which are assumed to exist. Otherwise, they can be replaced by the unique Moore–Penrose generalized inverse.

9.4 Performance Analysis and Computer Simulation

Stated above, we use "OBF" for the optimal batch fusion algorithm given in Theorem 9.1, "ORF" for the optimal recursive fusion given in Theorem 9.2, and "ODF" for the optimal distributed fusion algorithm given in Theorem 9.3, respectively. The

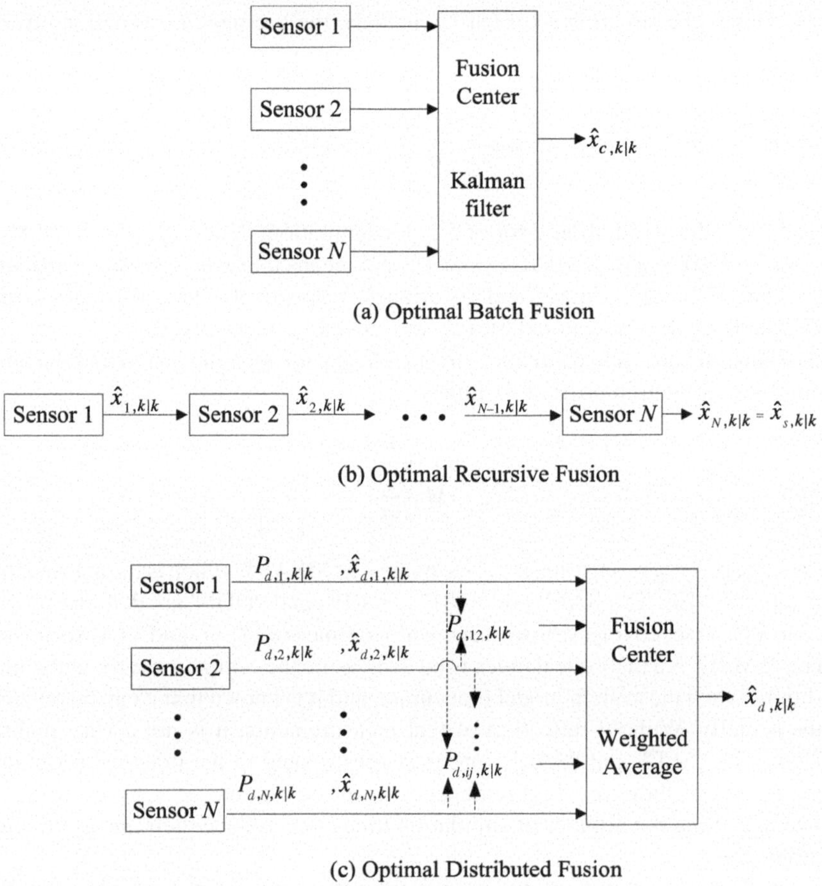

Fig. 9.1 Illustration of different fusion algorithms

architectures of the three algorithms are shown in Fig. 9.1. In this section, we will compare their estimation accuracy, computation complexity, and communication efficiency through an example.

We first introduce some measures for comparing estimation accuracy. In this context, all the estimation errors are zero mean.

(1) Sum of mean square error (SMSE), namely, the trace of the estimation error covariance matrix,

$$\operatorname{tr}(P) = E\{\tilde{x}'\tilde{x}\} \tag{9.91}$$

where $\tilde{x}$ is the estimation error. The SMSE of "OBF", "ORF", and "ODF" are denoted by $\operatorname{tr} P_{b,k|k}$, $\operatorname{tr} P_{r,k|k}$, and $\operatorname{tr} P_{d,k|k}$, respectively. Their averages along time are $\operatorname{tr} P_b$, $\operatorname{tr} P_r$, and $\operatorname{tr} P_d$, respectively.

(2) Statistical estimation error (SEE), namely, the average of estimation error of M times,

$$\tilde{x}_{k|k} = \frac{1}{M} \sum_{j=1}^{M} \tilde{x}_{j,k|k} \tag{9.92}$$

where $\tilde{x}_{j,k|k}$ is the estimation error of the jth simulation. When $\tilde{x}_{j,k|k} \sim \mathcal{N}(0, P_{k|k})$, then $\tilde{x}_{k|k} \sim \mathcal{N}(0, \frac{1}{M} P_{k|k})$, where $x \sim \mathcal{N}(\mu, \Sigma)$ means that x is Gaussian distributed with μ and Σ being its mean and covariance, respectively. The SEEs of "OBF", "ORF", and "ODF" are denoted by $\tilde{x}_{b,k}$, $\tilde{x}_{r,k}$, and $\tilde{x}_{d,k}$, respectively.

(3) Statistical absolute estimation error (SAEE), namely, the average of the absolution of the estimation error of M times,

$$\tilde{x}_{a,k|k} = \frac{1}{M} \sum_{j=1}^{M} |\tilde{x}_{j,k|k}| \tag{9.93}$$

The average of $\tilde{x}_{a,k|k}$ along time is $\tilde{x}_a$; we name it as **MSAEE** which is the abbreviation for mean of SAEE. The SAEEs of "OBF", "ORF", and "ODF" are denoted by $\tilde{x}^a_{b,k}$, $\tilde{x}^a_{r,k}$, and $\tilde{x}^a_{d,k}$, respectively, whose average along time are $\tilde{x}^a_b$, $\tilde{x}^a_r$, and $\tilde{x}^a_d$, respectively.

The "SMSE" is a theoretical index used to measure the estimation error under ideal conditions when the system model is accurate, and it is known that it can be obtained offline actually. Without reflection of real measurements, it is not a very realistic measure. The "SEE" and "SAEE" are measures related to the practical estimation errors. However, they may highly dependent on the random of $\tilde{x}_{j,k|k}$ even though the average along the number of simulation times may reduce their uncertainty to a certain degree.

To combine the theoretical index with the practical ones, we introduce another measure, which is the normalized estimation error square (NEES), also known as Mahalanobis distance [14, 16, 17]:

$$\varepsilon = (x - \hat{x})' P^{-1} (x - \hat{x}) = ||\tilde{x}||^2_{P^{-1}} \tag{9.94}$$

where x, $\hat{x}$, $\tilde{x}$, and P are the estimand, estimate, estimation error, and the assessed MSE matrix. NEES is the distance squared between x and $\hat{x}$ in terms of the P^{-1}-norm. To remove its dependence on the dimension of x and to reduce the fluctuation, define the average normalized estimation error square (ANEES) by

$$\bar{\varepsilon} = \frac{1}{nM} \sum_{j=1}^{M} \varepsilon_j \tag{9.95}$$

where ε_j denotes the NEES from the jth simulation, and M is the number of simulation times.

If $\tilde{x}$ is normal distributed with zero mean, namely, $\tilde{x} \sim \mathcal{N}(0, P^*)$, then it can be proven that

$$E\{\bar{\varepsilon}\} = \frac{\text{tr}(\tilde{P})}{n}, \quad \text{var}\{\bar{\varepsilon}\} = \frac{2\text{tr}(\tilde{P}\tilde{P})}{n^2 M} \tag{9.96}$$

where $\tilde{P} = P^{-1/2} P^* P^{-1/2}$ is the normalized MSE matrix [17, 18].

Notice that $\text{tr}(\tilde{P}) = \text{tr}(P^* P^{-1})$, so under ideal conditions when $P = P^*$, one shall have $E\{\bar{\varepsilon}\} = 1$. Denote

$$\text{tr}\, P_{br,k} = \log_{10}\left[\frac{\text{tr}(P_{b,k|k} P_{r,k|k}^{-1})}{n}\right] \tag{9.97}$$

$$\text{tr}\, P_{bd,k} = \log_{10}\left[\frac{\text{tr}(P_{b,k|k} P_{d,k|k}^{-1})}{n}\right] \tag{9.98}$$

It can be seen that $\text{tr}\, P_{br,k}$ or $\text{tr}\, P_{bd,k}$ are expected to be zero when $P_{r,k|k} = P_{b,k|k}$ or $P_{d,k|k} = P_{b,k|k}$, respectively. Hence, they can be used to measure the difference of the estimation performance between "ORF" and "OBF" or "ODF" and "OBF". We call these indices the expected difference of the normalized error square (EDNEES).

When $\tilde{x}_i \sim \mathcal{N}(0, P), \varepsilon_i = ||\tilde{x}_i||^2_{P^{-1}} \sim \chi^2(n)$ is standard chi-square with n degrees of freedom. So, the mean and covariance of ε_i/n are 1 and $2/n$, respectively. Therefore, we introduce ε_{br} and ε_{bd} to measure the performance difference between "ORF" to "OBF" or "ODF" to "OBF" by

$$\begin{aligned} \varepsilon_{br,k} &= \log_{10}\left[\prod_{j=1}^{M} \frac{\varepsilon^r_{j,k}/n}{\varepsilon^b_{j,k}/n}\right]^{1/M} \\ &= \frac{1}{M}\sum_{j=1}^{M}\left[\log_{10}\varepsilon^r_{j,k} - \log_{10}\varepsilon^b_{j,k}\right] \end{aligned} \tag{9.99}$$

$$\varepsilon_{bd,k} = \frac{1}{M}\sum_{j=1}^{M}\left[\log_{10}\varepsilon^d_{j,k} - \log_{10}\varepsilon^b_{j,k}\right] \tag{9.100}$$

where

$$\varepsilon^b_{j,k} = ||\tilde{x}^b_{j,k|k}||^2_{P_{b,k|k}} \tag{9.101}$$

$$\varepsilon^r_{j,k} = ||\tilde{x}^r_{j,k|k}||^2_{P_{r,k|k}} \tag{9.102}$$

$$\varepsilon^d_{j,k} = ||\tilde{x}^d_{j,k|k}||^2_{P_{d,k|k}} \tag{9.103}$$

are the NEES of "OBF", "ORF", and "ODF", respectively, and $\tilde{x}^b_{j,k|k}, \tilde{x}^r_{j,k|k}$, and $\tilde{x}^d_{j,k|k}$ are the corresponding estimation error of the jth simulation.

To avoid the positive and negative terms cancellations in Eqs. (9.99) and (9.100), we define

$$\varepsilon_{br,k}^{a} = \frac{1}{M}\sum_{j=1}^{M}[|\log_{10}\varepsilon_{j,k}^{r} - \log_{10}\varepsilon_{j,k}^{b}|] \tag{9.104}$$

$$\varepsilon_{bd,k}^{a} = \frac{1}{M}\sum_{j=1}^{M}[|\log_{10}\varepsilon_{j,k}^{d} - \log_{10}\varepsilon_{j,k}^{b}|] \tag{9.105}$$

The average of $\mathrm{tr}\, P_{br,k}$, $\mathrm{tr}\, P_{bd,k}$, $\varepsilon_{br,k}$, $\varepsilon_{bd,k}$, $\varepsilon_{br,k}^{a}$, and $\varepsilon_{bd,k}^{a}$ along time are denoted by $\mathrm{tr}\, P_{br}$, $\mathrm{tr}\, P_{bd}$, ε_{br}, ε_{bd}, ε_{br}^{a}, and ε_{bd}^{a}, respectively. Under ideal conditions, when the three algorithms are equivalent, they should all be zero.

We will use an example to compare the proposed algorithms by use of the above measures. Considering a target observed by three sensors, which is described by Eqs. (9.1) and (9.2) with

$$A_k = \begin{bmatrix} 1 & T \\ 0 & 1 \end{bmatrix} \tag{9.106}$$

$$Q_{k:} = Q = \begin{bmatrix} \frac{T^4}{4} & \frac{T^3}{2} \\ \frac{T^3}{2} & T^2 \end{bmatrix} . \times q \tag{9.107}$$

where $T = 1s$ is the sampling period, and $q = \begin{bmatrix} 10 & 0.8 \\ 0.8 & 0.1 \end{bmatrix}$ is the disturbance parameter. *Dot multiply* ".×" results in a matrix whose elements are the multiplication of the corresponding elements of two matrices with the same dimensions, namely,

$$\begin{bmatrix} a_{11} & a_{12} \\ a_{21} & a_{22} \end{bmatrix} . \times \begin{bmatrix} b_{11} & b_{12} \\ b_{21} & b_{22} \end{bmatrix} = \begin{bmatrix} a_{11}b_{11} & a_{12}b_{12} \\ a_{21}b_{21} & a_{22}b_{22} \end{bmatrix} \tag{9.108}$$

Therefore, the commutative law is suitable for *dot multiply* of two matrices.

Sensors 1 and 2 observe the position, and Sensor 3 observes the velocity, that is, the measurement matrices are

$$C_{1,k} = [1 \quad 0] \tag{9.109}$$

$$C_{2,k} = [1 \quad 0] \tag{9.110}$$

$$C_{3,k} = [0 \quad 1] \tag{9.111}$$

The measurement noise covariance is given by

$$R_k = \begin{bmatrix} Q_{1,1} & Q_{1,1} & Q_{1,2} \\ Q_{1,2} & Q_{1,1} & Q_{1,2} \\ Q_{1,2} & Q_{1,2} & Q_{2,2} \end{bmatrix} . \times \begin{bmatrix} r_1 & r_2 & r_2 \\ r_2 & r_1 & r_2 \\ r_2 & r_2 & r_3 \end{bmatrix} \tag{9.112}$$

The covariance between the system noise and the measurement noise is given by

$$S = [S_1 \quad S_2 \quad S_3] = \begin{bmatrix} Q_{1,1} & Q_{1,1} & Q_{1,2} \\ Q_{2,1} & Q_{2,1} & Q_{2,2} \end{bmatrix} . \times r_4 \tag{9.113}$$

where $r_1 = 100, r_2 = 0.1, r_3 = 10, r_4 = 0.25$, and $Q(i, j), i, j = 1, 2$ are the (i, j)th element of Q in (9.107). The initial condition is

$$x_0 = \begin{bmatrix} 10 \\ 0.5 \end{bmatrix}, \quad P_0 = \begin{bmatrix} 100 & 0 \\ 0 & 100 \end{bmatrix} \tag{9.114}$$

The Monte Carlo simulation results are shown in Figs. 9.2, 9.3, 9.4, 9.5, 9.6, 9.7, 9.8, 9.9, 9.10, and Table 9.1.

In Fig. 9.2, the measurements of Sensors 1, 2, 3 are shown in red dotted lines, and the original signal is shown in blue solid lines. It can be seen that the measurements are corrupted by noises.

Figure 9.3 shows the state estimates of a single simulation run. The red dotted line in (a), (c), and (e) are the position estimation by OBF, ORF, and ODF, respectively, and (b), (d), and (f) are the estimations of velocity by the three algorithms, respectively. The original signal is shown in blue solid line in each subfigure. From this figure, it can be seen that all algorithms generate good estimations.

Figure 9.4 shows the statistical estimation errors, namely, $\tilde{x}_{b,k}$, $\tilde{x}_{r,k}$, and $\tilde{x}_{d,k}$, respectively. Figure 9.5 shows the statistical estimation error with absolution, that is, $\tilde{x}^a_{b,k}$, $\tilde{x}^a_{r,k}$, and $\tilde{x}^a_{d,k}$, respectively. In Figs. 9.4 and 9.5, the statistical errors by ORF and ODF are shown by red dotted line, and the errors by OBF are shown by blue solid line. From Fig. 9.4, it can be seen that the three algorithms are comparable. However, from Fig. 9.5, it can be seen that the OBF is slightly better than the ODF, and much better than the ORF.

The traces of OBF, ORF, and ODF are shown in Fig. 9.6 with solid blue line, dotted red line, and dashed green line, respectively. Figure 9.7 shows the expected difference of the normalized estimation error square (EDNEES), namely, $\operatorname{tr} P_{br,k}$ and $\operatorname{tr} P_{bd,k}$ with solid blue and red dotted line, respectively. From these two figures, we can draw the same conclusion as that from Fig. 9.5.

Figures 9.8 and 9.10 show the illustration by ANEES or Mahalanobis distance-related measures. Figure 9.8 shows the average of the logarithm of Mahalanobis distance (denoted as "LDS" in the sequel) of OBF (blue solid line), ORF (red dotted line), and ODF (green dashed line), respectively. Figure 9.9 shows the difference of LDS between ORF and OBF (blue solid line) or ODF and OBF (red dotted line), namely, $\varepsilon_{br,k}$ and $\varepsilon_{bd,k}$, respectively. Figure 9.10 shows the difference of LDS with absolution, that is, $\varepsilon^a_{br,k}$ (blue solid line) and $\varepsilon^a_{bd,k}$ (red dotted line), respectively. Roughly, from Fig. 9.8, one can see that the three algorithms are comparable. From Figs. 9.9 to 9.10, however, one can find that the ODF is better than the ORF.

Briefly, from Figs. 9.2, 9.3, 9.4, 9.5, 9.6, 9.7, 9.8, 9.9, and 9.10, the three algorithms of OBF, ORF, and ODF are all effective. Roughly, they are comparable in precision.

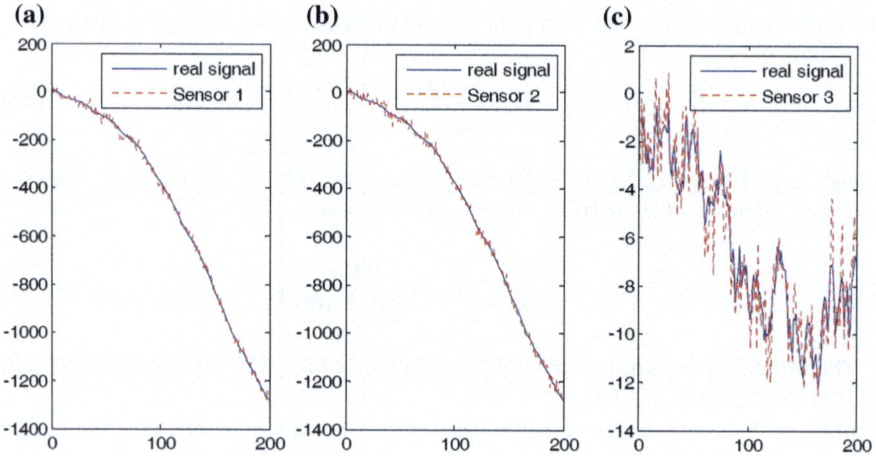

Fig. 9.2 The original signal and the corresponding measurements

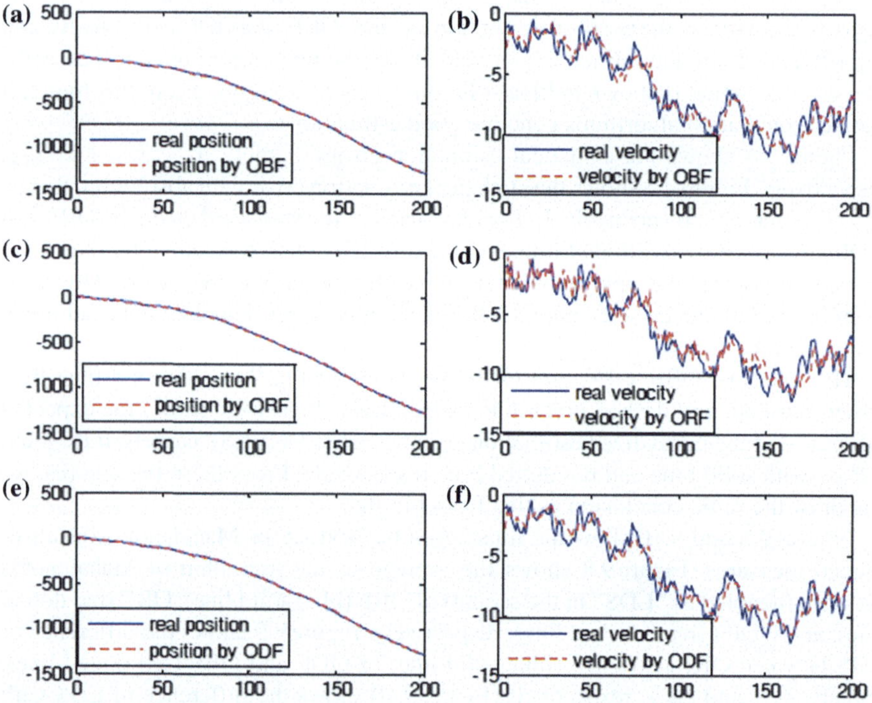

Fig. 9.3 The signal and the estimations

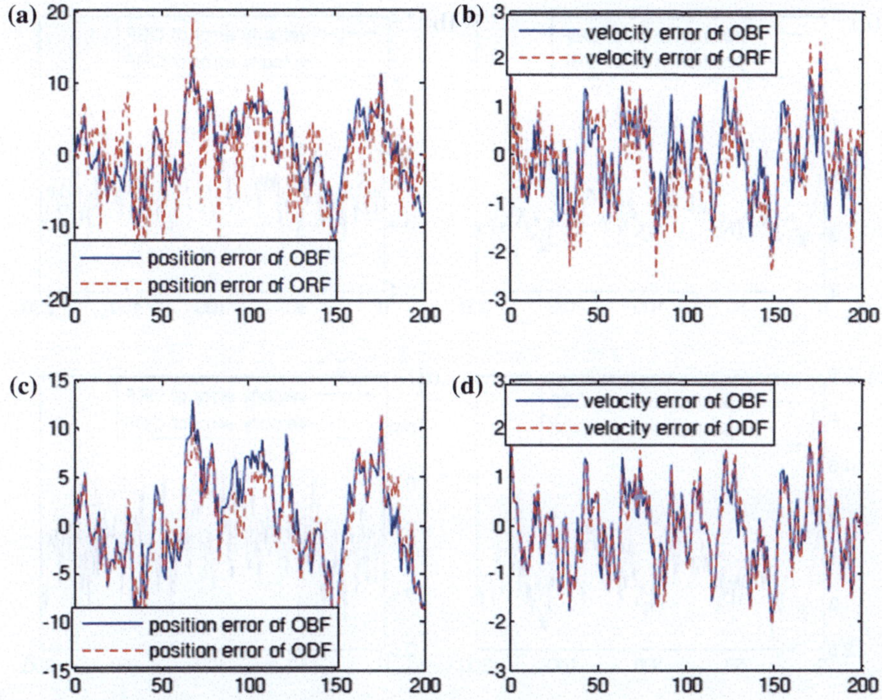

Fig. 9.4 The statistical estimation errors

Under more finer scale observation, OBF has the best estimation effect, then is the ODF, and followed by ORF.

In case 2, we consider the situation when the measurement noise of different sensors, and the measurement noise and the system noise are uncorrelated, namely, we set $q = 0.01$, $r_1 = 1$, $r_2 = 0$, $r_3 = 1$ and $r_4 = 0$ in this example.

The simulation results are shown in Figs. 9.11, 9.12, 9.13, 9.14, 9.15, and 9.16, which are the statistical absolute estimation errors, the trace of the estimation error covariances, the expected difference of the normalized estimation error square ($\text{tr}\, P_{rb,k}$ with red dotted line, $\text{tr}\, P_{db,k}$ with blue solid line), the average LDS of OBF (blue solid line), ORF (red dotted line) and ODF (greed dashed line), the difference of average LDS, and the difference of the average absolute LDS, respectively.

From Figs. 9.11, 9.12, 9.13, 9.14, 9.15, and 9.16, one can find that when the noise are uncorrelated, all the algorithms are effective. The ORF and the OBF are barely the same when position estimation are concerned, but from each figure we can see that the ORF is more effective than the ODF in this case.

The quantitative comparison of the three algorithms is shown in Table 9.1, from which we can draw the same conclusions as those from the figures.

To sum up, from the simulation in this section, one can see that the presented ORF and ODF are effective. When the measurement noise is uncorrelated and independent

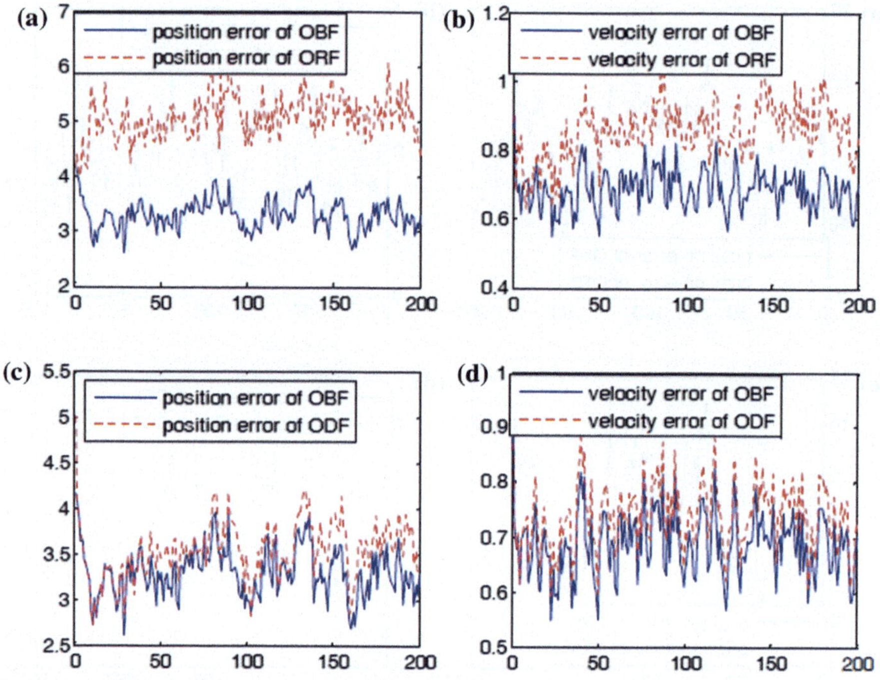

Fig. 9.5 The statistical absolute estimation errors

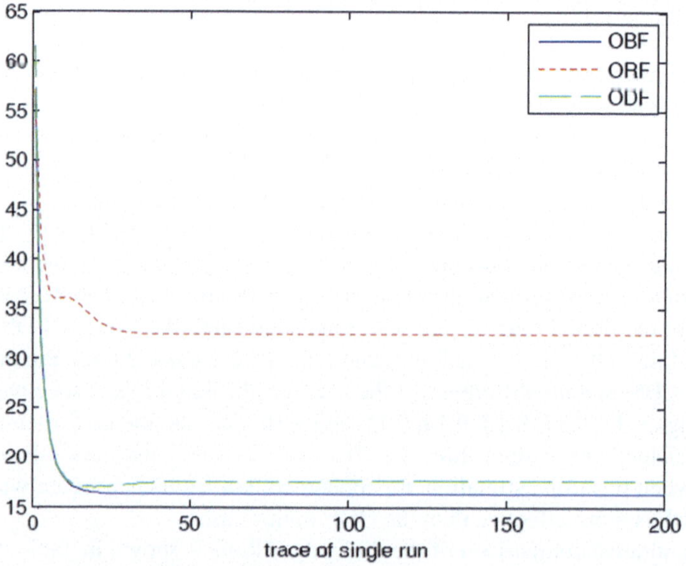

Fig. 9.6 Trace of the estimation error covariances

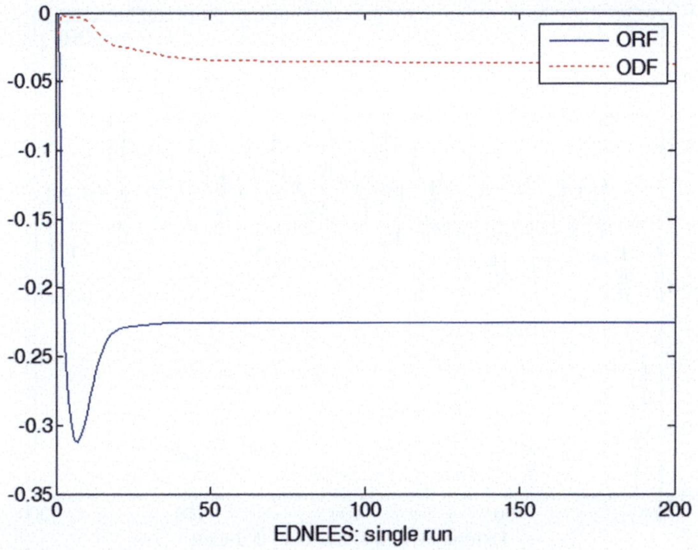

Fig. 9.7 The expected difference of the normalized estimation error square

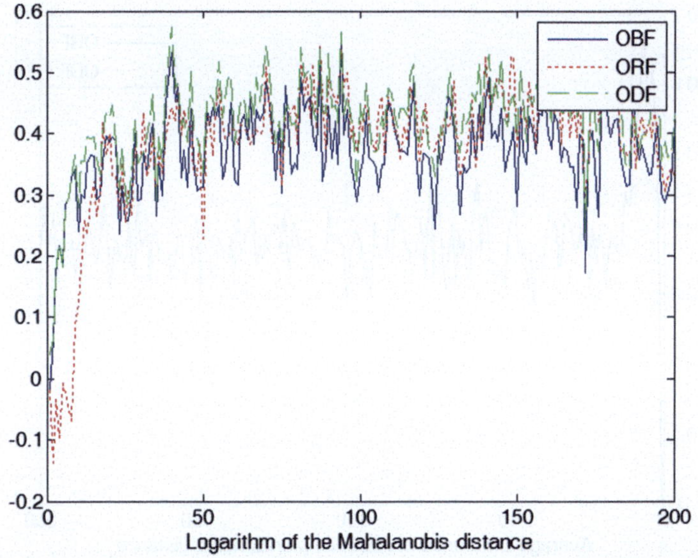

Fig. 9.8 The average of the normalized estimation error square after taking logarithm

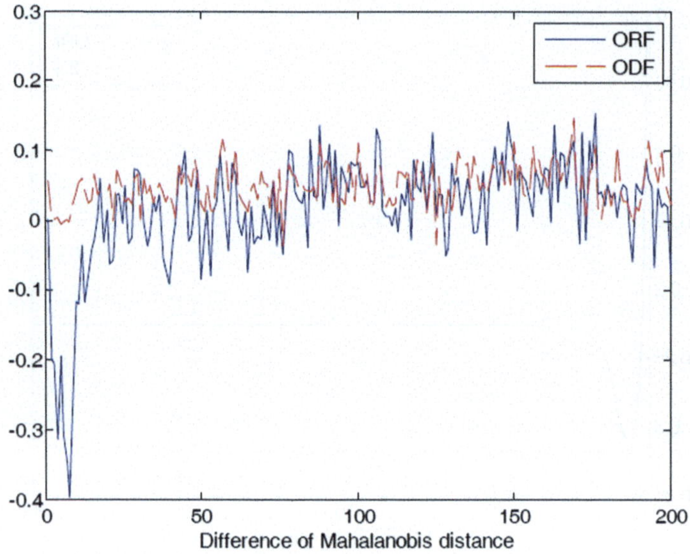

Fig. 9.9 Difference of the logarithm of Mahalanobis distance

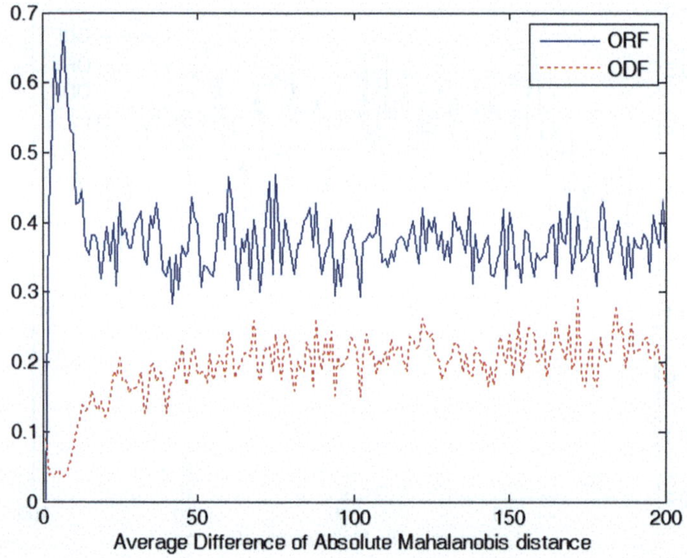

Fig. 9.10 Difference of the logarithm of Mahalanobis distance with absolution

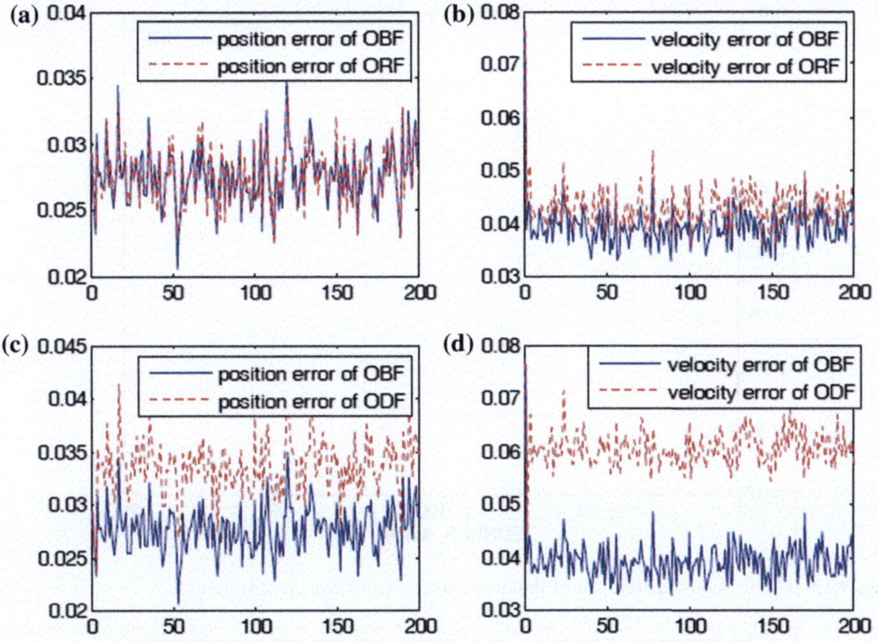

Fig. 9.11 The statistical absolute estimation errors

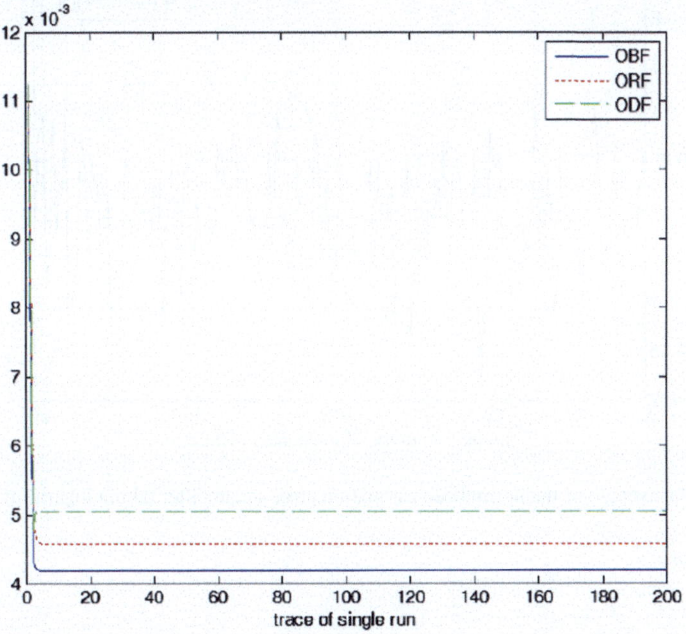

Fig. 9.12 Trace of the estimation error covariances

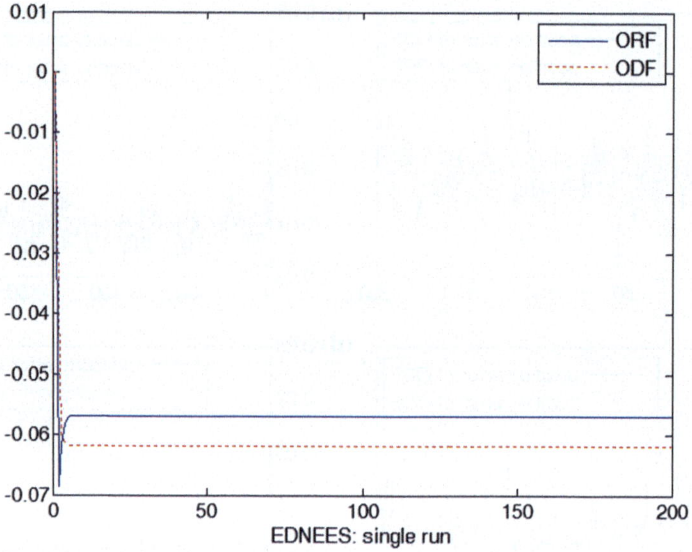

Fig. 9.13 The expected difference of the normalized estimation error square

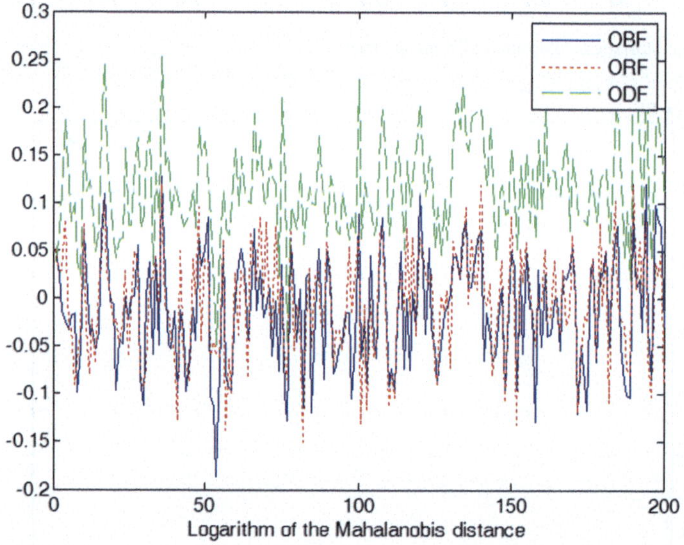

Fig. 9.14 The average of the normalized estimation error square after taking logarithm

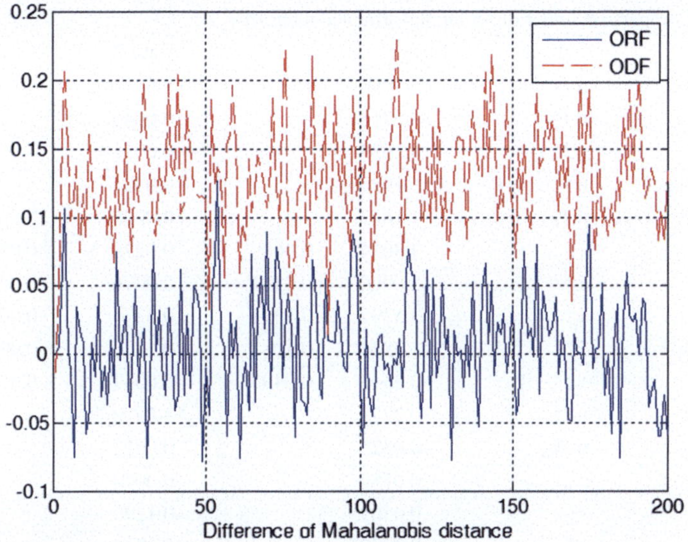

Fig. 9.15 Difference of the logarithm of Mahalanobis distance

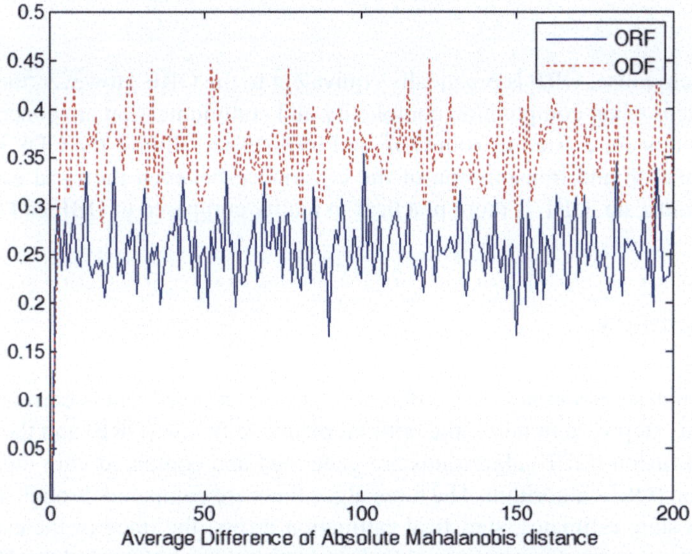

Fig. 9.16 Difference of the logarithm of Mahalanobis distance with absolution

Table 9.1 Comparison under different performance measures

Measure	Indices	Case 1		Case 2	
Trace	$\mathrm{tr}\, P_b$	16.9789		0.0042	
	$\mathrm{tr}\, P_r$	33.1718		0.0046	
	$\mathrm{tr}\, P_d$	18.1633		0.0051	
Trace difference	$\mathrm{tr}\, P_r - \mathrm{tr}\, P_b$	16.1929		0.0004	
	$\mathrm{tr}\, P_d - \mathrm{tr}\, P_b$	1.1843		0.0008	
MSAEE	$\tilde{x}_b^a$	3.3054	0.0689	0.0276	0.0393
	$\tilde{x}_r^a$	5.0900	0.8581	0.0276	0.0429
	$\tilde{x}_d^a$	3.5397	0.7391	0.0336	0.0606
MSAEE difference	$\tilde{x}_r^a - \tilde{x}_b^a$	1.7846	0.1084	0	0.0037
	$\tilde{x}_d^a - \tilde{x}_b^a$	0.2343	0.0493	0.0059	0.0213
ANEES	$\mathrm{tr}\, P_{br}$	−0.2275		−0.0567	
	$\mathrm{tr}\, P_{bd}$	−0.0325		0.0613	
Related	ε_{br}	0.0159		0.0070	
	ε_{bd}	0.0509		0.1290	
measure	ε_{br}^a	0.3772		0.2549	
	ε_{bd}^a	0.1938		0.3576	

of the system noise, ORF is practically equivalent to the OBF. However, the ORF is most efficient when computation complexity and communication are concerned, as augmentations are existed in both OBF and ODF, especially for the ODF in which the computation and transmission of the covariance between the local estimation errors are intricate. ORF is more practical in reality compared to OBF and ODF.

9.5 Summary

When the measurement noise of different sensors is coupled and is correlated with the previous step system noise, the optimal recursive fusion (ORF) and the optimal distributed fusion (ODF) algorithms are generated and compared with the optimal batch fusion (OBF) algorithm. The three algorithms are compared through an example, where state estimates, statistical estimation errors, the trace of the estimation error covariances, the average normalized estimation error square, and the difference of the logarithm of the Mahalanobis distance are included. From performance analysis, we conclude that the ORF and ODF are effective, which are slight worse than the OBF when estimation accuracy is concerned, but roughly comparable. When computation complexity and communication are concerned, ORF is the most efficient one among the three algorithms and therefore most practical.

References

1. X.R. Li, Y.M. Zhu, J. Wang, C.Z. Han, Optimal linear estimation fusion-part 1: unified fusion rules. IEEE Trans. Inf. Theory **49**(9), 2192–2208 (2003)
2. Y. Bar-Shalom, *Multitarget-Multisensor Tracking: Advanced Applications* (Artech House, Norwood, 1990)
3. C.Y. Chong, K.C. Chang, S. Mori, Distributed tracking in distributed sensor networks (1986), pp. 1863–1868
4. Z.S. Duan, X.R. Li, Lossless linear transformation of sensor data for distributed estimation fusion. IEEE Trans. Signal Process. **59**(1), 362–372 (2011)
5. E.B. Song, Y.M. Zhu, J. Zhou, Z.S. You, Optimal Kalman filtering fusion with cross-correlated sensor noises. Automatica **43**(8), 1450–1456 (2007)
6. X.R. Li, Optimal linear estimation fusion-part vii: dynamic systems (2003), pp. 455–462
7. Z.S. Duan, C.Z. Han, T. Tao, Optimal multi-sensor fusion target tracking with correlated measurement noises. IEEE Int. Conf. Syst. Man Cybern. **2**, 1272–1278 (2004)
8. Z.S. Duan, X.R. Li, The optimality of a class of distributed estimation fusion algorithm (2008), pp. 1–6
9. J.K. Uhlmann, General data fusion for estimates with unknown cross covariances (1996), pp. 536–547
10. C.Y. Chong, S. Mori, Convex combination and covariance intersection algorithms in distributed fusion (2001), pp. WeA2.11–WeA2.18
11. Y.M. Wang, X.R. Li, Distributed estimation fusion with unavailable cross-correlation. IEEE Trans. Aerosp. Electron. Syst. **48**, 259–278 (2012)
12. S. Sun, Z. Deng, Multi-sensor optimal information fusion Kalman filter. Automatica **40**(6), 1017–1023 (2004)
13. Z.S. Duan, C.Z. Han, Discrete-time linear estimation with correlated noises. Syst. Eng. Electron. **27**, 792–794 (2005)
14. Y. Bar-Shalom, X. Li, T. Kirubarajan, *Estimation with Applications to Tracking and Navigation: Theory Algorithms and Software* (Wiley, New York, 2001)
15. X.R. Li, Recursibility and optimal linear estimation and filtering, in *43rd IEEE Conference on Decision and Control*. Atlantis, Paradise Island, Bahamas, 14–17 Dec 2004, pp. 1761–1766
16. Y. Bar-Shalom, K. Birmiwal, Consistency and robustness of pdaf for target tracking in cluttered environments. Automatica **19**(4), 431–437 (1983)
17. X.R. Li, Z.L. Zhao, X.B. Li, Evaluation of estimation algorithms-II: credibility tests. IEEE Trans. Syst. Man Cybern. Part A Syst. Hum. **42**(1), 147–163 (2012)
18. J.H. Stapleton, *Linear Stochastical Models* (Wiley, New York, 1995)

Chapter 10
Optimal Estimation for Multirate Systems with Unreliable Measurements and Correlated Noise

In recent years, sensor networks have shown to be a persistent focus of research due to the rapid development of technology and its wildly use in multiple industries including military, law enforcement, agricultural and forestry-based projects, surveillance, and even information collection. Accordingly, considerable research attention has been devoted to state estimation techniques over sensor networks, not only due to a large number of potential applications but also because they provide more information than traditional communication systems with a single sensor. By using different research methods, a large number of research results on the design of fusion estimation algorithms in multisensor systems have been reported [1, 2].

For state estimation of multisensor systems, the existing researches rarely take into account the correlation of the noise [3, 4] which is widely existed in reality because different sensors usually observe the same target in a common noisy environment [5, 6]. The existed multisensor fusion estimation algorithms that consider the correlation of noises seldom consider the multirate sampling of sensors. For example, when the fusion center is not fed back to the local sensor and sensor noise is cross-correlated, a distributed fusion formula is proposed in [7] where the sensors are observing the single target with single sampling rate. When the error cross-correlation matrix between local estimates is unavailable, the distributed fusion approach is addressed in [8]. The performance of the Kalman filter with the mismatched process and measurement noise covariance is studied in [9]. For nonlinear systems with two kinds of noise correlations, the nonlinear information filters and the decentralized fusion algorithms on the basis of square-root cubature Kalman filter and cubature Kalman filter are researched in [10]. With correlated and autocorrelated noise, a distributed weighted robust Kalman filter fusion algorithm is derived for uncertain systems with multiple sensors in [11]. In consideration of the correlation of the process noise and the system noise, the optimal distributed fusion and sequential fusion algorithms are derived in [12].

The literature mentioned above seldom take consideration of the measurements randomly lose or unreliable that occur possibly in practical applications because of sensor intermittent faults or communication limitations. When the packet dropping

H. Ma et al., *Kalman Filtering and Information Fusion*,
https://doi.org/10.1007/978-981-15-0806-6_10

phenomenon or outlier of the sensor measurement occurs in multisensor fusion systems, the centralized and the distributed fusion estimation algorithms were developed in [13, 14]. To deal with the packet dropouts and delays simultaneously, an optimal distributed fusion Kalman filter is designed in [15]. Based on a packet dropout model, the optimal linear estimators for the discrete-time system are developed in [16], and a linear minimum variance (LMV) filter is proposed using the orthogonality principle in [17]. When the missing of data is Bernoulli distributed, the algorithms presented by the team of Wang [18], Sinopoli [19], and Kal [20] are promising in that they have proper computation complexity and can generate nearly optimal state estimations. Huang presents a Kalman filtering algorithm for a class of linear dynamic systems where the measurements' packet loss obeys Markov distribution [21]. When there are multistep transmission delays, multiple packet dropouts and correlated noises are concerned, a recursive nonlinear estimator is given in [22]. There are quite a few literature concerned about multisensor data fusion for the networked system. As an example, Sun et al. present optimal filtering algorithm for systems with multiple packet dropouts [23]. A networked federated filter algorithm with variable delays and packet losses is presented by Xia et al. in [24]. A multisensor distributed weighted Kalman filter with stochastic uncertainties, network delays, autocorrelated, and cross-correlated noises is put forward in [25].

For fusion of unreliable measurements, the problem of multirate or multiscale multisensor state estimation are seldom considered. The multirate sensor sampling problem is widely existed in applications, e.g., in GPS/INS integrated navigation. In fact, it is often unrealistic to sample all physical signals uniformly at one single rate in practice [26]. In [27], the multirate estimation model with four rates is constructed, and the state estimator in the multirate linear minimum mean square error is designed. In [28, 29], the multismart, multirate sensors' state estimation problem is studied. However, the above literature do not consider unreliable measurements. The optimal state estimation with unreliable measurements and multirate sensors is researched in [4, 30] without consideration of the noise correlation. For multirate dynamic systems, an optimal sequential fusion estimation algorithm with correlated noise and unreliable measurements is presented in [31], where the correlation of the measurement noise and the system noise at the same time step is not considered, and the distributed fusion of the observations is not considered. The distributed filtering problem through sensor networks has gained an ever-increasing interest, and a great number of research results concerning distributed algorithms have been reported (see [3, 8, 11]). Compared with the centralized Kalman filter, the distributed fusion algorithm involves a recursive computation of the filtering error cross-covariance matrix between any two subsystems; consequently, it has stronger fault-tolerance ability, smaller computation complexity, and better flexibility. In [32, 33], the distributed fusion for the linear dynamic system with unreliable measurements or data dropout is studied, but the correlation of the noises is not concerned.

From the above discussion, this paper focuses on the distributed fusion estimation problem for the multirate multisensor systems with unreliable measurements, where the measurement noises are correlated and coupled with the system noise of the current time step and the previous time step. Assuming that we have known the dynamics, but only partially know the measurement system, unreliable measurements may occur stochastically. The optimal distributed fusion estimation algorithm is derived. The effectiveness of the proposed algorithms is proven, and the feasibility is shown by a numerical example on target tracking.

This paper is organized as follows. In Sect. 10.1, the multirate multisensor estimation problem is formulated. Section 10.2 presents the local estimation algorithms in case of total reliable measurements, and stochastic unreliable measurements occur occasions, respectively, followed by the optimal and the suboptimal distributed fusion algorithm. Section 10.3 gives the simulation results, and Sect. 10.4 draws the conclusion.

10.1 Problem Formulations

As the measurement is unreliable or faulty with a certain probability, the discrete-time linear dynamic system with N multirate sensors satisfies

$$x_k = A_k x_k + B_k u_k + w_k, \tag{10.1}$$

$$z_{i,k_i} = \gamma_{i,k_i}[C_{i,k_i} x_{i,k_i} + v_{i,k_i}] \tag{10.2}$$

where $i = 1, 2, \ldots, N$ denote sensors, whose sampling rates are B_i, respectively. $x_k \in \boldsymbol{R}^n$ indicates the system state at the kth time instant with the highest sampling rate. $A_k \in \boldsymbol{R}^{n\times n}$ is the system transfer matrix. u_k is the control variable, B_k is the control innovation. w_k is zero-mean Gaussian-distributed system noise and meet

$$E[w_k] = 0 \tag{10.3}$$

$$E[w_k w_l^T] = Q_k \delta_{kl} \tag{10.4}$$

where δ_{kl} is the Kronecker delta function, $Q_k \geq 0$; $z_{i,k_i} \in \mathbb{R}^{q_i\times 1} (q_i \leq n)$ is the measurement of sensor i at time k with sampling rate B_i, which satisfies

$$B_i = \frac{B_1}{l_i}, i = 1, 2, \ldots, N \tag{10.5}$$

where l_i is a positive integer and $l_1 = 1$; $C_{i,k_i} \in \boldsymbol{R}^{q_i\times n}$ is the measurement matrix. $v_{i,k_i} \in \boldsymbol{R}^{q_i\times 1}$ is the Gaussian distributed measurement noise with zero mean and

$$E[v_{i,k_i} v_{j,k_j}^T] = \begin{cases} R_{ij,k}, & if\ k_i l_i = k_j l_j = k \\ 0, & otherwise \end{cases} \tag{10.6}$$

$$E[w_{k-1} v_{i,k_i}^T] = \begin{cases} S_{i,k}^*, & if\ l_i k_i = k \\ 0, & otherwise \end{cases} \tag{10.7}$$

$$E[w_k v_{i,k_i}^T] = \begin{cases} S_{i,k}, & if\ l_i k_i = k \\ 0, & otherwise \end{cases} \tag{10.8}$$

where $i = 1, 2, \ldots, N$.

The initial state vector x_0 is independent of w_k and v_{i,k_i}, and its mean and estimation error covariance are x_0 and P_0, respectively.

γ_{i,k_i} represents the measurement of sensor i at each moment which is supposed to be unreliable, denoting that the measurement is faulty with probability $(1 - \bar{\gamma}_i)$ or normal with $\bar{\gamma}_i$. Assuming that γ_{i,k_i} is a stochastic sequence with Bernoulli distribution that takes on 1 with probability $\bar{\gamma}_i$ and not 1 with probability $(1 - \bar{\gamma}_i)$, and γ_{i,k_i} is independent of w_k, v_{i,k_i} and x_0, $i = 1, 2, \ldots, N$.

In this article, the linear system estimation is researched in case that there are multiple sensors at different sampling rates synchronously observing a single target. Sensors in this multirate multisensor dynamic system are sampled at different scales. The scale index and the sensor index are both indicated by i. Bigger i denote coarser scales. For instance, $i = 1$ denotes the finest scale, where the sensor samples fastest. $i = 2, 3, \ldots, N - 1$ represent the coarser scales, and $i = N$ represents the coarsest scale, in which the sensor samples slowest. It is assumed that each sensor samples uniformly. k denotes time, and k_i denotes the k_ith measurement of sensor i and its sampling time is $l_i k_i$, we have

$$x_{i,k_i} = x_{l_i k_i}, i = 1, 2, \ldots, N \tag{10.9}$$

where $l_1 = 1, k_1 = k, x_{1,k_1} = x_k$.

Assume there are N_k sensors that are observing at time $k (k = 1, 2, \ldots)$. Without losing generality, we denote the sensors as $1, 2, \ldots, N_k$, where $1 \leq N_k \leq N$. Then system model (1) and (10.2) can be rewritten as

$$x_k = A_k x_k + B_k u_k + w_k \tag{10.10}$$

$$\begin{aligned} z_{i,\frac{k}{l_i}} &= \gamma_{i,\frac{k}{l_i}}[C_{i,\frac{k}{l_i}} x_k + v_{i,\frac{k}{l_i}}], \\ & i = 1, 2, \ldots, N_k \end{aligned} \tag{10.11}$$

where

$$z_{i,\frac{k}{l_i}} = z_{i,k_i} \tag{10.12}$$

$$\gamma_{i,\frac{k}{l_i}} = \gamma_{i,k_i} \tag{10.13}$$

$$C_{i,\frac{k}{l_i}} = C_{i,k_i} \tag{10.14}$$

$$v_{i,\frac{k}{l_i}} = v_{i,k_i} \tag{10.15}$$

In the sequel, we will obtain the optimal distributed Kalman filter fusion of x_k based on the multirate multisensor unreliable measurements that with coupled noises based on the above formulated system.

10.2 Optimal Distributed Fusion Algorithm

10.2.1 Local State Estimation with Normal Measurements

Theorem 10.1 *For system models (1) and (2), suppose that the measurements are all reliable, i.e.,* $\gamma_{i,k_i} = 1$ *for* $i = 1, 2, \ldots, N$*, then the local state estimation* $\hat{x}_{i,k|k}$ *and its filtering error covariance matrix are computed by*

$$\hat{x}_{i,k|k} = \begin{cases} \hat{x}_{i,k|k-1} + K_{i,\frac{k}{l_i}} \tilde{z}_{i,\frac{k}{l_i}}, & \text{if } k \ mod \ l_i = 0 \\ \hat{x}_{i,k|k-1}, & \text{if } k \ mod \ l_i \neq 0 \end{cases} \tag{10.16}$$

$$P_{i,k|k} = P_{i,k|k-1} - K_{i,\frac{k}{l_i}} K_{i,\frac{k}{l_i}} K_{i,\frac{k}{l_i}}^T \tag{10.17}$$

Where the filtering gain matrix for the state is obtained by

$$K_{i,\frac{k}{l_i}} = P_{i,k|k-1} C_{i,\frac{k}{l_i}}^T P_{\tilde{z}_i,\frac{k}{l_i}}^{-1} + S_{i,\frac{k}{l_i}}^* P_{\tilde{z}_i,\frac{k}{l_i}}^{-1} \tag{10.18}$$

The one-step state predictor and the corresponding error covariance matrix are computed by

$$\begin{aligned} \hat{x}_{i,k|k-1} &= A_{k-1}\hat{x}_{i,k-1|k-1} + \hat{w}_{i,k-1|k-1} \\ &\quad + B_{k-1}u_{k-1} \end{aligned} \tag{10.19}$$

$$\begin{aligned} P_{i,k|k-1} &= A_{k-1}P_{i,k-1|k-1}A_{k-1}^T \\ &\quad + A_{k-1}P_{x_i w_i,k-1|k-1} \\ &\quad + P_{x_i w_i,k-1|k-1}^T A_{k-1}^T \\ &\quad + P_{w_i,k-1|k-1} \end{aligned} \tag{10.20}$$

The white noise estimator is obtained by

$$\hat{w}_{i,k|k} = K_{w_i,\frac{k}{l_i}} \tilde{z}_{i,\frac{k}{l_i}} \tag{10.21}$$

$$P_{w_i,k|k} = Q_k - K_{w_i,\frac{k}{l_i}} S_{i,\frac{k}{l_i}}^T \tag{10.22}$$

$$K_{w_i,\frac{k}{l_i}} = S_{i,\frac{k}{l_i}} P_{\tilde{z}_i,\frac{k}{l_i}}^{-1} \tag{10.23}$$

and

$$P_{x_i w_i,k|k} = -K_{i,\frac{k}{l_i}} S_{i,\frac{k}{l_i}}^T \tag{10.24}$$

The innovation sequence $\tilde{z}_{i,\frac{k}{l_i}}$ with variance $P_{\tilde{z}_i,\frac{k}{l_i}}$ is computed by

$$\tilde{z}_{i,\frac{k}{l_i}} = z_{i,\frac{k}{l_i}} - C_{i,\frac{k}{l_i}}\hat{x}_{i,k|k-1} \tag{10.25}$$

$$\begin{aligned} P_{\tilde{z}_{i},\frac{k}{l_i}} &= C_{i,\frac{k}{l_i}} P_{i,k|k-1} C^T_{i,\frac{k}{l_i}} + C_{i,\frac{k}{l_i}} S_{i,\frac{k}{l_i}} * \\ &+ S^{*T}_{i,\frac{k}{l_i}} C^T_{i,\frac{k}{l_i}} + R_{i,\frac{k}{l_i}} \end{aligned} \tag{10.26}$$

Proof For any $i = 1, 2, \ldots, N$, define

$$Z_{i,k} \triangleq \{z_{i,1}, z_{i,2}, \ldots, z_{i,k}\} \tag{10.27}$$

$$Z^i_{1,k} \triangleq \{z_{1,k}, z_{2,k}, \ldots, z_{i,k}\} \tag{10.28}$$

$$\bar{Z}^i_{1,k} \triangleq \{Z^i_1(l)\}^k_{l=1} \tag{10.29}$$

where the measurements observed by sensor i in time $[0, k]$ are denoted by $Z_{i,k}$; $Z^i_{1,k}$ represents the measurements at time k sampled by sensors $1, 2, \ldots, i$; $\bar{Z}^i_{1,k}$ represents the measurements observed by sensors $1, 2, \ldots, i$ during time $[0, k]$; $\bar{Z}^N_{1,k}$ represents the measurements observed by all the sensors during time $[0, k]$.

Through the projection theorem, on the foundation of measurements of $\bar{Z}^N_1(k-1)$, the state prediction of x_k can be computed as

$$\hat{x}_{i,k|k-1} = A_{k-1}\hat{x}_{i,k-1|k-1} + \hat{w}_{i,k-1|k-1} + B_{k-1}u_{k-1} \tag{10.30}$$

and we derive

$$\hat{x}_{i,k|k} = \begin{cases} \hat{x}_{i,k|k-1} + K_{i,\frac{k}{l_i}}\tilde{z}_{i,\frac{k}{l_i}}, & \text{if } k \ mod\ l_i = 0 \\ \hat{x}_{i,k|k-1}, & \text{if } k \ mod\ l_i \neq 0 \end{cases} \tag{10.31}$$

The innovation sequence $\tilde{z}_{i,\frac{k}{l_i}}$ is computed by

$$\begin{aligned} \tilde{z}_{i,\frac{k}{l_i}} &= z_{i,\frac{k}{l_i}} - \hat{z}_{i,\frac{k}{l_i}} \\ &= C_{i,\frac{k}{l_i}}\tilde{x}_{i,k|k-1} + v_{i,\frac{k}{l_i}} \end{aligned} \tag{10.32}$$

Substituting (10.32) into $P_{\tilde{z}_{i},\frac{k}{l_i}} = E\{\tilde{z}_{i,\frac{k}{l_i}}\tilde{z}^T_{i,\frac{k}{l_i}}\}$, the innovation sequence covariance matrix can be computed as follows:

$$\begin{aligned} P_{\tilde{z}_{i},\frac{k}{l_i}} &= E\{\tilde{z}_{i,\frac{k}{l_i}}\tilde{z}^T_{i,\frac{k}{l_i}}\} \\ &= E\{[C_{i,\frac{k}{l_i}}\tilde{x}_{i,k|k-1} + v_{i,\frac{k}{l_i}}] \\ &\cdot [C_{i,\frac{k}{l_i}}\tilde{x}_{i,k|k-1} + v_{i,\frac{k}{l_i}}]^T\} \\ &= C_{i,\frac{k}{l_i}} P_{i,k|k-1} C^T_{i,\frac{k}{l_i}} + C_{i,\frac{k}{l_i}} S^*_{i,\frac{k}{l_i}} \\ &+ S^{*T}_{i,\frac{k}{l_i}} C^T_{i,\frac{k}{l_i}} + R_{i,\frac{k}{l_i}} \end{aligned} \tag{10.33}$$

Then, based on the projection theory, the following one-step prediction and filtering error equations for state x_k is computed by

$$\begin{aligned}\tilde{x}_{i,k|k-1} &= x_k - \hat{x}_{i,k|k-1} \\ &= A_{k-1}\tilde{x}_{i,k-1|k-1} + \tilde{w}_{i,k-1|k-1}\end{aligned} \tag{10.34}$$

$$\begin{aligned}\tilde{x}_{i,k|k} &= x_k - \hat{x}_{i,k|k} \\ &= \tilde{x}_{i,k|k-1} - K_{i,\frac{k}{l_i}}\tilde{z}_{i,\frac{k}{l_i}}\end{aligned} \tag{10.35}$$

Based on the projection theory, the white noise filter is computed by

$$\begin{aligned}\hat{w}_{i,k|k} &= \hat{w}_i(k\,|k-1) + K_{w_i,\frac{k}{l_i}}\tilde{z}_{i,\frac{k}{l_i}} \\ &= K_{w_i,\frac{k}{l_i}}\tilde{z}_{i,\frac{k}{l_i}}\end{aligned} \tag{10.36}$$

where the filtering gain for the white noise is

$$\begin{aligned}K_{w_i,\frac{k}{l_i}} &= E\{w_k\tilde{z}^T_{i,\frac{k}{l_i}}\}P^{-1}_{\tilde{z}_{i,\frac{k}{l_i}}} \\ &= E\{w_k[C_{i,\frac{k}{l_i}}\tilde{x}_{i,k|k-1} + v_{i,\frac{k}{l_i}}]^T\}P^{-1}_{\tilde{z}_{i,\frac{k}{l_i}}} \\ &= S_{i,\frac{k}{l_i}}P^{-1}_{\tilde{z}_i,\frac{k}{l_i}}\end{aligned} \tag{10.37}$$

Similarly, the filtering gain for the state is computed by

$$\begin{aligned}K_{i,k} &= E\{x_k\tilde{z}^T_{i,\frac{k}{l_i}}\}P^{-1}_{\tilde{z}_{i,\frac{k}{l_i}}} \\ &= E\{[\hat{x}_{i,k|k-1} + \tilde{x}_{i,k|k-1}][C_{i,\frac{k}{l_i}}\tilde{x}_{i,k|k-1} \\ &\quad + v_{i,\frac{k}{l_i}}]^T\}P^{-1}_{\tilde{z}_i,\frac{k}{l_i}} \\ &= P_{i,k|k-1}\,C^T_{i,\frac{k}{l_i}}P^{-1}_{\tilde{z}_{i,\frac{k}{l_i}}} + S^*_{i,\frac{k}{l_i}}P^{-1}_{\tilde{z}_{i,\frac{k}{l_i}}}\end{aligned} \tag{10.38}$$

By substituting (10.34) into $P_{i,k|k-1} = E\{\tilde{x}_{i,k|k-1}\tilde{x}^T_{i,k|k-1}\}$, the corresponding prediction error covariance matrix $P_{i,k|k-1}$ can be computed as

$$\begin{aligned}P_{i,k|k-1} &= E\{\tilde{x}_{i,k|k-1}\tilde{x}^T_{i,k|k-1}\} \\ &= E\{[A_{k-1}\tilde{x}_{i,k-1|k-1} \\ &\quad + \tilde{w}_{i,k-1|k-1}] \\ &\quad \cdot [A_{k-1}\tilde{x}_{i,k-1|k-1} \\ &\quad + \tilde{w}_{i,k-1|k-1}]^T\} \\ &= A_{k-1}P_{i,k-1|k-1}A^T_{k-1} \\ &\quad + A_{k-1}P_{x_iw_i,k-1|k-1}\end{aligned} \tag{10.39}$$

Substituting (10.35) into $P_{i,k|k} = E\{\tilde{x}_{i,k|k}\tilde{x}^T_{i,k|k}\}$, the state filtering error covariance matrix can be computed as

$$\begin{aligned} P_{i,k|k} &= E\{\tilde{x}_{i,k|k}\tilde{x}^T_{i,k|k}\} \\ &= E\{[\tilde{x}_{i,k|k-1} - K_{i,\frac{k}{l_i}}\tilde{z}_{i,\frac{k}{l_i}}] \\ &\cdot [\tilde{x}_{i,k|k-1} - K_{i,\frac{k}{l_i}}\tilde{z}_{i,\frac{k}{l_i}}]^T\} \\ &= P_{i,k|k-1} - K_{i,\frac{k}{l_i}} P^T_{\tilde{z}_{i,\frac{k}{l_i}}} K^T_{i,\frac{k}{l_i}} \end{aligned} \tag{10.40}$$

We have the white noise filtering error equation

$$\tilde{w}_{i,k|k} = w_k - \hat{w}_{i,k|k} = w_k - K_{w_i,\frac{k}{l_i}}\tilde{z}_{i,\frac{k}{l_i}} \tag{10.41}$$

Substituting (10.41) into $P_{w_i,k|k} = E\{\tilde{w}_{i,k|k}\tilde{w}^T_{i,k|k}\}$, we obtain the filtering error covariance matrix for the white noise

$$\begin{aligned} P_{w_i,k|k} &= E\{\tilde{w}_{i,k|k}\tilde{w}^T_{i,k|k}\} \\ &= E\{[w_k - K_{w_i,\frac{k}{l_i}}\tilde{z}_{i,\frac{k}{l_i}}] \\ &\cdot [w_k - K_{w_i,\frac{k}{l_i}}\tilde{z}_{i,\frac{k}{l_i}}]^T\} \\ &= Q_k - K_{w_i,\frac{k}{l_i}} S^T_{i,\frac{k}{l_i}} \end{aligned} \tag{10.42}$$

Therefore, the filtering error cross-covariance matrix for the state and the white noise is computed by

$$\begin{aligned} P_{x_i w_i,k|k} &= E\{\tilde{x}_{i,k|k}\tilde{w}^T_{i,k|k}\} \\ &= E\{[\tilde{x}_{i,k|k-1} - K_{i,\frac{k}{l_i}}\tilde{z}_{i,\frac{k}{l_i}}] \\ &\cdot [w_k - K_{w_i,\frac{k}{l_i}}\tilde{z}_{i,\frac{k}{l_i}}]^T\} \\ &= -K_{i,\frac{k}{l_i}} S^T_{i,\frac{k}{l_i}} \end{aligned} \tag{10.43}$$

So, the optimal local state estimations of x_k are obtained, which are $\hat{x}_{i,k|k}$ and $P_{i,k|k}$.

10.2.2 Local State Estimation with Unreliable Measurements

Theorem 10.2 *For system models (1) and (2), the local state estimation $\hat{x}_{i,k|k}$ and its filtering error covariance matrix are computed by*

$$\hat{x}_{i,k|k} = \begin{cases} \hat{x}_{i,k|k-1} + K_{i,\frac{k}{l_i}} \tilde{z}_{i,\frac{k}{l_i}}, & \text{if } k \ mod \ l_i = 0 \\ \hat{x}_{i,k|k-1}, & \text{if } k \ mod \ l_i \neq 0 \end{cases} \tag{10.44}$$

$$P_{i,k|k} = P_{i,k|k-1} - K_{i,\frac{k}{l_i}} P^T_{\tilde{z}_{i,\frac{k}{l_i}}} K^T_{i\frac{k}{l_i}} \tag{10.45}$$

The one-step state predictor and the corresponding error covariance matrix are computed by

$$\begin{aligned} \hat{x}_{i,k|k-1} &= A_{k-1}\hat{x}_{i,k-1|k-1} \\ &\quad + \hat{w}_{i,k-1|k-1} + B_{k-1}u_{k-1} \end{aligned} \tag{10.46}$$

$$\begin{aligned} P_{i,k|k-1} &= A_{k-1}P_{i,k-1|k-1}A^T_{k-1} \\ &\quad + A_{k-1}P_{x_i w_i,k-1|k-1} \\ &\quad + P^T_{x_i w_i,k-1|k-1}A^T_{k-1} \\ &\quad + P_{w_i,k-1|k-1} \end{aligned} \tag{10.47}$$

The white noise estimator is obtained by

$$\hat{w}_{i,k|k} = K_{w_i,\frac{k}{l_i}} \tilde{z}_{i,\frac{k}{l_i}} \tag{10.48}$$

$$P_{w_i,k|k} = Q_k - K_{w_i,\frac{k}{l_i}} S^T_{i,\frac{k}{l_i}} \tag{10.49}$$

$$K_{w_i,\frac{k}{l_i}} = S_{i,\frac{k}{l_i}} P^{-1}_{\tilde{z}_i,\frac{k}{l_i}} \tag{10.50}$$

and

$$P_{x_i w_i,k|k} = -K_{i,\frac{k}{l_i}} S^T_{i,\frac{k}{l_i}} \tag{10.51}$$

The innovation sequence $\tilde{z}_{i,\frac{k}{l_i}}$ *with variance* $P_{\tilde{z}_i,\frac{k}{l_i}}$ *is computed by*

$$\tilde{z}_{i,\frac{k}{l_i}} = z_{i,\frac{k}{l_i}} - C_{i,\frac{k}{l_i}} \hat{x}_{i,k|k-1} \tag{10.52}$$

$$\begin{aligned} P_{\tilde{z}_i,\frac{k}{l_i}} &= C_{i,\frac{k}{l_i}} P_{i,k|k-1} C^T_{i,\frac{k}{l_i}} \\ &\quad + C_{i,\frac{k}{l_i}} S^*_{i,\frac{k}{l_i}} + S^{*T}_{i,\frac{k}{l_i}} C^T_{i,\frac{k}{l_i}} \\ &\quad + R_{i,\frac{k}{l_i}} \end{aligned} \tag{10.53}$$

The filtering gain matrix for the state is computed by

$$K_{i,\frac{k}{l_i}} = \begin{cases} P_{i,k|k-1} C^T_{i,\frac{k}{l_i}} P^{-1}_{\tilde{z}_i,\frac{k}{l_i}} + S^*_{i,\frac{k}{l_i}} P^{-1}_{\tilde{z}_i,\frac{k}{l_i}}, & \text{if } |\rho_{i,\frac{k}{l_i}}| \leq \chi^2_\alpha(q_i) \\ 0, & \text{if } |\rho_{i,\frac{k}{l_i}}| > \chi^2_\alpha(q_i) \end{cases} \tag{10.54}$$

where

$$\rho_{i,\frac{k}{l_i}} = \tilde{z}^T_{i,\frac{k}{l_i}} P^{-1}_{\tilde{z}_i,\frac{k}{l_i}} \tilde{z}_{i,\frac{k}{l_i}} \tag{10.55}$$

In (10.54), q_i is the dimension of z_{i,k_i}, and $\chi^2_\alpha(q_i)$ is the one-sided χ^2 value with confidence α, $1 \leq i \leq N$.

Proof In case that unreliable measurements are existed, we apply the statistical characteristics of $\tilde{z}_{i,\frac{k}{l_i}}$ to judge whether measurement $z_{i,\frac{k}{l_i}}$ is faulty or normal. From the problem formulation and (10.32), one can see that when $z_{i,\frac{k}{l_i}}$ is reliable, $\tilde{z}_{i,\frac{k}{l_i}}$ is zero-mean Gaussian distributed with covariance $P_{\tilde{z}_i,\frac{k}{l_i}}$, i.e., $\tilde{z}_{i,\frac{k}{l_i}} \sim \mathcal{N}(0, P_{\tilde{z}_i,\frac{k}{l_i}})$, where $P_{\tilde{z}_i,\frac{k}{l_i}}$ is computed by (10.33).

Denote

$$\rho_{i,\frac{k}{l_i}} = \tilde{z}^T_{i,\frac{k}{l_i}} P^{-1}_{z_{i,\frac{k}{l_i}}} \tilde{z}_{i,\frac{k}{l_i}} \tag{10.56}$$

Then, if $\tilde{z}_{i,\frac{k}{l_i}}$ is normal, $\rho_{i,\frac{k}{l_i}} \sim \chi^2(q_i)$ should obey chi-square distribution with mean q_i and variance $2q_i$, respectively, where q_i denotes degrees of freedom and is the dimension of $z_{i,\frac{k}{l_i}}$.

Hence, $\rho_{i,\frac{k}{l_i}}$ is used as a measure to determine whether $z_{i,\frac{k}{l_i}}$ is faulty or normal. The null hypothesis in this hypothesis test problem is $H_0 : \gamma_{i,\frac{k}{l_i}} = 1$; the alternative hypothesis is $H_1 : \gamma_{i,\frac{k}{l_i}} \neq 1$. The rejection region is $(\chi^2_\alpha(q_i), +\infty)$, where $\chi^2_\alpha(q_i)$ is the boundary value of a unilateral chi-square distribution with confidence α, $1 \leq i \leq N$. Namely, if $|\rho_{i,\frac{k}{l_i}}| > \chi^2_\alpha(q_i)$, $z_{i,\frac{k}{l_i}}$ is deemed abnormal and won't be used for fusion estimation. If $|\rho_{i,\frac{k}{l_i}}| \leq \chi^2_\alpha(q_i)$, $z_{i,\frac{k}{l_i}}$ is taken as reliable and will be used for fusion estimation. Hence, we modified the innovation matrices as (10.54).

10.2.3 Optimal Distributed Fusion Estimation with Unreliable Measurements

Inspired by [34], which gives an optimal distributed fusion estimation for multisensor systems with the single sampling rate and uncorrelated noise, the optimal distributed Kalman filter for fusion of multirate multisensor systems with coupled noise for multirate multisensor systems with unreliable measurements shall be deduced in this section.

Theorem 10.3 *For system models (1) and (2), the optimal estimation of x_k in the form of distributed fusion at moment k can be computed by*

$$\begin{cases} \hat{x}_{k|k} = \sum_{i=1}^{N} \alpha_{i,k}\hat{x}_{i,k|k} \\ P_{k|k} = (e_k^T \Sigma^{-1}(k)e_k)^{-1} \\ \alpha_k = (e_k^T \Sigma^{-1}(k)e_k)^{-1} e_k^T \Sigma^{-1}(k) \end{cases} \tag{10.57}$$

where $\hat{x}_i(k|k)$ is computed by Theorem 10.2, and

$$\alpha_k = \begin{bmatrix} \alpha_{1,k} \; \alpha_{2,k} \cdots \alpha_{N,k} \end{bmatrix} \tag{10.58}$$

is an $n \times nN$ matrix.

$$e_k = [I_n \;\; I_n \;\; \ldots \;\; I_n]^T \tag{10.59}$$

is an $nN \times n$ matrix.

$$\Sigma(k) = (P_{x_i x_j,k|k}) \tag{10.60}$$

where

$$\begin{aligned} P_{x_i x_j,k|k} = & [I - K_{i,\frac{k}{l_i}} C_{i,\frac{k}{l_i}}]A_{k-1} \\ & \cdot P_{x_i x_j,k-1|k-1} A_{k-1}^T \\ & \cdot [I - K_{j,\frac{k}{l_j}} C_{j,\frac{k}{l_j}}]^T \\ & + [I - K_{i,\frac{k}{l_i}} C_{i,\frac{k}{l_i}}]P_{w_i x_j,k-1|k-1} \\ & \cdot A_{k-1}^T [I - K_{j,\frac{k}{l_j}} C_{j,\frac{k}{l_j}}]^T \\ & + [I - K_{i,\frac{k}{l_i}} C_{i,\frac{k}{l_i}}]P_{w_i w_j,k-1|k-1} \\ & \cdot [I - K_{j,\frac{k}{l_j}} C_{j,\frac{k}{l_j}}]^T \\ & + [I - K_{i,\frac{k}{l_i}} C_{i,\frac{k}{l_i}}]A_{k-1} \\ & \cdot P_{x_i w_j,k-1|k-1} [I - K_{j,\frac{k}{l_j}} C_{j,\frac{k}{l_j}}]^T \\ & + K_{i,\frac{k}{l_i}} R_{ij,k} K_{j,\frac{k}{l_j}}^T \end{aligned} \tag{10.61}$$

$$\begin{aligned} P_{x_i w_j,k|k} &= -[I - K_{i,\frac{k}{l_i}} C_{i,\frac{k}{l_i}}] A_{k-1} \\ &\cdot [P_{x_i x_j,k-1|k-1} A_{k-1}^T \\ &+ P_{x_i w_j,k-1|k-1}] C_{j,\frac{k}{l_j}}^T K_{w_j,\frac{k}{l_j}}^T \\ &- [I - K_{i,\frac{k}{l_i}} C_{i,\frac{k}{l_i}}][P_{w_i w_j,k-1|k-1} \\ &+ P_{w_i x_j,k-1|k-1} A_{k-1}^T] \\ &\cdot C_{j,\frac{k}{l_j}}^T K_{w_j,\frac{k}{l_j}}^T - K_{i,\frac{k}{l_i}} S_{i,k}^T \\ &+ K_{i,\frac{k}{l_i}} R_{ij,k} K_{w_j,\frac{k}{l_j}}^T \end{aligned} \tag{10.62}$$

$$\begin{aligned} P_{w_i w_j,k|k} &= Q_k - S_{j,k} K_{w_j,\frac{k}{l_j}}^T - K_{w_i,\frac{k}{l_i}} S_{i,k}^T \\ &+ K_{w_i,\frac{k}{l_i}} R_{ij,k} K_{w_j,\frac{k}{l_j}}^T \\ &+ K_{w_i,\frac{k}{l_i}} C_{i,\frac{k}{l_i}} A_{k-1} \\ &\cdot [P_{x_i x_j,k-1|k-1} A_{k-1}^T \\ &+ P_{x_i w_j,k-1|k-1}] C_{j,\frac{k}{l_j}}^T K_{w_j,\frac{k}{l_j}}^T \\ &+ K_{w_i,\frac{k}{l_i}} C_{i,\frac{k}{l_i}} [P_{w_i w_j,k-1|k-1} \\ &+ P_{w_i x_j,k-1|k-1} A_{k-1}^T] \\ &\cdot C_{j,\frac{k}{l_j}}^T K_{w_j,\frac{k}{l_j}}^T \end{aligned} \tag{10.63}$$

and

$$P_{k|k} \leq P_{x_i x_i,k|k},\ i = 1, 2, \ldots, N \tag{10.64}$$

Proof From (10.34) to (10.35), we have

$$\begin{aligned} \tilde{x}_{i,k|k} &= \tilde{x}_{i,k|k-1} - K_{i,\frac{k}{l_i}} \tilde{z}_{i,\frac{k}{l_i}} \\ &= \tilde{x}_{i,k|k-1} - K_{i,\frac{k}{l_i}} [C_{i,\frac{k}{l_i}} \tilde{x}_{i,k|k-1} + v_{i,\frac{k}{l_i}}] \\ &= A_{k-1} \tilde{x}_{i,k-1|k-1} + \tilde{w}_{i,k-1|k-1} \\ &- K_{i,\frac{k}{l_i}} \{C_{i,\frac{k}{l_i}} [A_{k-1} \tilde{x}_{i,k-1|k-1} \\ &+ \tilde{w}_{i,k-1|k-1}] + v_{i,\frac{k}{l_i}}\} \\ &= [I - K_{i,\frac{k}{l_i}} C_{i,\frac{k}{l_i}}] A_{k-1} \tilde{x}_{i,k-1|k-1} \\ &+ [I - K_{i,\frac{k}{l_i}} C_{i,\frac{k}{l_i}}] \tilde{w}_{i,k-1|k-1} \\ &- K_{i,\frac{k}{l_i}} v_{i,\frac{k}{l_i}} \end{aligned} \tag{10.65}$$

From (10.41), we have

$$\begin{aligned}\tilde{w}_{i,k|k} &= w_k - K_{w_i,\frac{k}{l_i}}\tilde{z}_{i,\frac{k}{l_i}}\\ &= w_k - K_{w_i,\frac{k}{l_i}}[C_{i,\frac{k}{l_i}}\tilde{x}_{i,k|k-1} + v_{i,\frac{k}{l_i}}]\\ &= w_k - K_{w_i,\frac{k}{l_i}}C_{i,\frac{k}{l_i}}A_{k-1}\tilde{x}_{i,k-1|k-1}\\ &- K_{w_i,\frac{k}{l_i}}C_{i,\frac{k}{l_i}}\tilde{w}_{i,k-1|k-1}\\ &- K_{w_{i,\frac{k}{l_i}}}v_{i,\frac{k}{l_i}}\end{aligned} \tag{10.66}$$

Then, we obtain the filtering error cross-covariance matrix between local filters of sensor i and sensor j by the following equation:

$$\begin{aligned}P_{x_ix_j,k|k} &= E\{\tilde{x}_{i,k|k}\tilde{x}^T_{j,k|k}\}\\ &= E\{[(I - K_{i,\frac{k}{l_i}}C_{i,\frac{k}{l_i}})A_{k-1}\\ &\cdot \tilde{x}_{i,k-1|k-1}\\ &+ (I - K_{i,\frac{k}{l_i}}C_{i,\frac{k}{l_i}})\tilde{w}_{i,k-1|k-1}\\ &- K_{i,\frac{k}{l_i}}v_{i,\frac{k}{l_i}}][(I - K_{j,\frac{k}{l_j}}C_{j,\frac{k}{l_j}})\\ &\cdot A_{k-1}\tilde{x}_{j,k-1|k-1}\\ &+ (I - K_{j,\frac{k}{l_j}}C_{j,\frac{k}{l_j}})\\ &\cdot \tilde{w}_{j,k-1|k-1}\\ &- K_{j,\frac{k}{l_j}}v_{j,\frac{k}{l_j}}]^T\}\\ &= [I - K_{i,\frac{k}{l_i}}C_{i,\frac{k}{l_i}}]A_{k-1}\\ &\cdot P_{x_ix_j,k-1|k-1}A^T_{k-1}\\ &\cdot [I - K_{j,\frac{k}{l_j}}C_{j,\frac{k}{l_j}}]^T\\ &+ [I - K_{i,\frac{k}{l_i}}C_{i,\frac{k}{l_i}}]P_{w_ix_j,k-1|k-1}\\ &\cdot A^T_{k-1}[I - K_{j,\frac{k}{l_j}}C_{j,\frac{k}{l_j}}]^T\\ &+ [I - K_{i,\frac{k}{l_i}}C_{i,\frac{k}{l_i}}]P_{w_iw_j,k-1|k-1}\\ &\cdot [I - K_{j,\frac{k}{l_j}}C_{j,\frac{k}{l_j}}]^T\\ &+ [I - K_{i,\frac{k}{l_i}}C_{i,\frac{k}{l_i}}]A_{k-1}P_{x_iw_j,k-1|k-1}\\ &\cdot [I - K_{j,\frac{k}{l_j}}C_{j,\frac{k}{l_j}}]^T\\ &+ K_{i,\frac{k}{l_i}}R_{ij,k}K^T_{j,\frac{k}{l_j}}\end{aligned} \tag{10.67}$$

where

$$\begin{aligned}
P_{x_i w_j,k|k} &= E\{\tilde{x}_{i,k|k}\tilde{w}_{j,k|k}^T\} \\
&= E\{[(I - K_{i,\frac{k}{l_i}}C_{i,\frac{k}{l_i}})A_{k-1} \\
&\quad \cdot \tilde{x}_{i,k-1|k-1} \\
&\quad + (I - K_{i,\frac{k}{l_i}}C_{i,\frac{k}{l_i}})\tilde{w}_{i,k-1|k-1} \\
&\quad - K_{i,\frac{k}{l_i}}v_{i,\frac{k}{l_i}}][w_k \\
&\quad - K_{w_{j,\frac{k}{l_j}}}C_{j,\frac{k}{l_j}}A_{k-1}\tilde{x}_{j,k-1|k-1} \\
&\quad - K_{w_{j,\frac{k}{l_j}}}C_{j,\frac{k}{l_j}} \\
&\quad \cdot \tilde{w}_{j,k-1|k-1} - K_{w_{j,\frac{k}{l_j}}}v_{j,\frac{k}{l_j}}]^T\} \\
&= -[I - K_{i,\frac{k}{l_i}}C_{i,\frac{k}{l_i}}]A_{k-1} \\
&\quad \cdot [P_{x_i x_j,k-1|k-1}A_{k-1}^T \\
&\quad + P_{x_i w_j,k-1|k-1}] \\
&\quad \cdot C_{j,\frac{k}{l_j}}^T K_{w_{j,\frac{k}{l_j}}}^T \\
&\quad - [I - K_{i,\frac{k}{l_i}}C_{i,\frac{k}{l_i}}][P_{w_i w_j,k-1|k-1} \\
&\quad + P_{w_i x_j,k-1|k-1}A_{k-1}^T]C_{j,\frac{k}{l_j}}^T K_{w_{j,\frac{k}{l_j}}}^T \\
&\quad + K_{i,\frac{k}{l_i}}R_{ij,k}K_{w_{j,\frac{k}{l_j}}}^T \\
&\quad - K_{i,\frac{k}{l_i}}S_{i,k}^T
\end{aligned} \tag{10.68}$$

and

$$\begin{aligned}
P_{w_i w_j,k|k} &= E\{\tilde{w}_{i,k|k}\tilde{w}_{j,k|k}^T\} \\
&= E\{[w_k - K_{w_{i,\frac{k}{l_i}}}C_{i,\frac{k}{l_i}}A_{k-1} \\
&\quad \cdot \tilde{x}_{i,k-1|k-1} - K_{w_{i,\frac{k}{l_i}}}C_{i,\frac{k}{l_i}}\tilde{w}_{i,k-1|k-1} \\
&\quad - K_{w_{i,\frac{k}{l_i}}}v_{i,\frac{k}{l_i}}][w_k \\
&\quad - K_{w_{j,\frac{k}{l_j}}}C_{j,\frac{k}{l_j}}A_{k-1}\tilde{x}_{j,k-1|k-1} \\
&\quad - K_{w_{j,\frac{k}{l_j}}}C_{j,\frac{k}{l_j}} \\
&\quad \cdot \tilde{w}_{j,k-1|k-1} - K_{w_{j,k}}v_{j,k}]^T\} \\
&= Q_k - S_{j,k}K_{w_{j,\frac{k}{l_j}}}^T
\end{aligned} \tag{10.69}$$

$$
\begin{aligned}
&- K_{w_i,\frac{k}{l_i}} S_{i,k}^T \\
&+ K_{w_i,\frac{k}{l_i}} R_{ij,k} K_{w_{j,\frac{k}{l_j}}}^T \\
&+ K_{w_i,\frac{k}{l_i}} C_{i,\frac{k}{l_i}} \\
&\cdot A_{k-1}[P_{x_ix_j,k-1|k-1} A_{k-1}^T \\
&+ P_{x_iw_j,k-1|k-1}] C_{j,\frac{k}{l_j}}^T K_{w_{j,\frac{k}{l_j}}}^T \\
&+ K_{w_i,\frac{k}{l_i}} C_{i,\frac{k}{l_i}} [P_{w_iw_j,k-1|k-1} \\
&+ P_{w_ix_j,k-1|k-1} A_{k-1}^T] C_{j,\frac{k}{l_j}}^T K_{w_{j,\frac{k}{l_j}}}^T
\end{aligned}
$$

Let

$$F_k = (y_k - e_k x_k)^T \Sigma^{-1}(k)(y_k - e_k x_k) \tag{10.70}$$

where

$$y_k = [\hat{x}_{1,k|k}^T \; \hat{x}_{2,k|k}^T \; \cdots \; \hat{x}_{N,k|k}^T]^T \tag{10.71}$$

$$e_k = [I_n \; I_n \; \ldots \; I_n]^T \tag{10.72}$$

$$\Sigma A_k = [P_{x_ix_j,k|k}] \tag{10.73}$$

The dimension of matrix e_k is $nN \times n$. ΣA_k is an $nN \times nN$ symmetric positive determined matrix.

F_k is a quadratic form of x_k for any $k = 1, 2, \ldots$, and is therefore a convex function of x_k.

By

$$\frac{dF_k}{dx} = 0 \tag{10.74}$$

we obtain the minimum point of F_x for

$$x_k = (e_k^T \Sigma^{-1}(k) e_k)^{-1} e_k^T \Sigma^{-1}(k) y_k \tag{10.75}$$

So, the optimal distributed fusion estimation can be computed by

$$\hat{x}_{k|k} = \alpha_k y_k = \sum_{i=1}^{N} \alpha_{i,k} \hat{x}_{i,k|k} \tag{10.76}$$

where

$$
\begin{aligned}
\alpha_k &= [\alpha_{1,k} \; \alpha_{2,k} \; \cdots \; \alpha_{N,k}] \\
&= (e_k^T \Sigma^{-1}(k) e_k)^{-1} e_k^T \Sigma^{-1}(k)
\end{aligned} \tag{10.77}
$$

By the unbiasedness property, we have

$$\tilde{x}_{k|k} = \sum_{i=1}^{N} \alpha_{i,k}\tilde{x}_{i,k|k} \tag{10.78}$$

So, we have the error covariance matrix

$$\begin{aligned} P_{k|k} &= \sum_{i=1}^{N} \alpha_{i,k} P_{x_i x_j,k|k} \alpha_{j,k}^T \\ &= \alpha^{T,k} \Sigma_k \alpha_k \\ &= (e_k^T \Sigma^{-1}(k) e_k)^{-1} \end{aligned} \tag{10.79}$$

Notice that by supposing $\alpha_{i,k} = I_n$, $\alpha_{j,k}^T = 0$ $(j = 1, 2, \ldots, N, j \neq i)$ in (10.79), then $P_{k|k} = P_{x_i x_i,k|k}$, so we have

$$P_{k|k} \leq P_{x_i x_i,k|k}, i = 1, 2, \ldots, N \tag{10.80}$$

Corollary 10.4 (The suboptimal distributed fusion estimation) *From Theorem 10.3, for system models (1) and (2), in the sense of linear minimum covariance, if the cross-correlation of the local estimation is neglected, namely, if we set* $P_{x_i x_j,k|k} = 0$ *for* $i, j = 1, 2, \ldots, N$*, then the kth suboptimal fusion estimation of* x_k *by distributed fusion can be computed as*

$$\begin{cases} \hat{x}_{k|k} = \sum_{i=1}^{N} \alpha_{i,k}\hat{x}_{i,k|k} \\ \alpha_{i,k} = \big[\sum_{i=1}^{N} P_{i,k|k}^{-1}\big]^{-1} P_i^{-1}(k|k) \\ P_{k|k} = \big[\sum_{i=1}^{N} P_{i,k|k}^{-1}\big]^{-1} \end{cases} \tag{10.81}$$

10.3 Numerical Example

An example is used to manifest the performance of the presented algorithm. A three-sensor tracking system is described by the following equations:

$$x_k = A_k x_k + B_k u_k + \Gamma_k w_k \tag{10.82}$$

$$z_{i,k_i} = \Gamma_{i,k_i}[C_{i,k_i} x_{i,k_i} + v_{i,k_i}] \tag{10.83}$$

where $i = 1, 2, 3$,

$$A_k = \begin{bmatrix} 1 & T_s & \frac{T_s^2}{2} \\ 0 & 1 & T_s \\ 0 & 1 & 1 \end{bmatrix} \tag{10.84}$$

$$Q_k = q\Gamma_k \Gamma^T(k) \tag{10.85}$$

$$\Gamma_k = [\frac{T_s^2}{2} \;\; T_s \;\; 1] \tag{10.86}$$

where $q = 0.01$ and $T_s = 0.01\,\mathbf{s}$ indicates the sampling period. w_k indicates the zero-mean Gaussian-distributed system noise with variance σ_w^2. The system noise is $\xi_k = \Gamma_k w_k$, and its covariance is $Q_k = \Gamma_k \Gamma_k^T \sigma_w^2$, where Γ_k is the process noise gain.

v_{i,k_i} represents the observation noise of the k_ith measurement of Sensor i with the sampling ratio of l_i, where $l_1 = 1$, $l_2 = 3$, $l_3 = 4$, and $v_{i,k}$ can be generated by

$$v_{i,k} = \beta_i w_{k-1} + \delta_i w_k + \eta(i,k) \tag{10.87}$$

Obviously, the measurement noises of different sensors are correlated, and they are also correlated with the system noise w_k and w_{k-1}, and the correlation strengths are decided by the value of β_i and δ_i. $\eta(i,k)$ is zero-mean Gaussian white noise with variance $\sigma_{\eta_i}^2$. Three sensors with different sampling rates observe the same target with the sample rates ratio $1 : \frac{1}{3} : \frac{1}{4}$. The measurement matrices are

$$\begin{cases} C_{1,k_1} = C_1 = [1 \;\; 0 \;\; 0] \\ C_{2,k_2} = C_2 = [1 \;\; 0 \;\; 0] \\ C_{3,k_3} = C_3 = [0 \;\; 1 \;\; 0] \end{cases} \tag{10.88}$$

where the position information is observed by Sensor 1 and Sensor 2 and velocity information is observed by Sensor 3.

The covariance matrix of the measurement noise meet

$$R_k = \begin{bmatrix} R_{11,k} & R_{12,k} & R_{13,k} \\ R_{21,k} & R_{22,k} & R_{23,k} \\ R_{31,k} & R_{32,k} & R_{33,k} \end{bmatrix} \tag{10.89}$$

$$\begin{cases} R_{11,k} = \beta_1^2 \sigma_w^2 + \delta_1^2 \sigma_w^2 + \sigma_{\eta_1}^2 \\ R_{12,k} = \beta_1 \beta_2 \sigma_w^2 + \delta_1 \delta_2 \sigma_w^2 \\ R_{13,k} = \beta_1 \beta_3 \sigma_w^2 + \delta_1 \delta_3 \sigma_w^2 \\ R_{21,k} = \beta_2 \beta_1 \sigma_w^2 + \delta_2 \delta_1 \sigma_w^2 \\ R_{22,k} = \beta_2^2 \sigma_w^2 + \delta_2^2 \sigma_w^2 + \sigma_{\eta_2}^2 \\ R_{23,k} = \beta_2 \beta_3 \sigma_w^2 + \delta_2 \delta_3 \sigma_w^2 \\ R_{31,k} = \beta_3 \beta_1 \sigma_w^2 + \delta_3 \delta_1 \sigma_w^2 \\ R_{32,k} = \beta_3 \beta_2 \sigma_w^2 + \delta_3 \delta_2 \sigma_w^2 \\ R_{33,k} = \beta_3^2 \sigma_w^2 + \delta_3^2 \sigma_w^2 + \sigma_{\eta_3}^2 \end{cases} \tag{10.90}$$

The covariance between v_{i,k_i} and w_{k-1} meet

$$S_k^* = [\beta_1 \sigma_w^2 \Gamma_{k-1} \;\; \beta_2 \sigma_w^2 \Gamma_{k-1} \;\; \beta_3 \sigma_w^2 \Gamma_{k-1}] \tag{10.91}$$

The covariance between v_{i,k_i} and w_k meet

$$S_k = [\delta_1 \sigma_w^2 \Gamma_k \;\; \delta_2 \sigma_w^2 \Gamma_k \;\; \delta_3 \sigma_w^2 \Gamma_k] \tag{10.92}$$

The initial condition is

$$x_0 = \begin{bmatrix} 10 \\ 0.5 \\ 0 \end{bmatrix}, P_0 = \begin{bmatrix} 1 & 0 & 0 \\ 0 & 1 & 0 \\ 0 & 0 & 1 \end{bmatrix} \tag{10.93}$$

We set $\sigma_w^2 = 0.01$, $\sigma_{\eta_1}^2 = 0.25$, $\sigma_{\eta_2}^2 = 0.16$, $\sigma_{\eta_3}^3 = 0.09$, $\beta_1 = \delta_1 = 5$, $\beta_2 = \delta_2 = 4$, $\beta_3 = \delta_3 = 1$. The unreliable probability of each sensor measurement is 0.1. We set confidence level $\alpha = 0.05$ when examining the reliability of the measurements.

In this section, information of three sensors is fused to estimate the target state. In the case of related noise, the discrimination between the estimated results gained by different estimation algorithms is compared, and the effect on fusion results when ignoring the noise correlation is analyzed.

We perform 30 Monte Carlo simulations at 600 sampling time, and the validity of the proposed algorithm is evaluated. Figures 10.1, 10.2, 10.3 and Tables 10.1, 10.2, and 10.3 demonstrate the simulation results.

Figure 10.1 demonstrates the RMSEs (position and velocity) of different algorithms, where the blue solid line demonstrates optimal distributed fusion algorithm with reliability test and with consideration of noise correlation, the red dotted line demonstrates the result by the suboptimal distributed fusion algorithm, and the green dashed line demonstrates single sensor filter result by the highest sampling rate sensor. It can be seen that the RMSEs of the solid line is the smallest, next is the dotted line, and the last one is the dashed line. It shows that the optimal distributed fusion algorithm is more effective than the suboptimal distributed fusion algorithm and the filter of one sensor. Table 10.1 demonstrates the position and velocity time-averaged RMSEs of different algorithms. The conclusion is the same as in Fig. 10.1. It demonstrates that the effectiveness of proposed optimal distributed fusion algorithm.

Figure 10.2a, b demonstrates the RMSEs (position and velocity), where the red dashed line and the blue solid line denote the results obtained by the distributed fusion algorithm ignoring noise correlation and considering noise correlation, respectively. It can be seen that the RMSEs of distributed fusion algorithm with consideration of noise correlation are smaller than those by the algorithm without consideration of noise correlation. It demonstrates that the distributed fusion algorithm considering noise correlation is efficacious and ignoring noise correlation will cut down the estimation accuracy when there is noise correlation in the system. The time-averaged RMSEs (position and velocity) of the distributed fusion algorithm considering noise correlation and ignoring noise correlation are shown in Table 10.2. The conclusion is the same as in Fig. 10.2.

Fig. 10.1 Comparison RMSEs of different algorithms

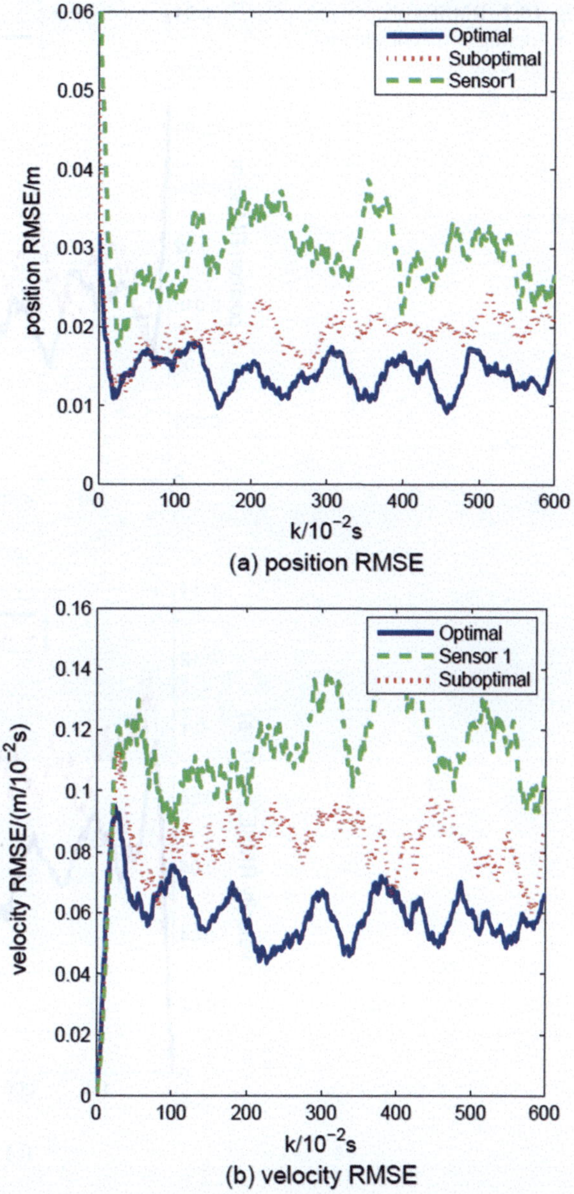

Figure 10.3 demonstrates the comparison of RMSEs between different algorithms, where blue solid line demonstrates the RMSEs of the distributed fusion algorithm with reliability test of the measurements, and the dashed line denotes the RMSEs of distributed fusion algorithm without test. It can be seen that RMSEs with reliability test are much smaller than those without test and signify that the reliability test arithmetic is more valid. Table 10.3 demonstrates the position and velocity time-

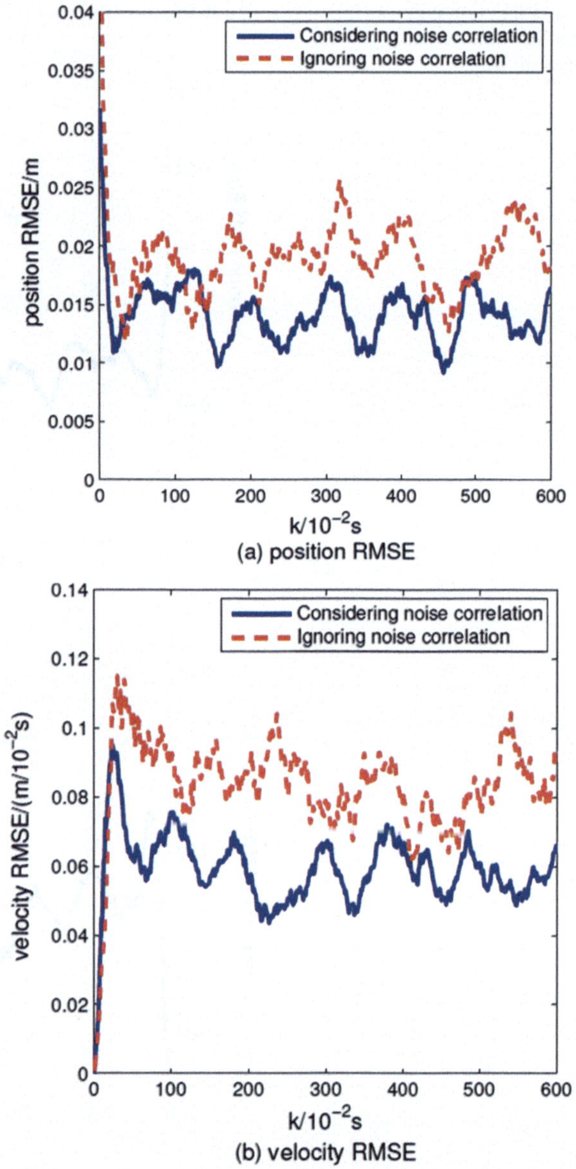

Fig. 10.2 RMSEs of considering noise correlation and ignoring noise correlation

averaged RMSEs with reliability test and without reliability test. The conclusion is the same as in Fig. 10.3.

In summary, the simulations in this section demonstrate that the optimal distributed fusion algorithm presented is effective. When unreliable measurements are available, and the measurement noises are autocorrelated and correlated with the system noise, ignoring the correlation or without unreliability test of the measurements will reduce the estimation accuracy.

Fig. 10.3 RMSEs with reliability test and without reliability test

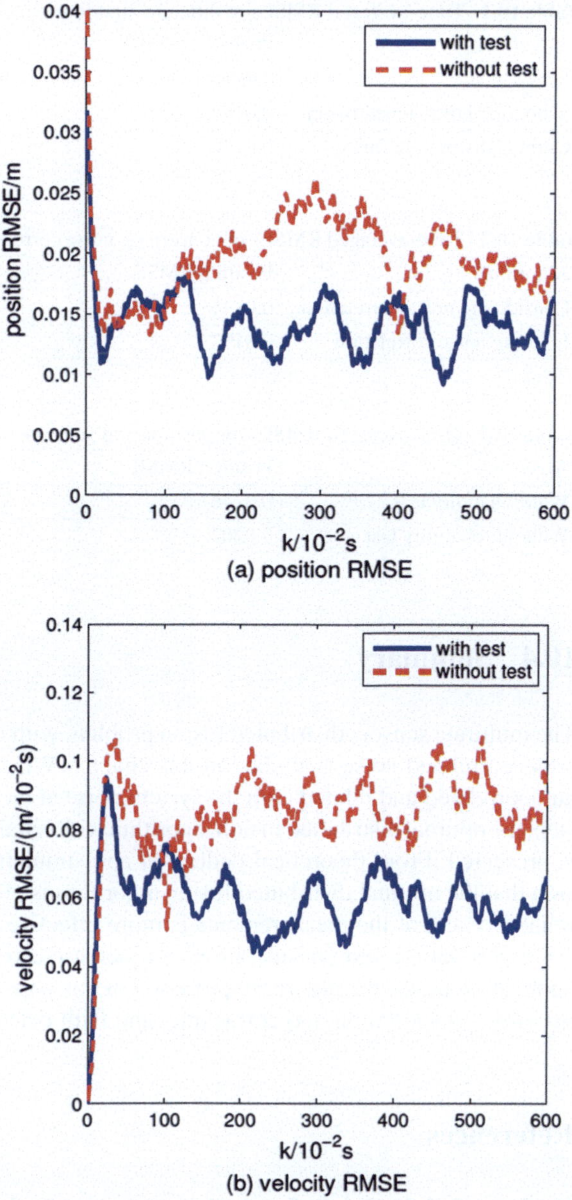

Table 10.1 Time-averaged RMSEs of different algorithms

Algorithm	Position RMSE	Velocity RMSE
Sensor 1	0.0286	0.1195
Suboptimal distributed fusion	0.0187	0.0855
Optimal distributed fusion	0.0146	0.0612

Table 10.2 Time-averaged RMSEs of concerning noise correlation

Algorithm	Position RMSE	Velocity RMSE
Considering noise correlation	0.0146	0.0612
Ignoring noise correlation	0.0197	0.0876

Table 10.3 Time-averaged RMSEs for position and velocity

Algorithm	Position RMSE	Velocity RMSE
With reliability test	0.0146	0.0612
Without reliability test	0.0202	0.0905

10.4 Summary

The multirate sensors distributed fusion problem with unreliable Measurements, and cross-correlated noise is studied in this chapter. When the measurement noises are autocorrelated and related with the system noise at the current step and the previous step, the optimal distributed fusion algorithm with reliability test of the measurements is presented. From theoretical deduction and simulation, we can draw the conclusion that the optimal distributed fusion algorithm is effective; fusion algorithm with reliability test of the measurements is more effective than that without test; taking noise correlation into consideration can increase estimation accuracy. In a host of application fields, the algorithm proposed in this paper has potential values, such as network transportation, target tracking, and fault detection.

References

1. S. Safari, F. Shabani, D. Simon, Multirate multisensor data fusion for linear systems using Kalman filters and a neural network. Aerosp. Sci. Technol. **39**, 465–471 (2014)
2. B. Chen, W. Zhang, L. Yu, Networked fusion Kalman filtering with multiple uncertainties. IEEE Trans. Aerosp. Electron. Syst. **51**(3), 2332C2349 (2015)
3. M.J.G. a Ligero, A. Hermoso-Carazo, J. Linares-Pérez, Distributed fusion estimation in networked systems with uncertain observations and markovian random delays. Signal Process. **106**, 114–122 (2015)

4. L.P. Yan, X.R. Li, Y.Q. Xia, M.Y. Fu, Modeling and estimation of asynchronous multirate multisensor system with unreliable measurements. IEEE Trans. Aerosp. Electron. Syst. **51**(3), 2012–2026 (2015)
5. J.X. Feng, M. Zeng, Optimal distributed Kalman filtering fusion for a linear dynamic system with cross-correlated noises. Int. J. Syst. Sci. **43**(2), 385–398 (2012)
6. R. Caballero-Águila, A. Hermoso-Carazo, J. Linares-Pérez, Optimal state estimation for networked systems with random parameter matrices, correlated noises and delayed measurements. Int. J. Gen. Syst. **44**(2), 142–154 (2015)
7. E.B. Song, Y.M. Zhu, J. Zhou, Z.S. You, Optimal Kalman filtering fusion with cross-correlated sensor noises. Automatica **43**(8), 1450–1456 (2007)
8. Y.M. Wang, X.R. Li, Distributed estimation fusion with unavailable cross-correlation. IEEE Trans. Aerosp. Electron. Syst. **48**(1), 259–278 (2012)
9. Q. Ge, T. Shao, Z. Duan, Performance analysis of the Kalman filter with mismatched noise covariances. IEEE Trans. Autom. Control. **61**(12), 4014–4019.1 (2016)
10. Q. Ge, D. Xu, C. Wen, Cubature information filters with correlated noises and their applications in decentralized fusion. Signal Process. **94**, 434–444 (2014)
11. J.X. Feng, Z.D. Wang, M. Zeng, Distributed weighted robust Kalman filter fusion for uncertain systems with autocorrelated and cross-correlated noises. Inf. Fusion **14**(1), 78–86 (2013)
12. L.P. Yan, X.R. Li, Y.Q. Xia, M.Y. Fu, Optimal state estimation with cross-correlated noises. Automatica **44**, 1135–1146 (2007)
13. A. Chiuso, L. Schenato, Information fusion strategies and performance bounds in packet-drop networks. Automatica **47**(7), 1304–1316 (2011)
14. R. Caballero-guila, I. Garca-Garrido, J. Linares-Prez, Information fusion algorithms for state estimation in multi-sensor systems with correlated missing measurements. Appl. Math. Comput. **226**(1), 548–563 (2014)
15. B. Chen, W.A. Zhang, L. Yu, Distributed fusion estimation with missing measurements, random transmission delays and packet dropouts. IEEE Trans. Autom. Control **59**(7), 1961–1967 (2014)
16. S.L. Sun, L.H. Xie, W.D. Xiao, Y.C. Soh, Optimal linear estimation for systems with multiple packet dropouts. Automatica **44**(5), 1333–1342 (2008)
17. Y. Liang, T.W. Chen, Q. Pan, Optimal linear state estimator with multiple packet dropouts. IEEE Trans. Autom. Control **55**(6), 1428–1433 (2010)
18. Z.D. Wang, D.W.C. Ho, X.H. Liu, Variance-constrained filtering for uncertain stochastic systems with missing measurements. IEEE Trans. Autom. Control **48**(7), 1254–1258 (2003)
19. B. Sinopoli, L. Schenato, M. Franceschetti, K. Poolla, M.I. Jordan, S.S. Sastry, Kalman filtering with intermittent observations. IEEE Trans. Autom. Control **49**(9), 1453–1464 (2004)
20. S. Kar, B. Sinopoli, J.M.F. Moura, Kalman filtering with intermittent observations: weak convergence to a stationary distribution. IEEE Trans. Autom. Control **57**(2), 405–420 (2012)
21. M.Y. Huang, S. Dey, Stability of Kalman filtering with Marikovian packet losses. Automatica **43**(4), 598–607 (2007)
22. S.Y. Wang, H.J. Fang, X.G. Tian, Recursive estimation for nonlinear stochastic systems with multi-step transmission delays, multiple packet dropouts and correlated noises. Signal Process. **115**, 164–175 (2015)
23. S. Sun, L. Xie, W. Xiao, N. Xiao, Optimal filtering for systems with multiple packet. IEEE Trans. Circuits Syst. II Express Briefs **55**(7), 695–699 (2008)
24. Y. Xia, J. Shang, J. Chen, G. Liu, Networked data fusion with packet losses and variable delays. IEEE Trans. Syst. Man Cybern. Part B Cybern. **39**(5), 1107–1120 (2009)
25. Z. Xing, Y. Xia, L. Yan, K. Lu, Q. Gong, Multi-sensor distributed weighted Kalman filter fusion with network delays, stochastic uncertainties, autocorrelated and cross-correlated noises. IEEE Trans. Syst. Man Cybern. Syst. (2016). https://doi.org/10.1109/TSMC.2016.2633283
26. R. Cristi, M. Tummala, Multirate, multiresolution, recursive Kalman filter. Signal Process. **80**(9), 1945–1958 (2000)
27. Y. Liang, T.W. Chen, Q. Pan, Multi-rate optimal state estimation. Int. J. Control **82**(11), 2059–2076 (2009)

28. M.S. Mahmoud, M.F. Emzir, State estimation with asynchronous multi-rate multi-smart sensors. Inf. Sci. **196**(1), 15–27 (2012)
29. L.P. Yan, L. Jiang, Y.Q. Xia, M.Y. Fu, State estimation and data fusion for multirate sensor networks. Int. J. Adapt. Control Signal Process. **30**(1), 3–15 (2016)
30. L.P. Yan, B. Xiao, Y.Q. Xia, M.Y. Fu, State estimation for asynchronous multirate multisensor nonlinear dynamic systems with missing measurements. Int. J. Adapt. Control Signal Process. **26**(6), 516–529 (2012)
31. L.P. Yan, J. Liu, L. Jiang, Y.Q. Xia, B. Xiao, Y. Liu, M.Y. Fu, Optimal sequential estimation for multirate dynamic systems with unreliable measurements and correlated noise, (Chengdu, China) (2016), pp. 4900–4905
32. L. Yan, D. Zhou, M. Fu, Y. Xia, State estimation for asynchronous multirate multisensor dynamic systems with missing measurements. IET Signal Process. **4**(6), 728–739 (2010)
33. C. Zhu, Y.Q. Xia, L.H. Xie, L.P. Yan, Optimal linear estimation for systems with transmission delays and packet dropouts. IET Signal Process. **7**(9), 814–823 (2012)
34. S. Sun, Z. Deng, Multi-sensor optimal information fusion Kalman filter. Automatica **40**(6), 1017–1023 (2004)

Chapter 11
CKF-Based State Estimation of Nonlinear System by Fusion of Multirate Multisensor Unreliable Measurements

In this chapter, the state estimation of a kind of nonlinear dynamic systems is studied, where multiple sensors observing the same target with different sampling rates, and the observations are randomly unreliable. Based on online checking of the reliability of the measurements, the cubature Kalman filter (CKF) is improved, which has better robustness and stability. Based on the improved CKF, an effective state estimation algorithm is presented, and the multirate multisensor information are effectively fused. The numerical simulation show the feasibility and the effectiveness of the presented algorithms.

11.1 Introduction

After several decades of development, the state estimation of linear systems has been relatively mature, but the research on the state estimation of nonlinear systems is not enough. In real life, most of the systems are nonlinear. Nonlinearity is the essential factor that affects the dynamic characteristics of the system. It brings difficulties to the analysis and design of the control systems. With the rapid development of networked control and sensor technologies [1, 2], the state estimation problem of data packet loss, network delay [2–4], and uncertainties in nonlinear systems [5] have become hot research issues. Therefore, developing some new methods to solve the state estimation problem of nonlinear system under network environment is a research topic, which has broad application prospects and challenges.

The optimal solution of the nonlinear state estimation problem needs to obtain the posterior probability distribution function of the system state [6]. Although Bayesian inference gives the optimal analytical solution for the state estimation of dynamic systems, it is almost impossible to propagate all the probability density characteristics [7, 8]. Therefore, in the past decades, people have put forward a large number of suboptimal filtering methods. The extended Kalman filter (EKF) [9, 10] is a typical

H. Ma et al., *Kalman Filtering and Information Fusion*,
https://doi.org/10.1007/978-981-15-0806-6_11

representative of the suboptimal filtering method. The basic idea is that the nonlinear system is linearized by first-order Taylor expansion. EKF needs to calculate the Jacobian matrix, not only it needs a large amount of calculation, and sometimes very difficult to achieve. In order to improve the filtering accuracy, some improved methods of EKF are proposed, such as the two-order EKF, the iterative EKF, and so on, but they are all at the expense of an increase in the amount of computation. In view of the deficiency of EKF, Julier and Uhlmann [11] first proposed the unscented Kalman filter (UKF) algorithm. By selecting the number of sigma points [12], UKF captures the mean and variance of the state vector, which overcomes the error caused by local linearization of EKF. Compared to the EKF, the advantage of UKF is that it can capture the high-order moments of the nonlinear transformation, and the filtering accuracy is also significantly improved [13]. However, the filtering performance of UKF is easily affected by the parameters. When the system state dimension is greater than or equal to 4, UKF may lead to nonpositive definite covariance matrix in the filtering process, and then affect the filtering process. In 2009, Arasaratnam and Haykin [14, 15] proposed cubature Kalman filter (CKF) algorithm. CKF uses a third-degree spherical–radial cubature rule to estimate the probability integral of the nonlinear filter. The weight of CKF in selecting the cubature points is positive, which avoid nonpositive definite property of the covariance matrix caused by weighted summation. CKF has better nonlinear approximation performance, numerical accuracy, and stability of the filter. Consequently, after the CKF algorithm is proposed, it has been widely accepted in many fields and been applied to a variety of estimation problems [16–18]. Particle filter (PF) [19], based on the idea of Monte Carlo simulation, randomly generates large numbers of particles to approximate the posterior probability density. PF requires a large amount of particles, so the calculation burden is huge. It is difficult to meet the real demand, and there exist problems such as particle degeneracy and impoverishment.

The above mentioned articles did not involve multisensor data fusion. The basic idea of multisensor data fusion is to make full use of the effective information of multisource sensors, in order to achieve the most ideal state estimation. Because of the advantages of high estimation precision, good fault-tolerance property, and strong complementarity, multisensor data fusion technology has been widely applied in the field of navigation, signal and image processing, target tracking, complex control system, and so on [20, 21]. For linear systems, when the system is in the presence of random packet loss or encounter unreliable observations, there are a lot of information fusion research results, including synchronous or asynchronous sampling, the same or different sampling rates, and so on [6, 22–26]. However, for such a case of nonlinear systems, the research results are less [27].

In this chapter, CKF-based state estimation algorithm by fusion of multirate multisensor unreliable measurements is presented. In simulation, it is compared to EKF and UKF, which show the effectiveness of the presented algorithm.

This chapter is organized as follows: Sect. 11.2 is the problem formulation. Section 11.3 is the multirate multisensor data fusion algorithm. Section 11.4 is the numerical simulation, and Sect. 11.5 reaches the conclusion.

11.2 Problem Formulation

A class of nonlinear dynamic system with multiple sensors observing a single target at different sampling rates with measurements randomly missing or faulty could be described by [27]:

$$x_{k_N+1} = f_{x_{k_N}} + w_{k_N} \tag{11.1}$$

$$y_{i,k_i} = \gamma_{i,k_i} h_{i,x_{k_i}} + v_{i,k_i},\ i = 1, 2, \ldots, N \tag{11.2}$$

where (11.1) is the system equation. $x_{k_N} \in R^n$ is the state vector at time $k_N T$, where T is the sampling period of sensor N. $f(*)$ is the system model, which is determined by nonlinear function f. $w_{k_N} \in R^{n\times 1}$ is supposed to be a zero-mean white Gaussian noise whose variance is Q_k. Equation (11.2) is the measurement equation. $i = 1, 2, \ldots, N$ denote sensors, whose sampling rates are S_i, respectively. $y_{i,k_i} \in R^{m_i\times 1}(m_i \leq n)$ is the k_ith measurement observed by sensor i, and $h_i(*)$ is the measurement model that can be determined by nonlinear function h_i. The sampling rates of different sensors meet $S_i = S_N/n_i$, $1 \leq i \leq N-1$, where n_i are known positive integers. The measurement noise $v_{i,k_i} \in R^{m_i\times 1}$ is supposed to be a zero-mean white Gaussian noise with variance being R_{i,k_i}, $i = 1, 2, \ldots, N$.

In this chapter, $i = N$ denotes the finest scale, where the sensor sampling is the fastest. $i = N-1, N-2, \ldots, 1$ denote coarser scales that have decreasing sampling rates. The higher the sampling rate of the sensor, the finer the scale. For $i = 1, 2, \ldots, N$, $x_{i,k} = x_{n_i,k}$, where $x_{i,k}$ is the state sampling value of moment k for the ith sensor.

The initial value of the state vector x_0 is a random vector, whose mean and estimation error covariance are x_0 and P_0, respectively. It is assumed that x_0, w_{k_N} and v_{i,k_i} are mutually independent.

The variable $\gamma_{i,k_i} \in R$ is a stochastic sequence that takes values on 0 and 1 with Bernoulli distribution, whose expectation is $\bar{\gamma}_i$, which is used to describe the faulty degree of measurements. It is assumed that γ_{i,k_i} is independent of w_{k_N}, v_{i,k_i}, and x_0, $i = 1, 2, \ldots, N$.

The objective of this chapter is to develop an algorithm, which can effectively fuse the measurements observed by different sensors with different sampling rates in case of stochastic data faulty, in order to generate the optimal estimation of state x_k.

Remark 11.1 For simiplicity and without lose of generosity, we denote $k_N = k$, and set $T = 1$ in the sequel.

11.3 Multirate Multisensor Data Fusion Algorithm

In the sequel, based on model (11.1), (11.2) and CKF algorithm, we will first generate the multirate multisensor data fusion algorithm under the assumption that all the observations are reliable. For the fusion of unreliable measurements, we will introduce the algorithm afterward.

Theorem 11.2 (CKF-based estimation by fusion of reliable measurements) *Suppose that the system model is formulated as in Sect. 11.2, and $\gamma_{i,k_i} = 1$ for $i = 1, 2, \ldots, N$ and $k_i = 1, 2, \ldots$. At time k, it is assumed that there are $p' + 1$ sensors meet mod $(k, n_{i_p}) = 0$, where $p = 0, 1, 2, \ldots, p'$ and $N = i_0 > N - 1 \geq i_1 \geq i_2 \geq \cdots \geq i_{p'} \geq 1$. The optimal state estimate $\hat{x}_{k-1|k-1}$ and the corresponding estimation error covariance $P_{k-1|k-1}$ at time $k-1$ have been generated. Then the optimal state estimate $\hat{x}_{k|k}$ and the estimation error covariance $P_{k|k}$ can be deduced based on CKF algorithm by the use of the following formulas (11.3) to (11.28):*

$$\hat{x}_{k|k} = \hat{x}_{i_{p'},k|k} \tag{11.3}$$

$$P_{k|k} = P_{i_{p'},k|k} \tag{11.4}$$

For $p = 1, 2, \ldots, p'$, we have

$$\begin{aligned}\hat{x}_{i_p,k|k} &= \hat{x}_{i_{p-1},k|k} + K_{i_p,k/n_{i_p}}[y_{i_p,k/n_{i_p}} \\ &\quad - \hat{y}_{i_p,k/n_{i_p}}]\end{aligned} \tag{11.5}$$

$$\begin{aligned}P_{i_p,k|k} &= P_{i_{p-1},k|k} \\ &\quad - K_{i_p,k/n_{i_p}} P_{\tilde{y}_{i_p},k/n_{i_p}} K^T_{i_p,k/n_{i_p}}\end{aligned} \tag{11.6}$$

where

$$X_{i_{p-1},l;k} = \hat{x}_{i_{p-1},k|k} + \sqrt{P_{i_{p-1},k|k}}\varepsilon_l \tag{11.7}$$

$$X^*_{i_p,l;k} = f(X_{i_{p-1},l;k}) \tag{11.8}$$

$$\hat{x}_{i_p,k|k} = w_l \sum_{l=1}^{2n} X^*_{i_p,l;k} \tag{11.9}$$

$$\begin{aligned}P_{i_p,k|k} &= w_l \sum_{l=1}^{2n} X^*_{i_p,l;k} X^{*T}_{i_p,l;k} \\ &\quad - \hat{x}_{i_p,k|k}\hat{x}^T_{i_p,k|k} \\ &\quad + Q_{i_p,k/n_{i_p}}\end{aligned} \tag{11.10}$$

$$X_{i_p,l;k} = \hat{x}_{i_p,k|k} + \sqrt{P_{i_p,k|k}}\varepsilon_l \tag{11.11}$$

$$Y_{i_p,l;k/n_{i_p}} = h_{i_p}(X_{i_p,l;k}) \tag{11.12}$$

$$\hat{y}_{i_p,k/n_{i_p}} = w_l \sum_{l=1}^{2n} Y_{i_p,l;k/n_{i_p}} \tag{11.13}$$

$$P_{\tilde{y}_{i_p},k/n_{i_p}} = w_l \sum_{l=1}^{2n} Y_{i_p,l;k/n_{i_p}} Y^T_{i_p,l;k/n_{i_p}}$$

$$
\begin{aligned}
&- \hat{y}_{i_p,k/n_{ip}} \hat{y}^T_{i_p,k/n_{ip}} \\
&+ R_{i_p,k/n_{ip}} &(11.14)
\end{aligned}
$$

$$
\begin{aligned}
P_{\tilde{x}\tilde{y}_N,k/n_{ip}} &= w_l \sum_{l=1}^{2n} X_{i_p,l;k} Y^T_{i_p,l;k/n_{ip}} \\
&- \hat{x}_{i_p,k|k} \hat{y}^T_{i_p,k/n_{ip}} &(11.15)
\end{aligned}
$$

$$
K_{i_p,k/n_{ip}} = P_{\tilde{x}\tilde{y}_{ip},k/n_{ip}} P^{-1}_{\tilde{y}_{ip},k/n_{ip}} \tag{11.16}
$$

For $p = 0$, the initial state estimate and the estimation error covariance are computed by

$$
\begin{aligned}
\hat{x}_{i_0,k|k} &= \hat{x}_{k|k-1} + K_{N,k}[y_{N,k} \\
&- \hat{y}_{N,k|k-1}] &(11.17) \\
P_{i_0,k|k} &= P_{k|k-1} \\
&- K_{N,k} P_{\tilde{y}_N,k|k-1} K^T_{N,k} &(11.18)
\end{aligned}
$$

where

$$
\begin{aligned}
X_{l;k-1} &= \hat{x}_{k-1|k-1} \\
&+ \sqrt{P_{k-1|k-1}} \varepsilon_l &(11.19) \\
X^*_{l;k|k-1} &= f(X_{l;k-1}) &(11.20) \\
\hat{x}_{k|k-1} &= w_l \sum_{l=1}^{2n} X^*_{l;k|k-1} &(11.21) \\
P_{k|k-1} &= w_l \sum_{l=1}^{2n} X^*_{l;k|k-1} X^{*T}_{l;k|k-1} \\
&- \hat{x}_{k|k-1} \hat{x}^T_{k|k-1} + Q_k &(11.22) \\
X_{l;k|k-1} &= \hat{x}_{k|k-1} + \sqrt{P_{k|k-1}} \varepsilon_l &(11.23) \\
Y_{N,l;k|k-1} &= h(X_{l;k|k-1}) &(11.24) \\
\hat{y}_{N,k|k-1} &= w_l \sum_{l=1}^{2n} Y_{N,l;k|k-1} &(11.25) \\
P_{\tilde{y}_N,k|k-1} &= w_l \sum_{l=1}^{2n} Y_{N,l;k|k-1} Y^T_{N,l;k|k-1} \\
&- \hat{y}_{N,k|k-1} \hat{y}^T_{N,k|k-1} + R_k &(11.26) \\
P_{\tilde{x}\tilde{y}_N,k|k-1} &= w_l \sum_{l=1}^{2n} X_{l;k|k-1} Y^T_{N,l;k|k-1}
\end{aligned}
$$

$$- \hat{x}_{k|k-1}\hat{y}^T_{N,k|k-1} \tag{11.27}$$

$$K_{N,k} = P_{\tilde{x}\tilde{y}_N,k|k-1}P^{-1}_{\tilde{y}_N,k|k-1} \tag{11.28}$$

The basic cubature points and the corresponding weights are obtained by using a cubature rule of degree three. The cubature point set: $\varepsilon_l = \sqrt{\frac{m^*}{2}}[1]_l$. *The corresponding weight,* $w_l = \frac{1}{m^*}, l = 1, 2, \ldots, m^*$. m^* *represents the number of cubature points.* $m^* = 2n$, *n means the state dimension of the system. Matrix* $[1] = [I_{n\times n}, -I_{n\times n}]$, *where* $I_{n\times n}$ *represents n-dimensional identity matrix and* $[1]_l$ *represents the lth line of matrix* [1].

Proof Define $Y_1^{l_i}(i) = \{y_{i,l}, l = 1, 2, \ldots, l_i\}$, which denotes the first to the l_ith observations by sensor i. By definition, the optimal state estimate $\hat{x}_{k-1|k-1}$ is obtained by use of all the information observed by sensors $i = 1, 2, \ldots, N$ sampled up to and include time $k-1$, that is,

$$\hat{x}_{k-1|k-1} = E\{x_{k-1}|Y^{l_i}_{1,i}; i = 1, 2, \ldots, N, l_i = [(k-1)/n_i]\} \tag{11.29}$$

where $[(k-1)/n_i]$ denotes the maximum integer that not large than $(k-1)/n_i$.

To generate $\hat{x}_{k|k}$, the information $y_{i_p}(k/n_{i_p})$ should be used in sequence of $p = 0, 1, 2, \ldots, p'$. For $p = 0$, $y_{N,k}$ is used to modify $\hat{x}_{k-1|k-1}$, and $\hat{x}_{i_0,k|k}$ could be obtained:

$$\begin{aligned}
\hat{x}_{i_0,k|k} &= E\{x_k|Y^{l_i-1}_{1,i}; i = 1, 2, \ldots, N, l_i = [k/n_i]; y_{N,k}\} \\
&= E\{x_k|Y^{l_i}_{1,i}; i = 1, 2, \ldots, N, l_i = [(k-1)/n_i]; y_{N,k}\} \\
&= E\{f_{x_{k-1}} + w_{k-1}|Y^{l_i}_{1,i}; \\
&\quad i = 1, 2, \ldots, N, l_i = [(k-1)/n_i]; y_{N,k}\} \\
&= E\{f_{x_{k-1}}|Y^{l}_{1,i}; i = 1, 2, \ldots, N, \\
&\quad l = [(k-1)/n_i]\} + P_{\tilde{x}\tilde{y}_N}(k|k-1)P^{-1}_{\tilde{y}_N}(k|k-1) \\
&\quad \cdot [y_{N,k} - \hat{y}_{N,k|k-1}] \\
&= \hat{x}_{k|k-1} + K_{N,k}[y_{N,k} - \hat{y}_{N,k|k-1}]
\end{aligned} \tag{11.30}$$

where, from CKF, we have

$$\begin{aligned}
\hat{x}_{k|k-1} = E\{f_{x_{k-1}}|Y^{l}_{1,i}; i = 1, 2, \ldots, N, \\
l = [(k-1)/n_i]\}
\end{aligned} \tag{11.31}$$

which is computed by use of (11.19) to (11.21), while $K_{N,k} = P_{\tilde{x}\tilde{y}_N,k|k-1}$ $P^{-1}_{\tilde{y}_N,k|k-1}$ is computed by (11.23) to (11.28).

The estimation error covariance of $\hat{x}_{i_0,k|k}$ is computed by

$$\begin{aligned} P_{i_0,k|k} &= E\{[x_k - \hat{x}_{i_0,k|k}][x_k - \hat{x}_{i_0,k|k}]^T\} \\ &= E\{[x_k - \hat{x}_{k|k-1} - K_{N,k}(y_{N,k} - \hat{y}_{N,k|k-1})] \\ &[x_k - \hat{x}_{k|k-1} - K_{N,k}(y_{N,k} - \hat{y}_{N,k|k-1})]^T\} \\ &= P_{k|k-1} - K_{N,k} P_{\tilde{y}_N,k|k-1} K_{N,k}^T \end{aligned} \tag{11.32}$$

where

$$P_{k|k-1} = E\{[x_k - \hat{x}_{k|k-1}][x_k - \hat{x}_{k|k-1}]^T\} \tag{11.33}$$

which is computed by Eq. (11.22), and

$$P_{\tilde{y}_N,k|k-1} = E\{[y_{N,k} - \hat{y}_{N,k|k-1}] \tag{11.34}$$

$$\cdot [y_{N,k} - \hat{y}_{N,k|k-1}]^T\} \tag{11.35}$$

is computed by use of (11.26).

Similarly, for $p = 1, 2, \ldots, p'$, we have (11.5) to (11.16).

Deductively, the optimal state estimate of x_k and the corresponding estimation error covariance are generated by (11.3) and (11.4), respectively.

For sensor i, if the system exists unreliable measurements and y_{i,k_i} is missing or fault, the data fusion algorithm of Theorem 11.2 should be modified.

Theorem 11.3 [23, 29] (CKF-based estimation by fusion of unreliable measurements) *Suppose that the system model is formulated as in Sect. 11.2. At time k, it is assumed that there are $p' + 1$ sensors meet mod$(k, n_{i_p}) = 0$, where $p = 0, 1, 2, \ldots, p'$ and $N = i_0 > N - 1 \geq i_1 \geq i_2 \geq \cdots \geq i_{p'} \geq 1$. The optimal state estimate $\hat{x}_{k-1|k-1}$ and the corresponding estimation error covariance $P_{k-1|k-1}$ at time $k - 1$ have been generated. Then the optimal state estimate $\hat{x}_{k|k}$ and the estimation error covariance $P_{k|k}$ can be deduced by modifying the CKF algorithm of Theorem 1 in that the $K_{i_p,k_{i_p}}$ in (11.16) and $K_{N,k}$ in (11.28) are set to be zeroes when $y_{i_p,k_{i_p}}$ or $y_{N,k}$ are deemed to be unreliable.*

The unreliability of the measurements is detected by use of the statistical property of the residuals. Namely, $y_{i_p,k_{i_p}}$ is claimed to be unreliable if $|\rho_{i_p,k}| > \chi^2_{\alpha^*}(m_{i_p})$, where

$$\begin{aligned} \rho_{i_p,k/n_{ip}} &= [y_{i_p,k/n_{ip}} \\ &- \hat{y}_{i_p,k/n_{ip}}]^T S^{-1}_{i_p,k/n_{ip}} [y_{i_p,k/n_{ip}} \\ &- \hat{y}_{i_p,k/n_{ip}}] \end{aligned} \tag{11.36}$$

$$S_{i_p,k/n_{ip}} = w_l \sum_{l=1}^{2n} Y_{i_p,l;k/n_{ip}} Y^T_{i_p,l;k/n_{ip}}$$

$$- y_{i_p,k/n_{ip}} y_{i_p,k/n_{ip}}^T$$
$$+ R_{i_p,k/n_{ip}} \tag{11.37}$$

From Theorems 11.2 and 11.3, we have the following theorem.

Theorem 11.4 *The state estimation algorithms presented by Theorems 11.2 and 11.3 are meaningful, and we have*

$$\lim_{k\to\infty} P_{k|k} \leq P_{S,k|k} \tag{11.38}$$

where $P_{S,k|k}$ is the estimation error covariance at time k by CKF algorithm based on model (1) and single sensor measurement, that is, when set $N = 1$ in (11.2).

11.4 Numerical Simulation

In this section, we will choose two nonlinear examples of dimension one and dimension five to illustrate the effectiveness of the presented algorithms.

11.4.1 Simple Example on Tracking of a Ship

A ship's propulsion system can be modeled by[28, 29]:

$$x_{k+1} = 0.1a{x_k}^2 + x_k + 0.1bu_k + w_k \tag{11.39}$$
$$y_{i,k_i} = x_{k_i} + v_{i,k_i}, i = 1, 2, 3, 4 \tag{11.40}$$

where the parameter "a" represents the resistance of the hull, and the power required to launch the ship is "b". "x" is the speed of the ship. The values of the parameters "a" and "b" are set to be $a = -0.58$ and $b = 0.2$. The control signal u_k is a square wave signal, its amplitude is from 0.9 to 1.1, and the period is 200. The system noise and the measurement noises are assumed to be Gaussian distributed with zero mean. The white system noise variance is $Q = 10^{-6}$, whose mean is zero. There are four sensors to observe the state variables, and the noise mean is zero; the covariances are $R_1 = R_2 = R_3 = R_4 = 10^{-4}$, respectively. During a period of time, the sampling points of Sensors 1, 2, 3, 4 are 300, 150, 100, and 75, respectively. The initial state and the corresponding error variance are $x_0 = 0$ and $P_0 = 100$. The variable parameter selection of UKF is $\alpha = 1$, $\beta = 0$, and $\kappa = 2$, which meet $n + k = 3$ [30].

To compare the estimation accuracy of different cases, we use the root mean square error (RMSE) and the time-averaged root mean square error (TARMSE).

Table 11.1 The TARMSE and the average CPU time of Monte Carlo simulation results

Algorithm	UKF	CKF	FUKF	FCKF
TARMSE ($\times 10^{-3}$)	0.3446	0.3434	0.3138	0.3158
TIME	0.0664	0.0499	0.1352	0.1042
Modified algorithm	UKF	CKF	FUKF	FCKF
TARMSE ($\times 10^{-3}$)	0.3660	0.3645	0.3378	0.3392
TIME	0.0778	0.0609	0.1604	0.1288

$$RMSE = \frac{1}{M}\sqrt{\sum_{l^*=1}^{M}(x_k - \hat{x}_{k|k}^{l^*})^2} \tag{11.41}$$

$$TARMSE = \frac{1}{TM}\sum_{k=1}^{T}\sqrt{\sum_{l^*=1}^{M}(x_k - \hat{x}_{k|k}^{l^*})^2} \tag{11.42}$$

where T is the length of the signal, M is the number of Monte Carlo simulation runs, x_k is the real state value, and $\hat{x_{k|k}}^{l^*}$ is the state estimate of moment k, which achieved across the l^*th Monte Carlo simulation run.

The Monte Carlo simulations of 50 runs are shown in Table 11.1 and Figs. 11.1, 11.2, 11.3, and 11.4. Table 11.1 is the TARMSE and the average CPU time of Monte Carlo simulations, whose upper part is the simulation result of the four algorithms without reliable data and lower part is the result of the four modified algorithms in the case of unreliable data. The fused UKF is called FUKF, and the fused CKF is called FUKF. Figures 11.1 and 11.2 show the measurements of sensors from 1 to 4 and RMSE of the above mentioned algorithms with reliable data. Figures 11.3 and 11.4 show the measurements of sensors from 1 to 4 and RMSE of the above mentioned algorithms with unreliable data at the probability of faulty at 10% for each sensor.

Analysis of simulation results for one-dimensional nonlinear case: For both cases (measurements all reliable or encounter 10% faulty), from the lowest to the highest, TARMSE values are presented with the following order: the value by fused UKF, the value by fused CKF, the value by CKF, and the value by UKF. It can be illustrated as follows: first, in the case of one dimension, the fused UKF estimation accuracy is the highest, slightly higher than the fused CKF; second, the sequential fused CKF algorithm is better than UKF and CKF algorithm, and the estimation accuracy is greatly improved; third, the modified algorithms by using Theorem 11.3 can effectively eliminate the influence of faulty measurements, and Figs. 11.2 and 11.4 are consistent to the above analysis; and fourth, in terms of the average CPU running time, the fused CKF algorithm saves about 23% of the running time compared with the fused UKF algorithm.

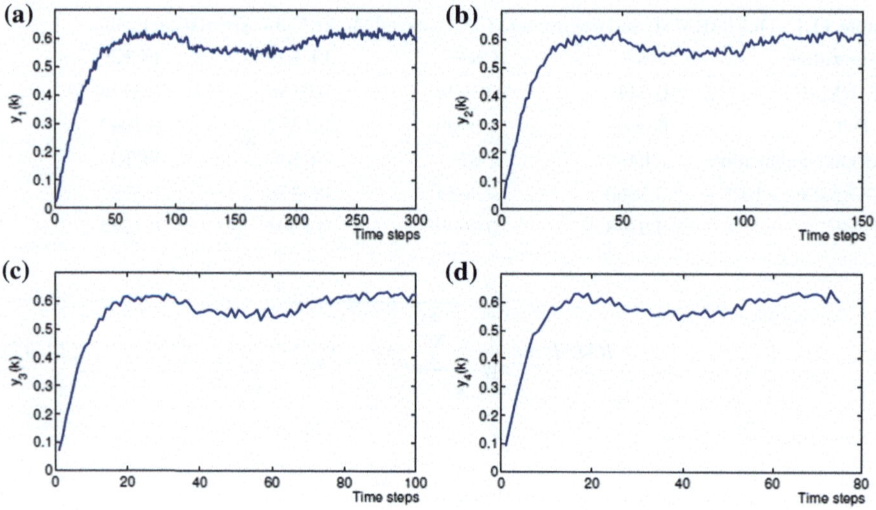

Fig. 11.1 The measurements of sensors from 1 to 4 without unreliable data

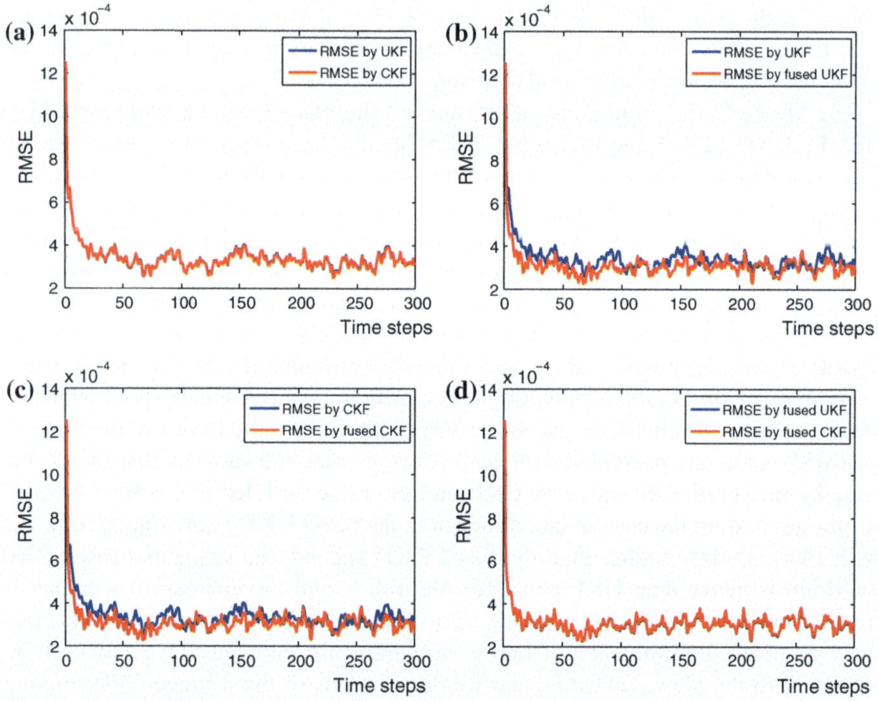

Fig. 11.2 RMSE of the above mentioned algorithms with reliable data fusion

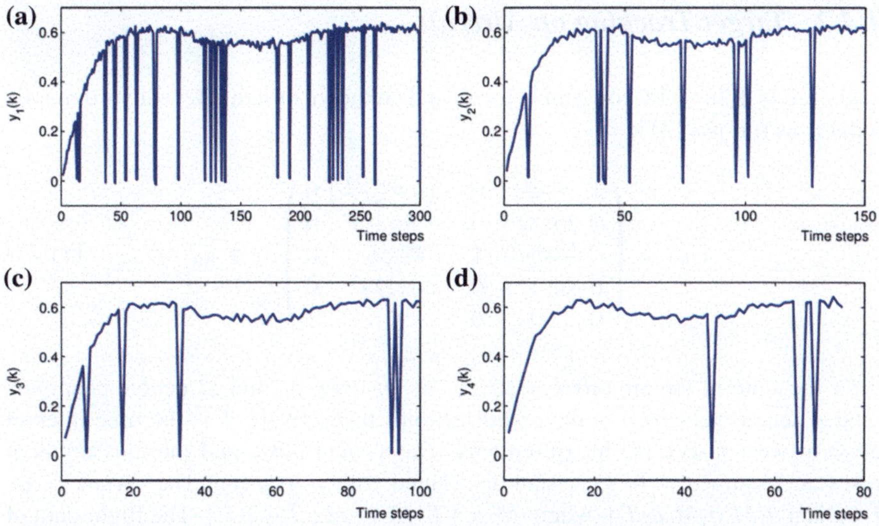

Fig. 11.3 The measurements of sensors from 1 to 4 with unreliable data

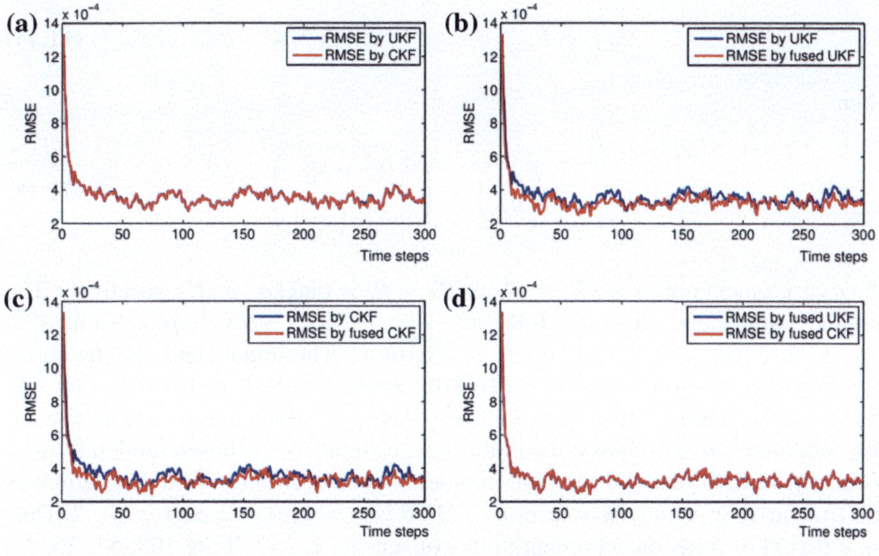

Fig. 11.4 RMSE of the above mentioned algorithms with unreliable data fusion

11.4.2 Target Tracking on Aircraft

An aircraft is flying with constant velocity and unknown turn rate Ω, and its dynamic model is as follows [31]:

$$x_{k+1} = \begin{pmatrix} 1 & \frac{\sin \Omega T}{\Omega} & 0 & -\left(\frac{1-\cos \Omega T}{\Omega}\right) & 0 \\ 0 & \cos \Omega T & 0 & -\sin \Omega T & 0 \\ 0 & \frac{1-\cos \Omega T}{\Omega} & 1 & \frac{\sin \Omega T}{\Omega} & 0 \\ 0 & \sin \Omega T & 0 & \cos \Omega T & 0 \\ 0 & 0 & 0 & 0 & 1 \end{pmatrix} x_k + w_k \tag{11.43}$$

where the state of the aircraft $x = [\xi \quad \dot{\xi} \quad \eta \quad \dot{\eta} \quad \Omega]^T$; ξ and Ω denote positions; $\dot{\xi}$ and $\dot{\eta}$ denote velocities in the and directions, respectively. T is the time interval between two consecutive measurements. The system noise and the measurement noises are assumed to be Gaussian distributed with zero mean. The variances are $Q = \text{diag}[\, q_1 M \; q_1 M \; q_2 T\,]$, where $M = [\, T^3/3 \; T^2/2;\; T^2/2 \; T\,]$. The flight data of the aircraft is measured by four radars, whose slant range is r, and azimuth is θ. The measurement equation is

$$y_{i,k_i} = h_{i,x_{k_i}} + v_{i,k_i},\, i = 1, 2, 3, 4 \tag{11.44}$$

where

$$h_{i,x_{k_i}} = \begin{pmatrix} r_{k_i} \\ \theta_{k_i} \end{pmatrix} = \begin{pmatrix} \sqrt{\xi_{k_i}^2 + \eta_{k_i}^2} \\ \tan^{-1}\left(\frac{\eta_{k_i}}{\xi_{k_i}}\right) \end{pmatrix} \tag{11.45}$$

The measurement noises are $R_1 = R_2 = R_3 = R_4 = \text{diag}[\, \sigma_r^2 \; \sigma_\theta^2\,]$, respectively. The system parameters are as follows: $T = 1\,\text{s}$, $\Omega = -3°\text{s}^{-1}$, $q_1 = 0.1\,\text{m}^2\,\text{s}^{-3}$, $q_2 = 1.75 \times 10^{-4}\,\text{s}^{-3}$, $\sigma_r = 10\,\text{m}$, $\sigma_\theta = \sqrt{10}\,\text{mrad}$. The initial state and the corresponding error covariance are $x_0 = [1000\,\text{m} \quad 300\,\text{m}\,\text{s}^{-1} \quad 1000\,\text{m}\,0\,\text{m}\,\text{s}^{-1} \quad -3°\text{s}^{-1}]^T$ and $P_0 = \text{diag}[100\,\text{m}^2 \quad 10\,\text{m}^2\,\text{s}^{-2} \quad 100\,\text{m}^2 \quad 10\,\text{m}^2\,\text{s}^{-2} \quad 100\,\text{mrad}^2\,\text{s}^{-2}]$. The simulation conditions are as follows: the initial state estimate is $\hat{x}_0$ chosen randomly from $N(x_0, P_0)$ in each run, and the total number of runs of Monte Carlo simulations is 50. The variable parameter selection of UKF is $\alpha = 1$, $\beta = 2$, and $\kappa = -2$. During a period of time, the sampling points of sensors 1, 2, 3, 4 are 108, 54, 36, 27, respectively.

Simulation results are shown in Table 11.2 and Figs. 11.5, 11.6 and 11.7. The upper part of Table 11.2 is the TARMSE of the first dimension and the third dimension of the four algorithms with data unreliable at the probability of 0.1 for each sensor, and the lower part is the corresponding average running time. Figure 11.5 shows the estimates of the above mentioned algorithms by fusion of reliable and unreliable data, respectively. Figures 11.6 and 11.7 show the RMSEs of the first dimension and the third dimension, respectively, by fusion of unreliable data.

Table 11.2 The TARMSE and average time

Algorithm	UKF	CKF	FUKF	FCKF
First dimension	7.7196	7.7911	7.1670	6.9283
Third dimension	5.6285	5.6673	5.1307	4.9672
TIME	0.0612	0.0424	0.1297	0.0894

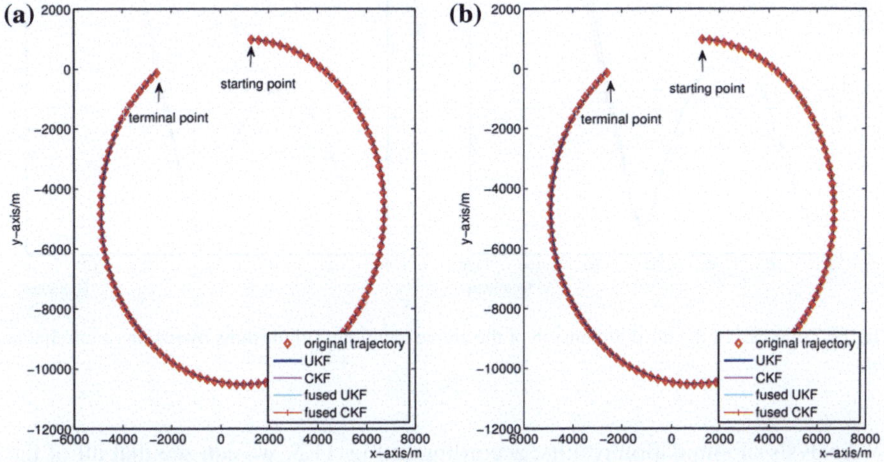

Fig. 11.5 The estimates of the above mentioned algorithms: **a** by fusion of reliable data; **b** by fusion of unreliable data

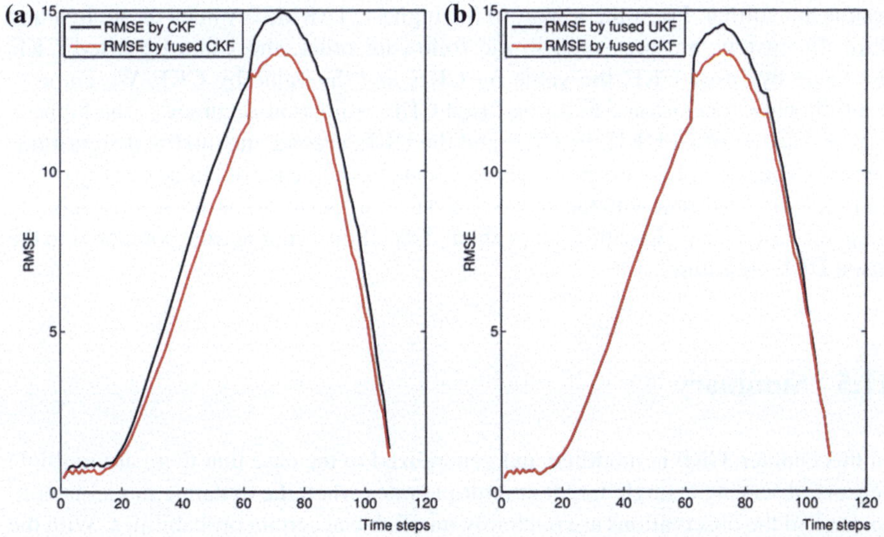

Fig. 11.6 RMSE of the first dimension of the above mentioned algorithms by fusion of unreliable data

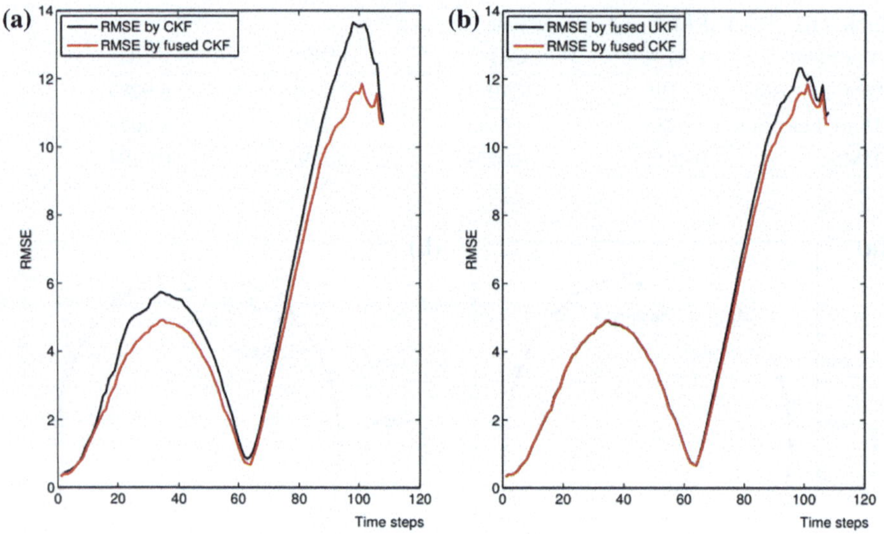

Fig. 11.7 RMSE of the third dimension of the above mentioned algorithms by fusion of unreliable data

Analysis of simulation results: according to Fig. 11.5, we can see that all of the above mentioned algorithms are able to track the target, but there is a difference between the estimated accuracy. In order to simplify the analysis process, this chapter only gives the first and third dimensional data analysis, and other dimensional analysis results are similar. From the lowest to the highest, TARMSE values of the first and third dimensions are presented in the following order: the value by fused CKF, the value by fused UKF, the value by UKF, and the value by CKF. We come to the following conclusions: first, the fused CKF estimation accuracy is the highest, higher than the fused UKF, the CKF, and the UKF; second, the modified algorithms by using Theorem 11.3 can effectively eliminate the influence of faults, and Figs. 11.6 and 11.7 are consistent with the above analysis; third, in terms of the average running time, the fused CKF algorithm saves about 32% of the running time compared to the fused UKF algorithm.

11.5 Summary

In this chapter, CKF is modified and generalized to the case that there are multiple sensors observing a single target at multiple rates when the dynamic model is nonlinear, and the observations are randomly unreliable at certain probabilities. With the increase of the state dimension of nonlinear system, the estimation accuracy of the presented sequential fusion CKF algorithm is higher than the fused UKF algorithm, and the running time of the algorithm is greatly reduced. The estimation accuracy of the presented algorithm is improved greatly compared to the CKF algorithm and the UKF algorithm. The modified algorithms by using Theorem 11.3 can effectively eliminate the influence of faulty measurements. Theoretical analysis and simulation results show the feasibility and effectiveness of the presented algorithm. The algorithm is applicable to many application fields, such as target tracking, integrated navigation, network control, fault diagnosis, and tolerance.

References

1. L. Wu, P. Shi, H. Gao, State estimation and sliding mode control of Markovian jump singular systems. IEEE Trans. Autom. Control **55**(5), 1213–1219 (2010)
2. C.L. Liu, F. Liu, Asynchronously-coupled consensus of second-order dynamic agents with communication delay. Int. J. Innov. Comput. Inf. Control **6**(11), 5035–5046 (2010)
3. Y.J. Zhang, S.S. Zhou, A.K. Xue, W.X. Zheng, Delay-dependent state estimation for time-varying delayed neural networks. Int. J. Innov. Comput. Inf. Control **5**(6), 1711–1724 (2009)
4. C.W. Liao, C.Y. Lu, K.Y. Zheng, C.C. Ting, A delay-dependent approach to design state estimator for discrete stochastic recurrent neural network with interval time-varying delays. ICIC Express Lett. **3**(3A), 465–470 (2009)
5. H.N. Wu, P. Shi, Adaptive variable structure state estimation for uncertain systems with persistently bounded disturbances. Int. J. Robust Nonlinear Control **20**(17), 2003–2015 (2010)
6. M.Y. Fu, Z.H. Deng, L.P. Yan, *Kalman Filter Theory and Its Application in Navigation System*, 2nd edn. (Science Publication, Beijing, 2010)
7. M.S. Arulampalam, S. Maskell, N. Gordon, T. Clapp, A tutorial on particle filters for online nonlinear/non-Gaussian Bayesian tracking. IEEE Trans. Signal Process. **50**(2), 174–188 (2002)
8. Y.X. Wu, D.W. Hu, M.P. Wu, X.P. Hu, *Quasi-Gaussian Particle Filtering* (Springer, Berlin, 2006)
9. R.E. Kalman, A new approach to linear filtering and prediction theory. J. Basic Eng. **82**(D), 35–46 (1960)
10. B.O.S. Teixeira, L.A.B. Torres, P. Iscold, L.A. Aguirre, Flight path reconstruction a comparison of nonlinear Kalman filter and smoother algorithms. Aerosp. Sci. Technol. **15**(1), 60–71 (2011)
11. S.J. Julier, J.K. Uhlmann, Unscented filtering and nonlinear estimation. Proc. IEEE **92**(3), 401–422 (2004)
12. R.V.D. Merwe, E.A. Wan, S.I. Julier, Sigma-point Kalman filters for nonlinear estimation and sensor fusion applications to integrated navigation **5120**, 1–30 (2004)
13. Y.X. Wu, D.W. Hu, M.P. Wu, X.P. Hu, A comparison of estimation accuracy by the use of KF, EKF and UKF filters, vol. 6 (Prague, Czech Republic) (2007), pp. 779–789
14. I. Arasaratnam, S. Haykin, Cubature Kalman filters. IEEE Trans. Autom. Control **54**(6), 1254–1269 (2009)
15. I. Arasaratnam, S. Haykin, T.R. Hurd, Cubature Kalman filtering for continuous-discrete systms: theory and simulations. IEEE Trans. Signal Process. **58**(10), 4977–4993 (2010)
16. Y. Zhang, G.S. Rui, J. Miao, Passive location of angle sensors using extend square-root cubature Kalman filter. Appl. Mech. Mater. (Electr. Inf. Mechatron. Appl.) **143–144**, 577–581 (2011)
17. C.Y. Sheng, J. Zhao, Y. Liu, W. Wang, Prediction for noisy nonlinear time series by echostate network based on dual estimation. Neurocomputing **82**, 186–195 (2012)
18. W.L. Li, Y.M. Jia, Location of mobile station with maneuvers using an IMM-based cubature Kalman filter. IEEE Trans. Ind. Electron. **59**(11), 4338–4348 (2012)
19. H.A.P. Blom, E.A. Bloem, Particle filtering for stochastic hybrid systems **3**, 3221–3226 (2004)
20. L. Yan, B. Liu, D. Zhou, An asynchronous multirate multisensor information fusion algorithm. IEEE Trans. Aerosp. Electron. Syst. **43**(3), 1135–1146 (2007)

21. S.S. Gao, Y.M. Zhong, X.Y. Zhang, B. Shirinzadeh, Multi-sensor optimal data fusion for INS/GPS/SAR integrated navigation system. Aerosp. Sci. Technol. **13**(4–5), 232–237 (2009)
22. L. Hong, Multi-resolutional distributed filtering. IEEE Trans. Autom. Control **39**(4), 853–856 (1994)
23. L. Yan, D. Zhou, M. Fu, Y. Xia, State estimation for asynchronous multirate multisensor dynamic systems with missing measurements. IET Signal Process. **4**(6), 728–739 (2010)
24. A. Censi, Kalman filtering with intermittent observations: convergence for semi-Markov chains and an intrinsic performance measure. IEEE Trans. Autom. Control **56**(2), 376–381 (2011)
25. S. Kar, B. Sinopoli, J.M.F. Moura, Kalman filtering with intermittent observations: weak convergence to a stationary distribution. IEEE Trans. Autom. Control **57**(2), 405–420 (2012)
26. S. Kluge, K. Reif, M. Brokate, Stochastic stability of the extended Kalman filter with intermittent observations. IEEE Trans. Autom. Control **55**(2), 514–518 (2010)
27. L.P. Yan, Z.H. Deng, M.Y. Fu, Study of asynchronous multirate data fusion estimation algorithm based on nonlinear systems. Acta Electron. Sin. **37**(12), 2735–2740 (2009)
28. C.L. Wen, D.H. Zhou, *Multiscale Estimation Theory and Application* (Publishing House of Tsinghua University, Beijing, 2002)
29. L.P. Yan, Y.Q. Xia, B.S. Liu, M.Y. Fu, *Multi-sensor Optimal Estimation Theory and Its Application* (Science Publishing House, Beijing, 2015)
30. S. Julier, J. Uhlmann, H.F. Durrant-Whyte, A new method for the nonlinear transformation of means and covariances in filters and estimators. IEEE Trans. Autom. Control **45**(3), 477–482 (2000)
31. Y. Bar-Shalom, X. Li, T. Kirubarajan, *Estimation with Applications to Tracking and Navigation: Theory Algorithms and Software* (Wiley, New York, 2001)

Part IV
Kalman Filtering for Multi-agent Systems

Chapter 12
Decentralized Adaptive Filtering for Multi-agent Systems with Uncertain Couplings

In this chapter, the problem of decentralized adaptive filtering for multi-agent systems with uncertain couplings is formulated and investigated. This problem is challenging due to the mutual dependency of state estimation and coupling estimation. First, the problem is divided into four typical types based on the origin of coupling relations and linearity of the agent dynamics. Then models of the four types are given, and the corresponding decentralized adaptive filtering algorithms are designed for the purpose of estimation of the unknown states and couplings which denotes the relations between agents and their neighbor agents in terms of states or outputs simultaneously, with preliminary stability analysis and discussions. For testing the effects of algorithm, with the so-called certainty equivalence principle, control signals are designed based on the results of state estimation and coupling estimation got by the proposed decentralized adaptive filtering algorithms. Extensive simulations are conducted to verify the effectiveness of considered algorithms.

12.1 Introduction

The concept of multi-agent systems [1–7] is proposed in 1980s with the development of complex system theory. Through the approach of dividing and conquering, decomposing a large complex system into small components or subsystems, which have intrinsic local coupling relations with each other, is a promising idea which can greatly simplify the control design of the whole system and reduce the total computational cost significantly. Hence, understanding how local interactions affect the global behavior plays a fundamental role either in theory or in practice.

A multi-agent system, as one simplified model for investigating complex systems, is composed of many agents which not only have local interactions among one another but also respond to others by making decisions and taking actions. In this way, the agents can perform one simple task independently or complete one difficult task by

H. Ma et al., *Kalman Filtering and Information Fusion*,
https://doi.org/10.1007/978-981-15-0806-6_12

cooperation and communication. Based on the above features, many applications of multi-agent systems can be found in vast areas such as robot systems, smart home systems, and so on.

There are many kinds of research about multi-agent systems in recent years. Main research topics may include the architecture of the multi-agent system, the communication among agents, coordination, and cooperation. The architecture discusses how the agents form the system and the communication discusses the way in which information transfers among agents. The coordination control of the multi-agent system is comprehensive, since this problem needs to consider the architecture and communication. And consensus control of multiple agents as one kind of coordination control has received more and more attentions in the recent decades. Consensus means the agents can make their states and velocities in agreement by cooperation and communication. This problem was first proposed by Olfati-Saber and Murray [8]. In the direction of distributed consensus in multi-agent cooperative control, Ren et al. have obtained numerous results [9–11]. These existing extensive work presented the consensus algorithms for single-integrator dynamics, double-integrator dynamics, and rigid body attitude dynamics.

However, many studies about multi-agent systems assume that the global team knowledge is available, that is, the relations among agents are known, and the group actions can be achieved in a centralized way. The situations where there are unknown or uncertain relations between agents are basically ignored for decades. However, these situations have important effects on the studies of multi-agent systems which often exist in the practical applications. In such situations, the agents must try to cope with the uncertainties occurred in local interactions with only limited available history information and local information.

In this chapter, we try to elaborate the decentralized filtering problem of one typical class of multi-agent systems, which focuses on dealing with coupling uncertainties. In such systems, there are unknown couplings among agents. As a preliminary study, in this chapter we use linear mappings to represent such unknown interactions briefly. Due to the unknown couplings among agents, common filter algorithms like standard Kalman filters [12–18] cannot be used directly. To address this challenging problem, in this contribution, the filtering process is achieved by adopting modifications to standard Kalman filter algorithm. This area is less explored in the existing literature despite that another challenging area, decentralized adaptive control of multi-agent systems, has been extensively studied by the authors [19]; hence, this chapter aims to lay a simple yet meaningful framework of decentralized adaptive filtering for multi-agent systems. Because of intrinsic differences between linear systems and nonlinear systems, we classify the considered system models as four different types according to the model linearity and the coupling type, and hence different decentralized filtering algorithms should be designed separately. And through estimated values of the states obtained by decentralized adaptive filtering, corresponding decentralized adaptive control law is designed according to the certainty equivalence principle. The stability of the filtering algorithm is also discussed with verification by simulations.

The rest of this chapter is organized as follows. First, Sect. 12.1 presents the system model and problem formulation. Then, in Sect. 12.2, the decentralized adaptive filtering algorithms are derived. Next in Sect. 12.3, based on the certainty equivalence principle, decentralized adaptive controller is designed to test the filter effects. In Sect. 12.4, some remarks on the stability analysis for the proposed decentralized filtering algorithms are given. And, in Sect. 12.5, simulations of decentralized adaptive filter and control algorithms are reported and the corresponding results are analyzed. Finally, we conclude this chapter with some concluding remarks in Sect. 12.6.

Notations. The following notations are used throughout this chapter. $\mathbf{R}^n$ and $\mathbf{R}^m$ stand for the set of real vectors with dimension n and m, respectively. I and $0_{m\times n}$ mean the identity matrix and zero matrix. The subscript of matrix represents the dimension of the matrix of dimension m by n. Wherever the dimensions of the matrices are not mentioned, they are assumed to be of compatible dimensions. $\mathrm{E}(\cdot)$ and $P()$ denote the mathematical expectation and variance. $[\cdot]^T$ denotes the transposition of a matrix and $(\cdot)^{-1}$ denotes the inversion of a matrix. $\widehat{*}$ and $\bar{*}$ on the top represent the corresponding posterior and prior terms, respectively.

12.2 Problem Statement

Since agents often have similar structures and the same state dimension, we assume $x_i \in \boldsymbol{R}^n$, $u_i \in \boldsymbol{R}^r$, $y_i \in \boldsymbol{R}^m$, $w_i \in \boldsymbol{R}^n$, and $v_i \in \boldsymbol{R}^m$ for every i in $\{1, 2, \ldots, N\}$.

12.2.1 Model 1: Linear Model with Output Coupling

Considering one class of multi-agent system where there are N agents and every agent can be described by the following dynamic model:

$$\begin{aligned} x_{i,k} &= A_{i,k-1}x_{i,k-1} + B_{i,k-1}u_{i,k-1} + \\ &\sum_{j_p\in\mathcal{N}_i} G_{i,j_p}y_{j_p,k-1} + w_{i,k-1}, \\ y_{i,k} &= C_{i,k}x_{i,k} + v_{i,k}, \\ \mathcal{N}_i &= \{j_1, j_2, \ldots, j_{n_i}\} \end{aligned} \tag{12.1}$$

where coefficient matrices $G_{i,j}$ are known with dimension $n \times m$, $A_{i,k}$ are known with dimension $n \times n$, $B_{i,k}$ are known with dimension $n \times r$, and $C_{i,k}$ are known with dimension $m \times n$ $(i = 1, 2, \ldots, N; k = 1, 2, 3, \ldots)$. If $m = n$, one typical case is $G_{i,j} = q_{i,j_p}$, i.e.,

$$\begin{aligned} x_{i,k} &= A_{i,k-1}x_{i,k-1} + B_{i,k-1}u_{i,k-1} + \\ &\sum_{j_p\in\mathcal{N}_i} q_{i,j_p}y_{j_p,k-1} + w_{i,k-1}. \end{aligned} \tag{12.2}$$

12.2.2 Model 2: Linear Model with State Coupling

The dynamic model of every agent is given below:

$$
\begin{aligned}
x_{i,k} &= A_{i,k-1}x_{i,k-1} + B_{i,k-1}u_{i,k-1} + \\
&\sum_{j_p \in \mathcal{N}_i} q_{i,j_p} x_{j_p,k-1} + w_{i,k-1}, \\
y_{i,k} &= C_{i,k}x_{i,k} + v_{i,k}, \\
\mathcal{N}_i &= \{j_1, j_2, \ldots, j_{n_i}\}
\end{aligned} \tag{12.3}
$$

where coefficient matrices $A_{i,k}$ are known with dimension $n \times n$, $B_{i,k}$ are known with dimension $n \times r$, and $C_{i,k}$ are known with dimension $m \times n$ ($i = 1, 2, \ldots, N$; $k = 1, 2, 3, \ldots$).

12.2.3 Model 3: Nonlinear Model with Output Coupling

Each agent i has the following nonlinear dynamics:

$$
\begin{aligned}
x_{i,k} &= f_i(x_{i,k-1}, w_{i,k-1}) + B_{i,k-1} \times \\
&\quad u_{i,k-1} + \sum_{j_p \in \mathcal{N}_i} q_{i,j_p} g_{i,j_p}(y_{j_p,k-1}), \\
y_{i,k} &= h_i(x_{i,k}, v_{i,k}), \\
\mathcal{N}_i &= \{j_1, j_2, \ldots, j_{n_i}\}
\end{aligned} \tag{12.4}
$$

where f_i, h_i, and $g_{i,j}$ are known nonlinear mappings.

12.2.4 Model 4: Nonlinear Model with State Coupling

The dynamic model of every agent is

$$
\begin{aligned}
x_{i,k} &= f_i(x_{i,k-1}, w_{i,k-1}) + B_{i,k-1} \times \\
&\quad u_{i,k-1} + \sum_{j_p \in \mathcal{N}_i} q_{i,j_p} g_{i,j_p}(x_{j_p,k-1}), \\
y_{i,k} &= h_i(x_{i,k}, v_{i,k}), \\
\mathcal{N}_i &= \{j_1, j_2, \ldots, j_{n_i}\}
\end{aligned} \tag{12.5}
$$

where f_i, h_i, and $g_{i,j}$ are known nonlinear mappings.

In the four models, the subscripts i and j_p denote the ith and j_pth agent. Here, $x_{i,k}$ and $x_{i,k-1}$ are the states of ith agent in the kth and $(k-1)$th moment, respectively. Accordingly, $x_{j_p,k-1}$ denotes the state of j_pth agent in the $(k-1)$th moment. And y, u, w, and v denote the output, input, process noise, and measurement noise, respectively. The inputs are written as the above since they are designed to test effects of the estimation. The difference is that agents are linear in Model 1 and Model 2, but agents are nonlinear in Model 3 and Model 4. And agents in Model 1 and Model 3 are coupled by outputs and in Model 2 and Model 4 they are coupled by states.

The coupling term $\sum_{j_p \in \mathcal{N}_i} q_{ij_p} \cdots$ shows influence between agents and their neighbor agents, where scalar q_{ij_p} denotes the corresponding coupling strength. To simplify the unknown relations, we use unknown parameters q_{ij_p} and known functions $g_{i,j_p,k-1}$ to present the coupling structure. And this term represents the influence of neighbor agent j_p on agent i. The index set $\mathcal{N}_i$ denotes the set of agents which have effects on agent i.

Suppose that the noises are white, zero mean, and uncorrelated, and the noises of different agents are also uncorrelated. If the noises are correlated, we can also use methods proposed in the book written by Simon to remove the correlation. For simplicity and clarity, we may suppose that the a priori statistical information on noises are given by

$$\begin{aligned} w_{i,k} &\sim \mathrm{N}(0, Q_{i,k}), \\ v_{i,k} &\sim \mathrm{N}(0, R_{i,k}). \end{aligned} \tag{12.6}$$

That is, $w_{i,k}$ and $v_{i,k}$ have known covariance matrices $Q_{i,k}$ and $R_{i,k}$ with dimension $n \times n$ and $m \times m$, respectively.

The whole description of the multi-agent dynamic system is shown above. Among the four models, Model 1 and Model 2 are basis of Model 3 and Model 4; hence, the understanding of the former two models will bring benefits to the latter two models. However, due to the nonlinearity of Model 3 and Model 4, they are more challenging in terms of closed-loop complexity and technical difficulties involved, and later we may see that different algorithms may be needed in general for resolving the decentralized adaptive filtering problems for different models. For achieving control objectives of agents, it would be desirable that the states need to be known. Since the system has unknown parameters, we cannot estimate the states only by standard Kalman filter; hence, new filter algorithm needs to be studied to estimate the states and coupling parameters in the linear state-space model.

12.3 Decentralized Adaptive Filter

First, the inputs are assumed to be zero for they are only testing terms and cannot affect the study of filter algorithm. And the first step to deal with the unknown coupling parameters q_{i,j_p} is to consider the coupling parameters as a part of states, that is, to extend the states. The new state vector is changed to

$$x^*_{i,k} = \begin{bmatrix} x_{i,k} \\ q_{ij_1} \\ q_{ij_2} \\ \vdots \\ q_{ij_{n_i}} \end{bmatrix}_{(n+n_i)\times 1} \tag{12.7}$$

Then the filtering algorithms are proposed for the different models introduced above.

12.3.1 Model 1

The algorithm is based on the standard Kalman filter and a decentralized one. With the augmented states, the dynamic model of original multi-agent system can be rewritten as follows:

$$\begin{aligned} x^*_{i,k} &= A^*_{i,k-1}x^*_{i,k-1} + w^*_{i,k-1}, \\ y^*_{i,k} &= y_{i,k} = C^*_{i,k}x^*_{i,k} + v^*_{i,k}, \end{aligned} \tag{12.8}$$

where (supposing $\mathcal{N}_i = \{j_1, j_2, \dots, j_{n_i}\}$)

$$A^*_{i,k-1} = \begin{bmatrix} A_{i,k-1} & G_{i,j_1}y_{j_1,k-1} & \cdots & G_{i,j_{n_i}}y_{j_{n_i},k-1} \\ 0_{1\times n} & 1 & \cdots & 0 \\ 0_{1\times n} & 0 & \cdots & 0 \\ \vdots & \vdots & \ddots & \vdots \\ 0_{1\times n} & 0 & \cdots & 1 \end{bmatrix} \tag{12.9}$$

$$\begin{aligned} C^*_{i,k} &= \left[C_{i,k}, 0_{m\times 1}, 0_{m\times 1}, \dots, 0_{m\times 1}\right]_{m\times(n+n_i)} \\ w^*_{i,k-1} &= \begin{bmatrix} I_{n\times n} \\ 0_{1\times n} \\ 0_{1\times n} \\ \vdots \\ 0_{1\times n} \end{bmatrix}_{(n+n_i)\times n} \times w_i(k-1) \\ v^*_{i,k} &= v_{i,k}, \end{aligned} \tag{12.10}$$

and

$$
\begin{aligned}
w^*_{i,k} &\sim \mathrm{N}(0, Q^*_{i,k}),\\
v^*_{i,k} &\sim \mathrm{N}(0, R^*_{i,k}),\\
Q^*_{i,k} &= \begin{bmatrix} Q_{i,k} & 0_{n\times n_i} \\ 0_{n_i\times n} & 0_{n_i\times n_i} \end{bmatrix}_{(n+n_i)\times(n+n_i)}\\
R^*_{i,k} &= R_{i,k}.
\end{aligned}
\tag{12.11}
$$

Then with this modified model, the standard Kalman filter can be applied directly. For clarity, the following equations show the filtering process:

$$
\begin{aligned}
\bar{P}^*_{i,k} &= A^*_{i,k-1}\widehat{P}^*_{i,k-1}[A^*_{i,k-1}]^{\mathrm{T}} + Q^*_{i,k-1},\\
K^*_{i,k} &= \bar{P}^*_{i,k}[C^*_{i,k}]^{\mathrm{T}}\{C^*_{i,k}\bar{P}^*_{i,k}[C^*_{i,k}]^{\mathrm{T}} + R^*_{i,k}\}^{-1}\\
\bar{x}^*_{i,k} &= A^*_{i,k-1}\widehat{x}^*_{i,k-1},\\
\widehat{x}^*_{i,k} &= \bar{x}^*_{i,k} + K^*_{i,k}(y^*_{i,k} - C^*_{i,k}\bar{x}^*_{i,k}),\\
\widehat{P}^*_{i,k} &= [I - K^*_{i,k}C^*_{i,k}]\bar{P}^*_{i,k}[I - K^*_{i,k}C^*_{i,k}]^{\mathrm{T}} +\\
&\quad K^*_{i,k}R^*_{i,k}K^*_{i,k},
\end{aligned}
\tag{12.12}
$$

where $\bar{P}^*_{i,k}$ and $\widehat{P}^*_{i,k}$ denote the prior error covariance matrix and the posterior error covariance matrix at time k, respectively. Here, $\bar{x}^*_{i,k}$ and $\widehat{x}^*_{i,k}$ denote the prior estimation and the posterior estimation of states at time k, respectively. The Kalman gain $K^*_{i,k}$ is the gain matrix to modify the prior estimation with outputs.

Algorithm 1 (DFL-decentralized filter for linear system).
Input: coefficient matrices, neighborhood indices, and outputs of every agent
Output: estimated values of states and coupled parameters
1 Extend dimensions of states
2 Update the state equations using the new extending states
3 Initialize estimation values and calculate the initial error covariance matrix
4 Initialize simulation steps T
5 **for** $k = 2$ to T **do**
6 **for** $i = 1$ to N **do**
7 Form matrix $A^*_{i,k}$ with neighbor outputs y_{j_l} via (9)
8 Calculate the next prior error covariance $\bar{P}^*_{i,k}$ via (12)
9 Calculate the Kalman gain matrix $K^*_{i,k}$ via (12)
10 Time update—calculate $\bar{x}^*_{i,k}$ from $\widehat{x}^*_{i,k-1}$ via (12)
11 Measurement update—calculate $\widehat{x}^*_{i,k}$ with $K^*_{i,k}$ via (12) as new measurement $y^*_{i,k}$ is available
12 Calculate the posterior error covariance $\widehat{P}^*_{i,k}$ via (12)
13 **end for**
14 **end for**
15 Output and display sequences of $\widehat{x}^*_{i,k}$ which contains estimation of coupled parameters and states.

12.3.2 Model 2

This algorithm is based on the standard Kalman filter and a decentralized one. With the augmented states, the dynamic model of original multi-agent system can be rewritten as follows:

$$\begin{aligned} x^*_{i,k} &= A^*_{i,k-1} x^*_{i,,k-1} + w^*_{i,,k-1}, \\ y^*_{i,k} &= y_{i,k} = C^*_{i,k} x^*_{i,k} + v^*_{i,k}, \end{aligned} \tag{12.13}$$

where

$$A^*_{i,k-1} = \begin{bmatrix} A_{i,k-1} \, x_{j_1,k-1} \, x_{j_2,k-1} \cdots x_{j_{n_i},k-1} \\ 0_{1\times n} & 1 & 0 & \cdots & 0 \\ 0_{1\times n} & 0 & 1 & \cdots & 0 \\ \vdots & \vdots & \vdots & \ddots & \vdots \\ 0_{1\times n} & 0 & 0 & \cdots & 1 \end{bmatrix} \tag{12.14}$$

Except for matrix $A^*_{i,k-1}$, other terms are the same with Model 1. The new problem is that some terms of $A^*_{i,k-1}$ are unknown; hence, the standard Kalman filter cannot be applied to the modified model directly. Since agents can communicate with each other, the estimated values of these unknown ones can be transmitted to replace the real values. That is, in the filtering process of Model 1, the matrix $A^*_{i,k-1}$ is changed to

$$A^*_{i,k-1} = \begin{bmatrix} A_{i,k-1} \, \widehat{x}_{j_1,k-1} \, \widehat{x}_{j_2,k-1} \cdots \widehat{x}_{j_{n_i},k-1} \\ 0_{1\times n} & 1 & 0 & \cdots & 0 \\ 0_{1\times n} & 0 & 1 & \cdots & 0 \\ \vdots & \vdots & \vdots & \ddots & \vdots \\ 0_{1\times n} & 0 & 0 & \cdots & 1 \end{bmatrix}, \tag{12.15}$$

where $\widehat{x}_{j_1,k-1}, \widehat{x}_{j_2,k-1}, \ldots$ are the parts of $\widehat{x}^*_{j_1,k-1}, \widehat{x}^*_{j_2,k-1}, \ldots$. With this algorithm, estimation of states and unknown parameters can be achieved.

Algorithm 2 (DFL2-decentralized filter for linear system 2).
Input: coefficient matrices, neighborhood indices, and outputs of every agent
Output: estimated values of states and coupled parameters
Extend dimensions of states
Update the state equations using the new extending states
Initialize estimation values and calculate the initial error covariance matrix
Initialize simulation steps T
for $k = 2$ to T **do**
  for $i = 1$ to N **do**
   Form matrix $A^*_{i,k}$ via (15) with neighbor outputs $\widehat{x}_{j_p}$ $(j_p \in \mathcal{N}_i)$
   Calculate the next prior error covariance $\bar{P}^*_{i,k}$ via (12)
   Calculate the Kalman gain matrix $K^*_{i,k}$ via (12)
   Time update—calculate $\bar{x}^*_{i,k}$ from $\widehat{x}^*_{i,k-1}$ via (12)
   Measurement update—calculate $\widehat{x}^*_{i,k}$ with $K^*_{i,k}$ via (12) as new measurement $y^*_{i,k}$ is available
   Get the posterior error covariance matrix $\widehat{P}^*_{i,k}$ via (12)
  end for
end for

15 Output and display sequences of $\widehat{x}^*_{i,k}$ which contains estimation of coupled parameters and states.

12.3.3 Model 3

Since the model is nonlinear, the extended Kalman filter can be considered as the basic algorithm. First, using the same notations $x^*_{i,k}$, $w^*_{i,k}$, $v^*_{i,k}$ as in Model 1, we obtain the modified model of agent dynamics

$$\begin{aligned} x^*_{i,k} &= f^*_i(x^*_{i,k}, w^*_{i,k-1}), \\ y^*_{i,k} &= y_{i,k} = h^*_i(x^*_{i,k}, v^*_{i,k-1}), \end{aligned} \tag{12.16}$$

where f^*_i and h^*_i are two functions defined by

$$x^*_{i,k} = \begin{bmatrix} f_i(\cdots) + \sum\limits_{j_p \in \mathcal{N}_i} q_{i,j_p} g_{i,j_p}(y_{j_p,k-1}) \\ q_{i,j_1} \\ q_{i,j_2} \\ \vdots \\ q_{i,j_{n_i}} \end{bmatrix} \tag{12.17}$$

$$y^*_{i,k} = y_{i,k} = h_i(x_{i,k}, v_{i,k}). \tag{12.18}$$

Then we find that the problem of simultaneously estimating the unknown states and the coupling parameters can be transformed to a filter problem of the new model, i.e., estimating the augmented states $x^*_{i,k}$. The new model is a nonlinear dynamic model; hence, the extended Kalman filter algorithm [20] can be applied in this modified model. For clarity, the filtering process is shown as follows.

The filtering process:
(1) Calculate partial derivatives matrices of functions

$$A^*_{i,k-1} = \frac{\partial f^*_i}{\partial x^*_i}|_{\hat{x}^*_{i,k-1}}, \tag{12.19}$$

$$G^*_{i,k-1} = \frac{\partial f^*_i}{\partial w_i}|_{\hat{x}^*_{i,k-1}}, \tag{12.20}$$

where $\hat{x}^*_{i,k-1}$ denotes posterior estimation of $x^*_{i,k-1}$.
(2) Time update

$$
\begin{aligned}
\bar{P}^*_{i,k} &= A^*_{i,k-1}\hat{P}^*_{i,k-1}[A^*_{i,k-1}]^{\mathrm{T}} + \\
&\quad G^*_{i,k-1}Q_{k-1}[G^*_{i,k-1}]^{\mathrm{T}} \\
\bar{x}^*_{i,k} &= f^*_i(\hat{x}^*_{i,k-1}, 0_{n\times n})
\end{aligned}
\tag{12.21}
$$

(3) Calculate partial derivatives matrices of functions

$$C^*_{i,k} = \frac{\partial h^*_i}{\partial x^*_i}|_{\bar{x}^*_{i,k}}, \tag{12.22}$$

$$J^*_{i,k} = \frac{\partial h^*_i}{\partial v_i}|_{\bar{x}^*_{i,k}}, \tag{12.23}$$

where $\bar{x}^*_{i,k}$ is the prior estimation.
(4) Measurement update

$$
\begin{aligned}
K^*_{i,k} &= \bar{P}^*_{i,k}[C^*_{i,,k}]^{\mathrm{T}}\{C^*_{i,k}\bar{P}^*_{i,k}[C^*_{i,k}]^{\mathrm{T}} + R_k\}^{-1} \\
\hat{x}^*_{i,k} &= \bar{x}^*_{i,k} + K^*_{i,k}[y^*_{i,k} - h^*_i(\bar{x}^*_{i,k}, 0_{m\times m})] \\
\hat{P}^*_{i,k} &= [I - K^*_{i,k}C^*_{i,k}]\bar{P}^*_{i,k}.
\end{aligned}
\tag{12.24}
$$

Algorithm (DFNL1-decentralized filter for nonlinear system 1).
Input: functions, neighborhood indices, and outputs of every agent
Output: estimated values of states and coupled parameters
1 Extend dimension of states
2 Update the state equations using the new extending states
3 Initialize estimation values and calculate the initial error covariance matrix
4 Initialize simulation steps T
5 **for** $k = 2$ to T **do**
6 **for** $i = 1$ to N **do**
7 From $A^*_{i,k}$ and $G^*_{i,k-1}$ via (19) and (20) with $\hat{x}^*_{i,k-1}$ and y_{j_p} $(j_p \in \mathcal{N}_i)$
8 Calculate the next prior error covariance $\bar{P}^*_{i,k}$ via (21)
9 Time update—calculate $\bar{x}^*_{i,k}$ from $\hat{x}^*_{i,k-1}$ via (21)
10 Calculate matrices $C^*_{i,k}$ and $J^*_{i,k}$ via (22) and (23)
11 Calculate the Kalman gain matrix $K^*_{i,k}$ via (24)
12 Measurement update—calculate $\widehat{x}^*_{i,k}$ with $K^*_{i,k}$ via (24) as new measurement $y^*_{i,k}$ is available
13 Get the posterior error covariance matrix $\widehat{P}^*_{i,k}$ via (24)
14 **end for**
15 **end for**
16 Output and display sequences of $\widehat{x}^*_{i,k}$ which contains estimation of coupled parameters and states.

12.3.4 Model 4

Rewriting the dynamic model, we can obtain a similar model as given above. The difference is the state equation as follows:

$$x_{i,k}^* = \begin{bmatrix} f_i(\cdots) + \sum\limits_{j_p \in \mathcal{N}_i} q_{i,j_p} g_{i,j_p}(x_{j_p,k-1}) \\ q_{i,j_1} \\ q_{i,j_2} \\ \vdots \\ q_{i,j_{n_i}} \end{bmatrix} \tag{12.25}$$

There are terms which are different from output coupling of Model 3, such as $g_{i,j_1}(x_{j_1,k-1})$. That is, when partial derivative matrices of functions are calculated, these terms are unknown which are different from terms only including outputs. Hence, some modifications are necessary to extend the algorithm. Since posterior estimation values are close to the true values, we use the posterior estimation instead.

$$A_{i,k-1}^* = \frac{\partial f_i^*}{\partial x_i^*}|_{\hat{x}_{i,k-1}^*}, \tag{12.26}$$

$$G_{i,k-1}^* = \frac{\partial f_i^*}{\partial w_i}|_{\hat{x}_{i,k-1}^*}. \tag{12.27}$$

Note that unknown terms should be replaced by their posterior estimates. For example, $g_{i,j_1}(x_{j_1,k-1})$ should be replaced by $g_{i,j_1}(\hat{x}_{j_1,k-1})$. The rest of estimating process is similar to that in Model 3.

Algorithm (DFNL2-Decentralized Filter for Nonlinear System 2).
Input: functions, neighborhood indices, and outputs of every agent
Output: estimated values of states and coupled parameters

Extend dimensions of states
Update the state equations using the new extending states
Initialize estimation values and calculate the initial error covariance matrix
Initialize simulation steps T
for $k = 2$ to T **do**
  for $i = 1$ to N **do**
    From $A_{i,k}^*$ and $G_{i,k-1}^*$ via (26) and (27) with $\hat{x}_{i,k-1}^*$
    Calculate the next prior error covariance $\bar{P}_{i,k}^*$ via (21)
    Time update—calculate $\bar{x}_{i,k}^*$ from $\hat{x}_{i,k-1}^*$ via (21)
    Calculate matrices $C_{i,k}^*$ and $J_{i,k}^*$ via (22) and (23)
    Calculate the Kalman gain matrix $K_{i,k}^*$ via (24)
    Measurement update—calculate $\hat{x}_{i,k}^*$ with $K_{i,k}^*$ via (24) as new measurement $y_{i,k}^*$ is available
    Get the posterior error covariance matrix $\widehat{P}_{i,k}^*$ via (24)
  end for
end for
Output and display sequences of $\hat{x}_{i,k}^*$ which contains estimation of coupled parameters and states.

12.4 Decentralized Controller Design

Controller design will be easy if the coupling parameters are exactly known a priori. However, since the coupling parameters are unknown in our settings, corresponding control parameters cannot be decided for the purpose of compensating the unknown coupling effects. To resolve this problem, one natural way is to regard the estimation of those parameters instead as their true values. This idea is in fact the so-called certainty equivalence principle, which is widely used in adaptive controller design. One can imagine that controllers based on this principle may yield satisfactory control performance if the parameter estimates are very close to the true parameters.

For testing estimation effects of filter algorithms, decentralized adaptive regulators are designed hereinafter. That is, control inputs are designed to regulate the states to zero based on the parameter estimates. For simplicity and clarity, coefficient matrices of inputs are assumed to be identity matrices, that is, $B_{i,k-1}$ is equal to I.

For Model 1, the control input of agent i can be designed as follows:

$$u_{i,k-1} = -A_{i,k-1}\hat{x}_{i,k-1} - \sum_{j_p \in \mathcal{N}_i} \widehat{q}_{i,j_p} y_{j_p,k-1}, \tag{12.28}$$

where $\hat{x}_{i,k-1}$ and $\widehat{q}_{i,j_p}$ are terms of $\hat{x}^*_{i,k-1}$.

For Model 2, we have

$$u_{i,k-1} = -A_{i,k-1}\widehat{x}_{i,k-1} - \sum_{j_p \in \mathcal{N}_i} \widehat{q}_{i,j_p} \hat{x}_{j_p,k-1}. \tag{12.29}$$

For Model 3, we have

$$u_{i,k-1} = \\ - f_i(\hat{x}_{i,k-1}, 0_{n\times 1}) - \sum_{j_p \in \mathcal{N}_i} \widehat{q}_{i,j_p} g_{i,j_p}(y_{j_p,k-1}), \tag{12.30}$$

where $0_{n\times 1}$ replaces the noise $w_{i,k-1}$ since the noise $w_{i,k-1}$ cannot be known when control is designed.

For Model 4, we have

$$u_{i,k-1} = \\ - f_i(\hat{x}_{i,k-1}, 0_{n\times 1}) - \sum_{j_p \in \mathcal{N}_i} \widehat{q}_{i,j_p} g_{i,j_p}(\hat{x}_{j_p,k-1}). \tag{12.31}$$

If the estimation results are good enough, then control inputs designed as above can regulate states to zero approximately. In this way, the closed-loop signals may not diverge; consequently, the estimation results of filtering algorithms introduced before can be tested via simulations.

12.5 Remarks on Stability Analysis

12.5.1 Linear Situations

The filtering process applied in the new Model 1 is essentially the standard Kalman filter for the augmented system. Though there are some unknown coupling parameters, the new system obtained by augmenting states can be considered as a normal multi-agent system (8) without uncertain couplings. In the new augmented system model, the system matrix of agent i is

$$A^*_{i,k-1} = \begin{bmatrix} A_{i,k-1}\, y_{j_1,k-1}\, y_{j_2,k-1} \cdots y_{j_{n_i},k-1} \\ 0_{1\times n} \quad 1 \quad 0 \quad \cdots \quad 0 \\ 0_{1\times n} \quad 0 \quad 1 \quad \cdots \quad 0 \\ \vdots \quad \vdots \quad \vdots \quad \ddots \quad \vdots \\ 0_{1\times n} \quad 0 \quad 0 \quad \cdots \quad 1 \end{bmatrix} \tag{12.32}$$

That is, all elements of matrix A_i^* are essentially known for each agent, despite that it is time-varying and random according to the practical measurements. So conditions under which this algorithm is stable can be given by the conditions to stabilize a standard Kalman filter with time-varying system matrix.

Theorem 1 *For the dynamical system (1), Algorithm 1 (DFL) is optimal in the sense that it provides best linear unbiased estimation for the unknown coupling parameters and the unknown states, provided that the initial conditions are unbiased.*

We shall remark that closed-loop stability analysis of Kalman filter with time-varying matrices is a challenging theoretical problem, which is still open up to now to the best knowledge of the authors since sufficient and necessary conditions to guarantee closed-loop stability of time-varying Kalman filter are not reported in the literature, although the counterpart problem has been extensively studied for Kalman filter with time-invariant matrices.

According to [21], we have the following results by regarding the signals $y_{i,k}$ in the deterministic sense:

Theorem 2 *If the new dynamical system (8) is uniformly completely observable and uniformly completely controllable, and $\hat{P}^*_{i,0} \geq 0$, then the discrete-time linear filter given in Algorithm 1 (DFL) is uniformly asymptotically stable.*

See [21] for the mathematical definitions of uniformly completely observable, uniformly completely controllable, and uniformly asymptotically stable. Theorem 2 is one straightforward conclusion of Theorem 7.5 of [21], whose proof is rather involved; hence, we do not repeat the details of the proof. From this theorem, we may have the following results (for each bounded sample of $\{y_k\}$).

Corollary 1 *For the dynamical system (1), the decentralized filter given in Algorithm 1 (DFL) is uniformly asymptotically stable if the following conditions hold for the dynamical system (8):*

For a positive integer N_1, there exist constants $\alpha_1 > 0$ and $\beta_1 > 0$ such that

$$\begin{aligned}\alpha_1 I \le &\sum_{m_1=k-N_1+1}^{k} A_i^*(k \sim m_1) Q_{i,m_1-1}^* \times \\ &[A_i^*(k \sim m_1)]^{\mathrm{T}} \le \beta_1 I,\end{aligned} \tag{12.33}$$

where $k \ge N_1$ and

$$A_i^*(k \sim m_1) = A_{i,k-1}^* A_{i,k-2}^* \cdots A_{i,m_1}^*. \tag{12.34}$$

For a positive integer N_2, there exist constants $\alpha_2 > 0$ and $\beta_2 > 0$ such that

$$\begin{aligned}\alpha_2 I \le &\sum_{m_2=k-N_2+1}^{k} (A_i^*(m_2 \sim k))^{\mathrm{T}} (R_{i,m_2}^*)^{-1} \times \\ &A_i^*(m_2 \sim k) \le \beta_2 I,\end{aligned} \tag{12.35}$$

where $k \ge N_2$ and

$$A_i^*(m_2 \sim k) = (A_{i,k-1}^* A_{i,k-2}^* \cdots A_{i,m_2}^*)^{-1}. \tag{12.36}$$

$\hat{P}_{i,0}^* \ge 0$ can be obtained by applying $\hat{x}_{i,0}^*$ appropriately.

Corollary 2 *For the dynamical system (1), let $P_{i,k}^{1*}$ and $P_{i,k}^{2*}$ be any solutions obtained from the filtering process of Algorithm 1 (DFL) starting from different initial conditions $P_{i,0}^{1*}$ and $P_{i,0}^{2*}$. Then, we can conclude that*

$$\| P_{i,k}^{1*} - P_{i,k}^{2*} \| \le c_1 \exp^{-c_2 k} \| P_{i,0}^{1*} - P_{i,0}^{2*} \| \to 0. \tag{12.37}$$

Note that Theorem 2 regards the matrices $A_{i,k}^*$ as time-varying matrices, which in fact form a stochastic process. Thus it would be desirable to reflect the stochastic nature of $A_{i,k}^*$ in theoretical studies, which would be much more challenging since statistical properties like probabilistic distributions are not available or even impossible to establish explicitly for those involved stochastic processes. As far as we know, there are very limited researches along this direction, partially due to the intrinsic mathematical difficulties involved. However, with the concepts of stochastic boundedness of signals, it is possible to establish the so-called stochastic BIBO stability, under some conditions of weakly stochastically boundedness and stochastic stationary properties for the involved stochastic processes, via the techniques developed in [22]. To save space, we do not make such detailed analysis.

As to linear Model 2 with state couplings, since each agent has no access to the states of its neighbor agents, the matrices $A^*_{i,k}$ can only be approximated by using posterior estimates of neighbor agents. This subtle difference from Model 1 makes the closed-loop stability analysis in this case more difficult than that for Model 1. More efforts are needed to establish some rigorous theoretical results in this case.

12.5.2 Nonlinear Situations

Though there are unknown coupling parameters among agents, the filtering process applied in the modified Model 3 relies on the extended Kalman filter. Note that in this process, we need to calculate the partial derivatives matrices of functions:

$$G^*_{i,k-1} = \frac{\partial f^*_i}{\partial w_i}|_{\hat{x}^*_{i,k-1}}, \tag{12.38}$$

$$A^*_{i,k-1} = \begin{bmatrix} \frac{\partial f_i}{\partial x_i}|_{\hat{x}_{i,k-1}} & G_{i,j,k-1} \\ 0_{n_i \times n} & I_{n_i \times n_i} \end{bmatrix}, \tag{12.39}$$

where

$$G_{i,j} = \begin{bmatrix} g_{i,j_1}(y_{j_1,k-1}) \\ g_{i,j_2}(y_{j_2,k-1}) \\ \vdots \\ g_{i,j_{n_i}}(y_{j_{n_i},k-1}) \end{bmatrix}^{\mathrm{T}}. \tag{12.40}$$

We can find that these partial derivatives matrices of Model 3 are all available to agent i; hence, stability analysis of extended Kalman filter [23] may be applied directly, and we have the following results.

Theorem 3 *For the dynamical system (3) with decentralized filtering Algorithm 3 (DFNL1), the state estimation error of agent i is exponentially bounded in mean square sense if the following conditions hold for each agent i:*

(1) There are positive real numbers $\bar{a}$, $\bar{c}$, $\bar{p}$, $\underline{p}$, $\underline{q}$, $\underline{r}$ such that the following bounds on various matrices are fulfilled for every k:

$$\|A^*_{i,k}\| \leq \bar{a}, \tag{12.41}$$

$$\|C^*_{i,k}\| \leq \bar{c}, \tag{12.42}$$

$$\underline{p}I \leq P^*_{i,k} \leq \bar{p}I, \tag{12.43}$$

$$\underline{q}I \leq Q^*_{i,k}, \tag{12.44}$$

$$\underline{r}I \le R^*_{i,k}. \tag{12.45}$$

(2) $A^*_{i,k}$ is nonsingular for every $k \ge 0$.

(3) There are positive real numbers $\epsilon_\phi, \epsilon_\psi, \kappa_\phi, \kappa_\psi$ such that the nonlinear functions ϕ, ψ

$$\phi(z, \hat{z}, w) = f_i^*(z, w) - f_i^*(\hat{z}, w) - \frac{\partial f_i^*}{\partial z}\mid_{\hat{z}} (z - \hat{z}), \tag{12.46}$$

$$\psi(z, \hat{z}, v) = h_i^*(z, v) - h_i^*(\hat{z}, v) - \frac{\partial h_i^*}{\partial z}\mid_{\hat{z}} (z - \hat{z}), \tag{12.47}$$

satisfy

$$\|\phi(z, \hat{z}, w)\| \le \kappa_\phi \|z - \hat{z}\|^2, \tag{12.48}$$

$$\|\psi(z, \hat{z}, v)\| \le \kappa_\psi \|z - \hat{z}\|^2, \tag{12.49}$$

for any $z, \hat{z}$ with $\|z - \hat{z}\| \le \epsilon_\phi$ and $\|z - \hat{z}\| \le \epsilon_\psi$, respectively.

Theorem 4 *For the dynamical system (3) with decentralized filtering Algorithm 3 (DFNL1), the state estimation error of agent i is exponentially bounded in mean square if the following conditions hold:*

(1) Matrices $Q^*_{i,k}$ and $R^*_{i,k}$ are uniformly bounded below, i.e., there exist positive constants $\underline{q}$ and $\underline{r}$ such that

$$\underline{q}I \le Q^*_{i,k}, \tag{12.50}$$

$$\underline{r}I \le R^*_{i,k}, \tag{12.51}$$

(2) The nonlinear system satisfies the observability rank condition, i.e.,

$$U(z_k) = \begin{bmatrix} & \frac{\partial h_i^*}{\partial z}\mid_{z_k} & \\ \frac{\partial h_i^*}{\partial z}\mid_{z_{k+1}} & \frac{\partial f_i^*}{\partial z}\mid_{z_k} & \\ & \vdots & \\ \frac{\partial h_i^*}{\partial z}\mid_{z_{k+\bar{n}-1}} & \frac{\partial f_i^*}{\partial z}\mid_{z_{k+\bar{n}-2}} \cdots & \frac{\partial f_i^*}{\partial z}\mid_{z_k} \end{bmatrix}, \tag{12.52}$$

has full rank $\bar{n}$ at z_k, for every $z_k \in \mathcal{K}$, where $\mathcal{K} \in \mathbf{R}^{\bar{n}}$ is a compact set. Here $\bar{n} = n + n_i$ is the dimension of the augmented state space.

(3) The nonlinear functions f_i^* and h_i^*, $k = 1, 2, \ldots$ are twice differentiable and $\frac{\partial f_i^*}{\partial z}\mid_{z_k} \ne 0$. (For brevity, here z_k denotes $x^*_{i,k}$.)

12.6 Simulation Studies

12.6.1 Model 1

(1) Simulation 1 (Sim l)

We choose a multi-agent system with three agents, each of which has the following model:

$$
\begin{aligned}
x_{i,k} &= A_i x_{i,k-1} + w_{i,k-1} + \\
&\sum_{j_p \in \mathcal{N}_i} q_{i,j_p}(y_{j_p,k-1}), \\
y_{i,k} &= C_i \times x_{i,k} + v_{i,k},
\end{aligned} \tag{12.53}
$$

where

$$
\begin{aligned}
A_i &= \begin{bmatrix} 0.8 & 0 \\ 0 & 0.8 \end{bmatrix}, \\
C_i &= \begin{bmatrix} 0.4 & 0 \\ 0 & 0.4 \end{bmatrix}, \\
q_{i,j_p} &= 0.6, \\
Q_{i,k} = R_{i,k} &= \begin{bmatrix} 0.04 & 0 \\ 0 & 0.04 \end{bmatrix},
\end{aligned} \tag{12.54}
$$

$$
\mathcal{N}_i = \begin{cases} \{i-1\}, & \text{if } i = 2, 3, \\ \{3\}, & \text{if } i = 1. \end{cases} \tag{12.55}
$$

Filtering process is initialized as

$$
\hat{x}^*_{i,1} = \mathrm{E}[x^*_{i,1}], \tag{12.56}
$$

where

$$
\hat{q}_{i,j_p,1} = 6. \tag{12.57}
$$

Then, error covariance matrix of the first step can be calculated:

$$
\hat{P}^*_{i,1} = \mathrm{E}[(x^*_{i,1} - \hat{x}^*_{i,1})[x^*_{i,1} - \hat{x}^*_{i,1}]^{\mathrm{T}}]. \tag{12.58}
$$

Simulation results for this model with Algorithm 1 (DFL) are given in Figs. 12.1, 12.2, and 12.3.

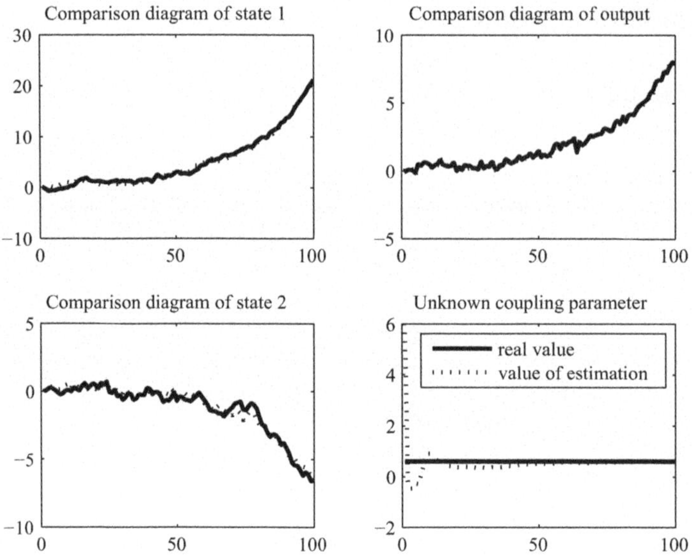

Fig. 12.1 Simulation (Sim1) of Model 1: the first agent

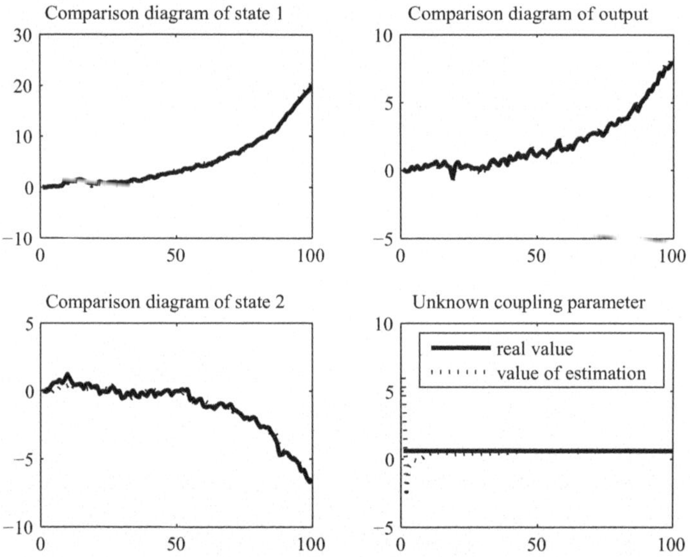

Fig. 12.2 Simulation (Sim1) of Model 1: the second agent

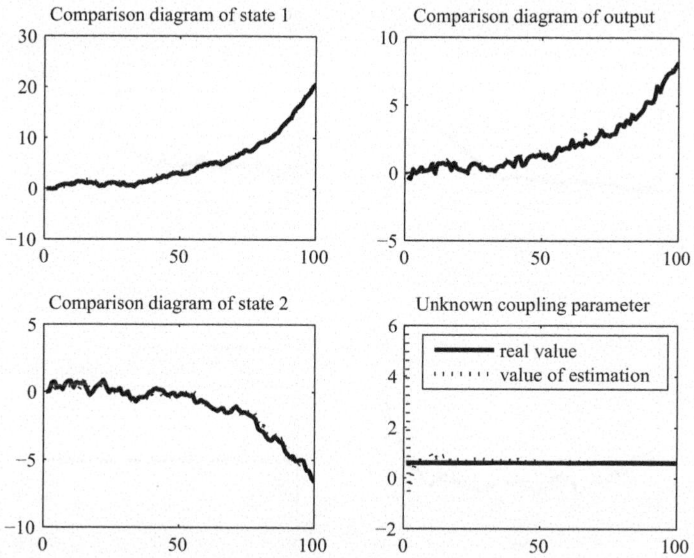

Fig. 12.3 Simulation (Sim1) of Model 1: the third agent

In each figure, the subfigure on top left shows the first element of states and the subfigure at bottom left shows the second. The top right subfigure shows the first element of outputs.

From these figures, one can see that the real values and the estimated values are approximately the same.

(2) Simulation 2 (Sim2)

We choose a collection of 100 agents, each of which is the same with the above model, to verify the algorithm. Since there are many agents to be shown, for brevity, only one agent is shown in Fig. 12.4 and others are similar to this agent.

For displaying the effectiveness of algorithm, two numerical indices are introduced:

$$
\begin{aligned}
P(q) &= \frac{1}{N}\sum_{i=1}^{N}(agent(i).qpre(T) - agent(i).q)^2, \\
P(x) &= \frac{1}{2N \times T}\sum_{i=1}^{N}\sum_{j=1}^{T}[(agent(i).xpre(1, j)- \\
&\quad agent(i).x(1, j))^2 + (agent(i).xpre(2, j)- \\
&\quad agent(i).x(2, j)]^2. \qquad (12.59)
\end{aligned}
$$

That is, $P(q)$ represents mean value of error covariance estimation about q in the last step (at time T) and $P(x)$ denotes mean value of error covariance estimation about

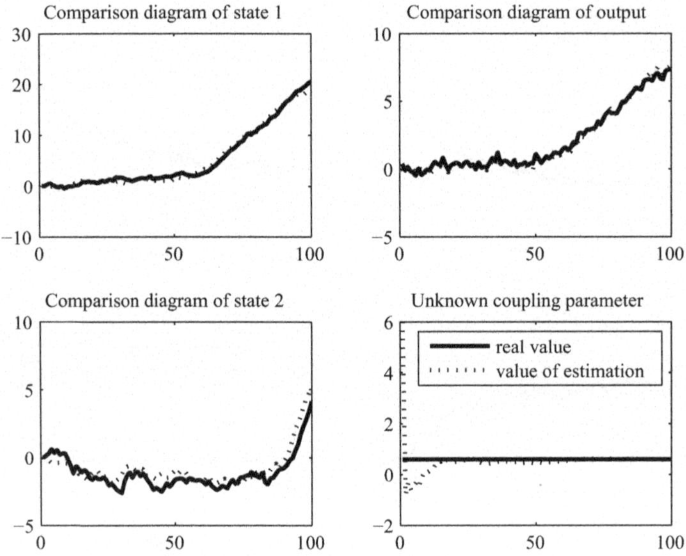

Fig. 12.4 Simulation (Sim2) of Model 1: the first agent

Table 12.1 Indices of DFL (Sim2)

$P(q)$	$P(x)$
0.0001	0.0228

x of every step. Since states of every agent are two-dimensional, error covariance of every element of states is added.

As we can see, the indices in Table 12.1 are small compared with values of states and coupling parameters.

(3) Control Simulation (ContrSim)

From the above simulations, we may notice that the system without control may not be stable due to the coupling effects even if matrix A_i is stable. Now we use the corresponding system with control inputs to test the estimation effects. By the certainty equivalence principle, control signals which regulate states to zero can be calculated to test the estimation results. To show the effectiveness of control design based on estimated values, we take the following coefficient matrices and parameters:

$$A_i = \begin{bmatrix} 1 & 0 \\ 0 & 1 \end{bmatrix},$$
$$C_i = \begin{bmatrix} 0.5 & 0 \\ 0 & 0.5 \end{bmatrix},$$
$$q_{i,j_p} = 0.6, \tag{12.60}$$

and dimension of states, inputs, output is 2.

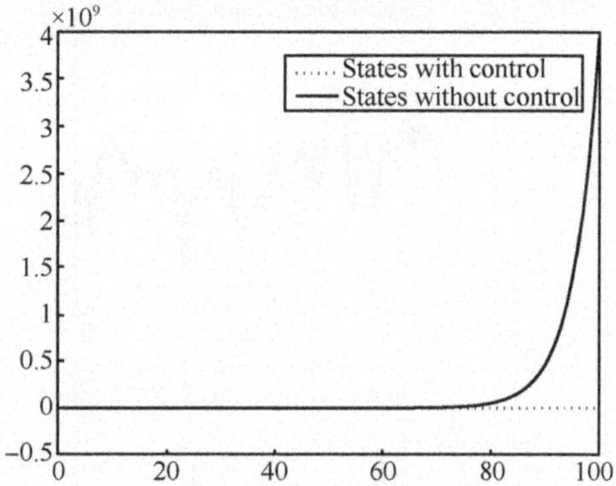

Fig. 12.5 States comparison with or without control

Filtering process is initialized as

$$\hat{x}^*_{i,1} = E[x^*_{i,1}], \tag{12.61}$$

where

$$\hat{q}_{i,j_p,1} = 6. \tag{12.62}$$

By designing control according to the certainty equivalence, the estimation effects can be shown in Fig. 12.5. Here, only one state of agent 1 is shown for simplicity and others are similar to Fig. 12.5. From Fig. 12.5, it is clearly shown that by control signals based on the estimated values, states can be regulated to be a signal close to zero signal.

12.6.2 *Model 2*

(1) Simulation 1 (Sim1)

Consider another multi-agent system where there are 5 agents and every agent can be described as the following two-dimensional dynamic model:

$$\begin{aligned} x_{i,k} &= A_{i,k-1}x_i + q_{i,j_p}x_{i-1,k-1} + w_{i,k-1}, \\ y_{i,k} &= C_{i,k}x_{i,k} + v_{i,k}. \end{aligned} \tag{12.63}$$

As we expect that this model is stable without inputs, here A_i, q_i and C_i can be chosen as follows:

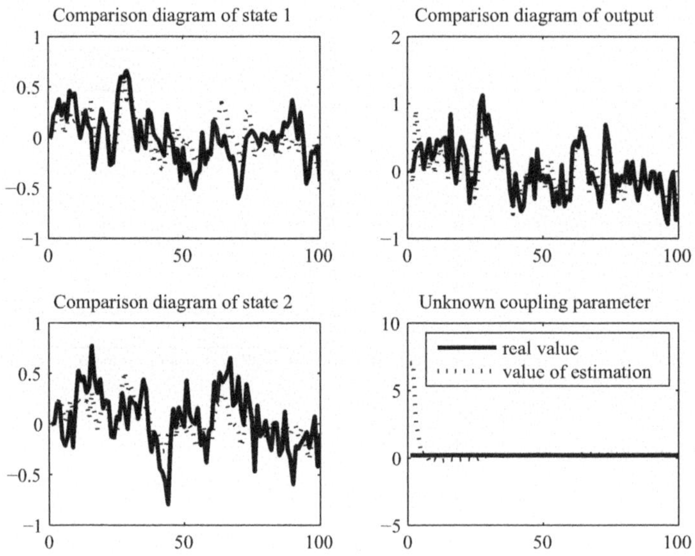

Fig. 12.6 Simulation (Sim1) of Model 2: the third agent

$$
\begin{aligned}
A_i &= \begin{bmatrix} 0.6 & 0 \\ 0 & 0.6 \end{bmatrix}, \\
C_i &= \begin{bmatrix} 0.8 & 0.7 \end{bmatrix}, \\
q_{i,j_p} &= 0.2,
\end{aligned} \tag{12.64}
$$

$$
\mathcal{N}_i = \begin{cases} \{i-1\}, & \text{if } i = 2, 3, 4, 5, \\ \{5\}, & \text{if } i = 1. \end{cases} \tag{12.65}
$$

For comparison of different methods, all methods are initialized as follows:

$$
\hat{x}^*_{i,1} = \mathrm{E}[x_{i,1}], \tag{12.66}
$$

where

$$
\hat{q}_{i,j_p,1} = 7. \tag{12.67}
$$

Simulation results with Algorithm 2 (DFL2) are depicted in Figs. 12.6, 12.7, and 12.8. Here, three agents are shown for conciseness.

In every figure, the subfigure in top left corner shows the first element of states and the subfigure in bottom left corner shows the second.

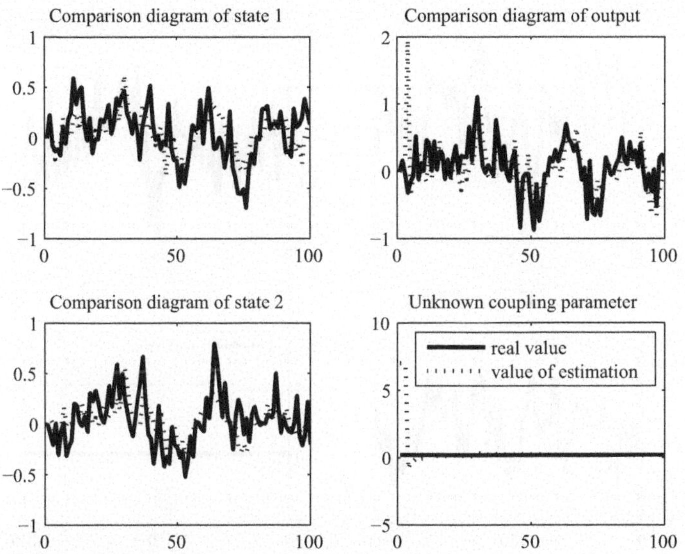

Fig. 12.7 Simulation (Sim1) of Model 2: the fourth agent

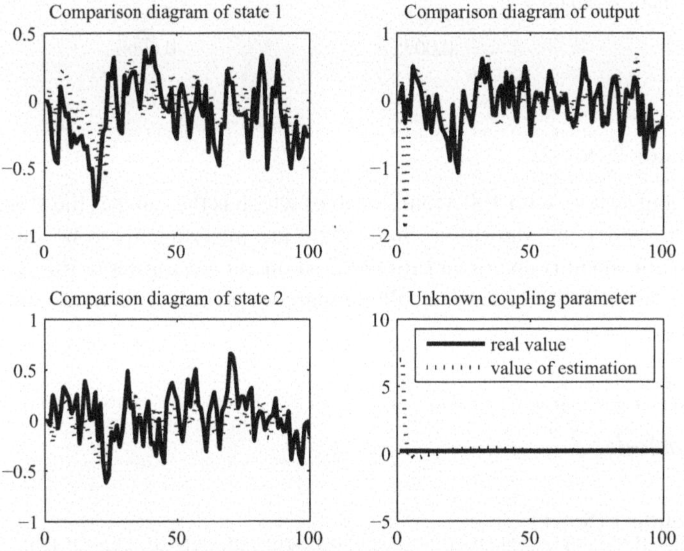

Fig. 12.8 Simulation (Sim1) of Model 2: the fifth agent

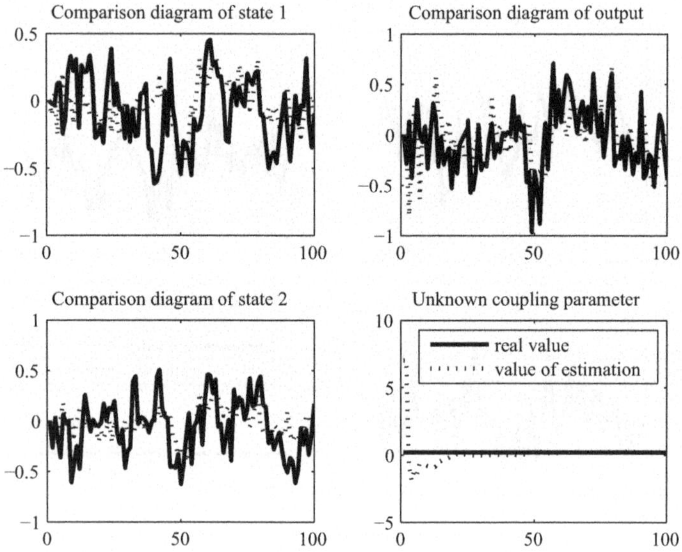

Fig. 12.9 Simulation (Sim2) of Model 1: the first agent

Table 12.2 Indices of DFL2 (Sim2)

$P(q)$	$P(x)$
0.0095	0.0066

(2) Simulation 2 (Sim2)

A multi-agent system with 100 agents, each of which is the same as the above model, is chosen to verify the algorithm. Since there are many agents to be shown, to be brief, only one agent is shown in Fig. 12.9 and others are similar to this agent.

The indices in Table 12.2 are small compared with values of states and coupling parameters.

12.6.3 Model 3

(1) Simulation 1 (Sim1)
A multi-agent one-dimensional system with 3 agents is chosen and the dynamic model of every agent is given as follows:

$$\begin{aligned} x_{i,k} &= -(x_{i,k-1})^2 + w_{i,k-1} + \\ &\sum_{j_p \in \mathcal{N}_i} q_{i,j_p} y_{j_p,k-1}, \\ y_{i,k} &= Cons \times (x_{i,k})^2 + v_{i,k}, \end{aligned} \tag{12.68}$$

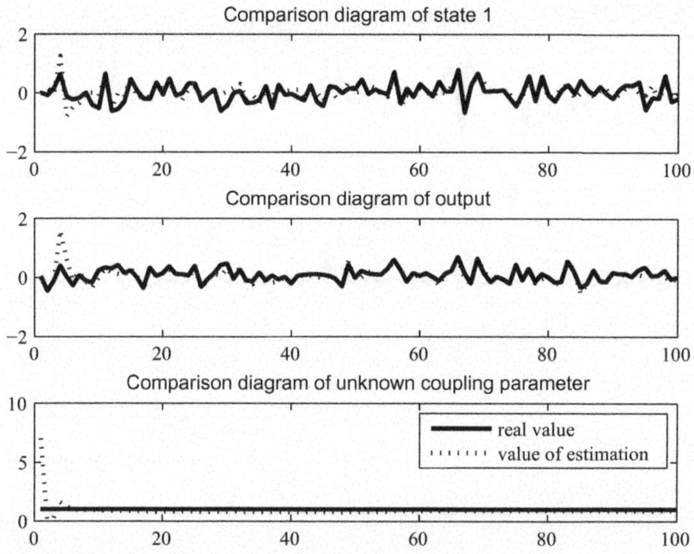

Fig. 12.10 Simulation (Sim1) of Model 3: the first agent

where we take constants $Cons = 1$ and $q_{i,j_p} = 1$:

$$\mathcal{N}_i = \begin{cases} \{i-1\}, & \text{if } i = 2, 3, \\ \{3\}, & \text{if } i = 1, \end{cases} \tag{12.69}$$

$$\begin{aligned} w_{i,k} &\sim \mathrm{N}(0, Q_{i,k}), \\ v_{i,k} &\sim \mathrm{N}(0, R_{i,k}), \end{aligned} \tag{12.70}$$

where $Q_{i,k} = 0.04$ and $R_{i,k} = 0.04$.

Filtering process is initialized as

$$\hat{x}^*_{i,1} = E[x^*_{i,1}], \tag{12.71}$$

where

$$\hat{q}_{i,j_p,1} = 7. \tag{12.72}$$

Then error covariance matrix of the first step can be calculated:

$$\hat{P}^*_{i,1} = \mathrm{E}[(x^*_{i,1} - \hat{x}^*_{i,1})[x^*_{i,1} - \hat{x}^*_{i,1}]^{\mathrm{T}}]. \tag{12.73}$$

The simulation results with Algorithm 3 (DFNL1) are depicted in Figs. 12.10, 12.11, and 12.12.

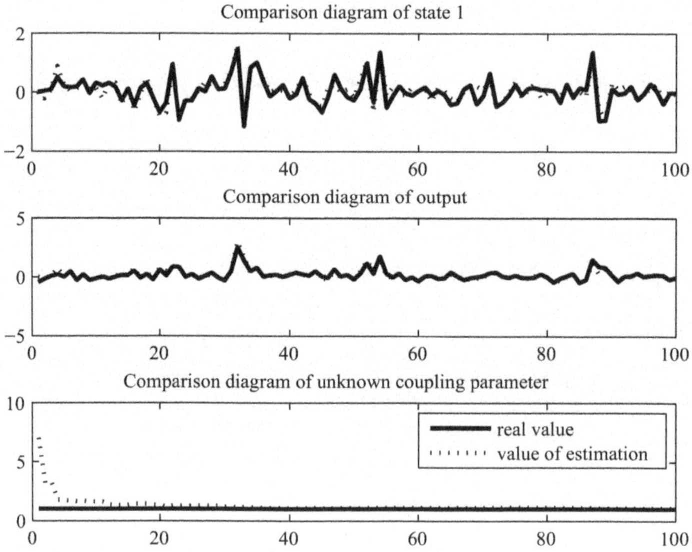

Fig. 12.11 Simulation (Sim1) of Model 3: the second agent

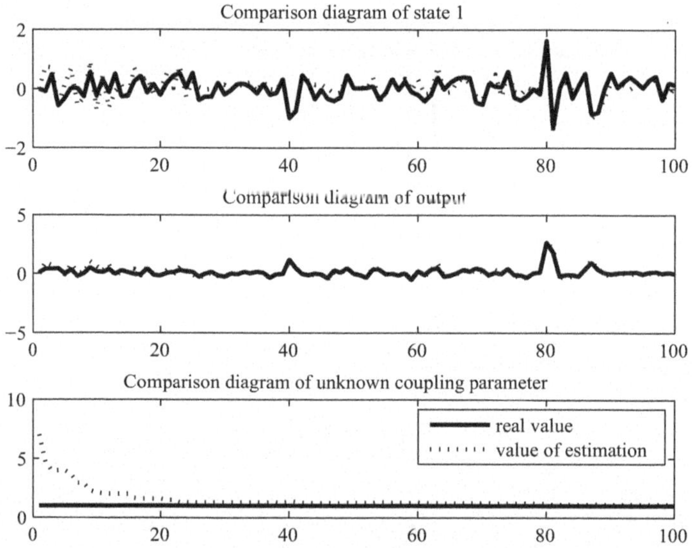

Fig. 12.12 Simulation (Sim1) of Model 3: the third agent

Table 12.3 Indices of DFNL1 (Sim2)

$P(q)$	$P(x)$
0.0590	0.0228

Through these figures, one can see that the estimated values and the real values are nearly equal.
(2) Simulation 2 (Sim2)
A multi-agent system with 100 agents, each of which is the same as the above model, is chosen to verify the algorithm.

The indices in Table 12.3 are small compared with values of states and coupling parameters.

12.6.4 Model 4

(1) Simulation 1 (Sim1)
A multi-agent one-dimensional system with 3 agents is chosen and the dynamic model of every agent is given as follows:

$$\begin{aligned} x_{i,k} &= -(x_{i,k-1})^2 + w_{i,k-1} + \\ &\sum_{j_p \in \mathcal{N}_i} q_{i,j_p} x_{j_p,k-1}, \\ y_{i,k} &= Cons \times (x_{i,k})^2 + v_{i,k}, \end{aligned} \tag{12.74}$$

where we take constants $Cons = 1.2$ and $q_{i,j_p} = 0.3$:

$$\mathcal{N}_i = \begin{cases} \{i-1\}, & \text{if } i = 2, 3, \\ \{3\}, & \text{if } i = 1, \end{cases} \tag{12.75}$$

$$\begin{aligned} w_{i,k} &\sim \mathrm{N}(0, Q_{i,k}), \\ v_{i,k} &\sim \mathrm{N}(0, R_{i,k}), \end{aligned} \tag{12.76}$$

where $Q_{i,k} = 0.04$ and $R_{i,k} = 0.04$.

Filtering process is initialized as

$$\hat{x}^*_{i,1} = \mathrm{E}[x^*_{i,1}], \tag{12.77}$$

where

$$\hat{q}_{i,j_p,1} = 6. \tag{12.78}$$

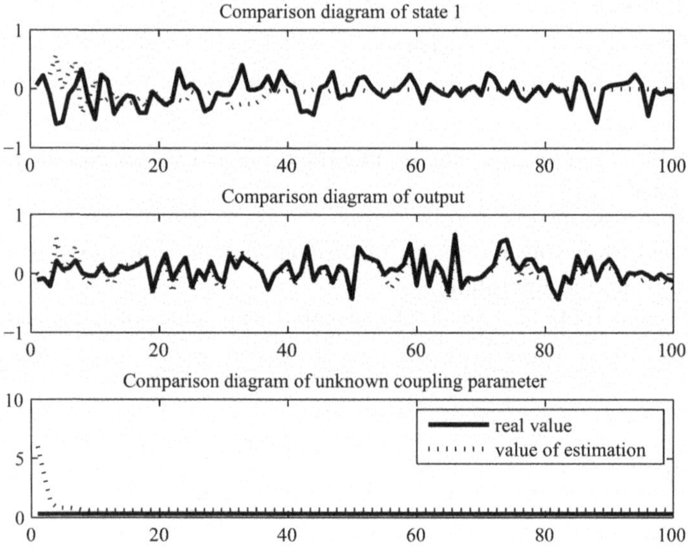

Fig. 12.13 Simulation (Sim1) of Model 4: the first agent

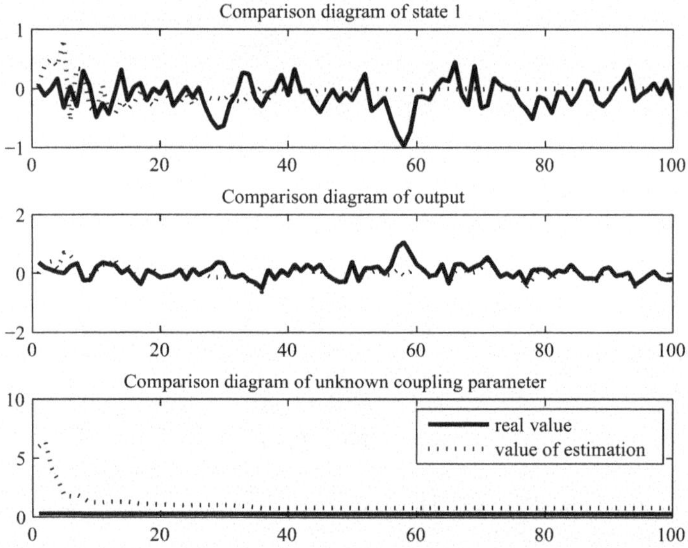

Fig. 12.14 Simulation (Sim1) of Model 4: the second agent

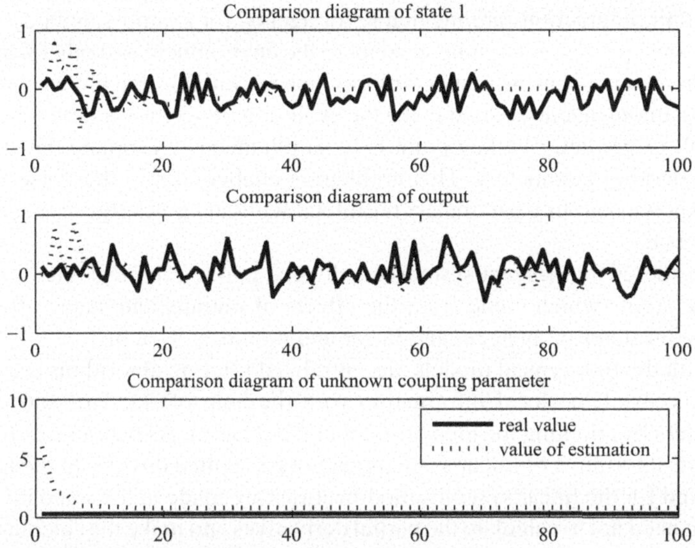

Fig. 12.15 Simulation (Sim1) of Model 4: the third agent

Table 12.4 Indices of DFNL2 (Sim2)

$P(q)$	$P(x)$
0.0160	0.0707

Then error covariance matrix of the first step can be calculated:

$$\hat{P}_{i,1}^{*} = \mathrm{E}\{[x_{i,1}^{*} - \hat{x}_{i,1}^{*}][x_{i,1}^{*} - \hat{x}_{i,1}^{*}]^{\mathrm{T}}\}, \tag{12.79}$$

The simulation results with Algorithm 4 (DFNL2) are depicted in Figs. 12.13, 12.14, and 12.15.

(2) Simulation 2 (Sim2)

A multi-agent system with 100 agents, each of which is the same as the above model, is chosen to verify the algorithm.

The indices in Table 12.4 are small compared with values of states and coupling parameters.

12.7 Conclusion

In this contribution, we have introduced a decentralized adaptive filtering problem for a class of multi-agent systems which contain uncertain coupling parameters. This problem is important and fundamental for further studies on controlling multi-agent systems. In practice, many situations of complex systems involve such a problem since uncertain interactions are inevitable for agents, for example, nobody can successfully grow up without learning to adapt to the environment and other people. For every agent, only noisy measurement information is available, but there are uncertain interactions among agents, which make the evolution of agents become complicated and difficult to predict. The key point is to simultaneously estimate the states and unknown coupling parameters. This problem is challenging in the sense that estimating unknown coupling parameters is difficult without knowledge of precise states and vice versa.

The considered multi-agent systems are divided into four typical types, including coupling relations which come from the effects of outputs and states of neighbor agents have on states of an agent and linear and nonlinear systems.

Then with the four typical models, decentralized filtering algorithms are deduced based on the available local information with the help of standard and extended Kalman filters by adopting the idea of states augmentation. As we can see, the modified filtering algorithms of linear systems cannot be applied directly in the nonlinear systems. And for the linear systems, modifications are made in the standard Kalman filter which need not to calculate the partial derivatives and make the calculation simple. But for the nonlinear systems, modifications are made in the extended Kalman filter. So the linear systems are not considered as the special case of the nonlinear systems. For testing the effects of estimation, decentralized adaptive controllers are designed based on certainty equivalence principle. And we provide some remarks to address the challenging issue of closed-loop stability analysis of the proposed algorithms. Then by extensive simulations, the introduced decentralized filtering algorithms have illustrated satisfactory estimation accuracy.

In future, in-depth studies and more methods are still needed to cope with more complex situations in which influence relation among agents needs to be represented by coupling nonlinear functions. And theoretic closed-loop analysis of these methods should be conducted in detail. There is still a long way ahead along this new direction of filtering for multi-agent systems with coupling uncertainties, especially decentralized adaptive filtering of multi-agent systems with parametric coupling uncertainties as well as nonparametric coupling uncertainties.

References

1. W.S. Yan, J.B. Li, Y.T. Wang, Consensus for damaged multi-agent system. Acta Autom. Sin. **38**(11), 1880–1884 (2012)
2. H.W. Yu, Y.F. Zheng, Dynamic behavior of multi-agent systems with distributed sampled control. Acta Autom. Sin. **38**(3), 357–365 (2012)
3. R. Lukomski, K. Wilkosz, Modeling of multi-agent system for power system topology verification with use of petri nets, in *2010 Modern Electric Power Systems* (2010), pp. 1–6
4. H. Ma, S.S. Ge, K.-Y. Lum, WLS-based partially decentralized adaptive control for coupled ARMAX multi-agent dynamical system, in *American Control Conference* (2009), pp. 3543–3548
5. H.B. Ma, Decentralized adaptive synchronization of a stochastic discrete-time multi-agent dynamic model. SIAM J. Control Optim. **48**(2), 859–880 (2009). Published 25 Feb 2009

6. H.B. Ma, C.G. Yang, M.Y. Fu, Decentralized adaptive control of discrete-time multi-agent systems, in *Discrete Time Systems*, ed. by M.A. Jordan (I-Tech Education and Publishing, Vienna, 2011), pp. 229–254
7. T. Al-Kanhal, M. Abbod, Multi-agent system for dynamic manufacturing system optimization. Lect. Notes Comput. Sci. **5103**(3), 634–643 (2008)
8. R. Olfati-Saber, R. Murray, Consensus problems in networks of agents with switching topology and time-delays. IEEE Trans. Autom. Control **49**(9), 1520–1533 (2004)
9. W. Ren, Decentralization of coordination variables in multi-vehicle systems, in *IEEE International Conference on Networking* (2006), pp. 550–555
10. W. Ren, I.K. Moore, I.Y. Chen, High-order consensus algorithms in cooperative vehicle systems, in *IEEE International Conference on Networking* (2006), pp. 475–462
11. Y. Cao, D. Stuart, R. Wei, Z. Meng, Distributed containment control for multiple autonomous vehicles with double-integrator dynamics: algorithms and experiments. IEEE Trans. Control Syst. Technol. **19**(4), 929–938 (2011)
12. R.E. Kalman, A new approach to linear filtering and prediction problems. Trans. ASME - J. Basic Eng. **82** (Series D), 35–45 (1960)
13. D. Simon, *Optimal State Estimation* (Wiley, New York, 2006)
14. J. Wang, Q.H. Liang, K. Liang, A new extended Kalman filter based carrier tracking loop, in *Proceedings - 2009 3rd IEEE International Symposium on Microwave, Antenna, Propagation and EMC Technologies for Wireless Communications* (IEEE, 2009), pp. 1181–1184
15. W. Yao, L. Xiong-Lin, Robustness analysis of Kalman filtering algorithm for multi-rate systems. Acta Electron. Sin. **38**(2), 156–174 (2012)
16. X. Jie, G. Wei, S. Yu-Xing, Research on real-time block Kalman filtering estimation methods for the un-modeled system based on output measurements. Acta Electron. Sin. **40**(10), 1958–1964 (2012)
17. X. Jie, G. Wei, S. Yu-Xing, Metro transfer passenger forecasting based on Kalman filter. J. Beijing Jiaotong Univ. **37**(3), 112–116 (2013)
18. L. Zhou-Zhou, Multi-sensor information fusion and its application. Electron. Des. Eng. **21**(11), 116–123 (2013)
19. H.B. Ma, Y.L. Zhao, M.Y. Fu, C.G. Yang, Decentralized adaptive control for a class of semi-parametric uncertain multi-agent systems, in *Proceedings of the 10th World Congress on Intelligent Control and Automation*, pp. 2060–2065
20. B.M. Quine, A derivative-free implementation of the extended Kalman filter. Automatica **42**(11), 1927–1934 (2006)
21. A.H. Jazwinski, *Stochastic Process and Filter Theory* (Academic, New York, 1970)
22. V. Solo, Stability of the Kalman filter with stochastic timevarying parameters 57–61 (1996)
23. K. Reif, S. Gunther, E. Yaz, R. Unbehauen, Stochastic stability of the discrete-time extended Kalman filter. Autom. Control **44**(4), 714–728 (1999)

Chapter 13
Comparison of Several Filtering Methods for Linear Multi-agent Systems with Local Unknown Parametric Couplings

In this chapter, several filtering methods for multi-agent systems with coupling uncertainties are investigated. The considered multi-agent system is composed of many agents, each of which evolves with a discrete-time stochastic linear time-varying dynamics, and each agent will be locally influenced by its neighbor agents. So the states evolution of each agent is not only related with its previous states but also related with its neighbors' previous states. Every agent can only observe its own measurements (outputs) and its neighbor agents' outputs, while the states are invisible to any agent because of communication limitations existing in the considered multi-agent system. We suppose that each agent has a priori knowledge on the structure of itself and its neighbors, that is to say, the coefficient matrices occurred in the individual dynamic system are available for use by the agent itself and those agents who have interactions with the agent. Because of the information constraints, without knowing the coupling gains of the local interactions, it is not easy for each agent to estimate its states by traditional Kalman filter or other state observers. For the wide applications of multi-agent systems and the existence of coupling uncertainties in many practical applications, the problem of decentralized filtering of multi-agent systems is more important; this chapter introduces one general framework of multi-agent systems for convenience of filtering studies. For the considered coupled linear discrete-time multi-agent system with uncertain linear local couplings, based on the key idea of *state augmentation* and *the certainty equivalence principle* borrowed from the area of adaptive control, we introduce several filtering methods to resolve the fundamental problem considered in this chapter. By conducting extensive simulations, the consuming time and estimation errors of every method are compared for one typical example, which suggests which method is more precise and fast.

13.1 Introduction

During the process of recognizing the world, it is more and more difficult to describe a whole system by using only a single dynamic equation. The concept of complex systems [1–3] was proposed to refer to those large-scale systems which can exhibit

H. Ma et al., *Kalman Filtering and Information Fusion*,
https://doi.org/10.1007/978-981-15-0806-6_13

certain complex behaviors or complexities. One complex system consists of many parts, and almost every part has relations with some others. Due to the local interactions, the whole system can exhibit complicated or ordered collective phenomena, sometimes called *emergence*, which may not be seen or predicted directly from the form of local interactions. Due to the wide background of complex systems, many scholars began to study the complex systems. One important point of studying complex systems is how to model them. By transforming complex systems to mathematical models, it is possible to investigate control algorithms quantitatively for the purpose of controlling the whole system to achieve certain expected goal. After decades of research, theories of complex systems [4] have been successfully applied in understanding the nature of complex systems and describing complex systems. Complex systems have potential applications in many fields, including physics, biology, computer science, and so on, such as intelligent robots [5, 6], traffic control [7], aerospace field [8], etc.

Theory of multi-agent systems, as a useful conceptual tool to model and analyze complex systems, has been developed [9–13] in 1980s and the past decades. The basic idea of multi-agent system is to describe the complex system by a collection of many agents, each of which abstracts an individual component of the whole system and evolves with time and local interactions with other neighbor agents. When the multi-agent system is used to describe an actual system, structural characteristics and behavior of the actual system can be represented by communication and mutual influence among agents. They have certain advantages over single-agent systems. In multi-agent systems, every agent can make its own decision and evolution just like in single-agent systems; however, one unique feature of multi-agent systems is that the agents in multi-agent systems may also use the local information from the neighbor agents rather their own information; hence, in this way, it is possible to finish many jobs, some of which may not be imagined by any single agent, distributedly, and decentralized by the coordination or cooperation of agents.

Based on these global and collective characteristics, the multi-agent system theory [14] has been widely used in many fields, such as intelligent robot [15, 16], intelligent transportation system (ITS) [17, 18], expert system [19], software development [20], and so on. But, at the same time, these characteristics bring out many challenges which need to be solved. For instance, structural complexity and information uncertainty, they incur many difficulties in analysis of the whole system. When the whole system is composed of agents which have influence on their neighbor agents, the problem is that though agents can recognize its own structure and outputs, they may not have access to full knowledge of the relations with its neighbor agents. Usually, relations among agents can be represented by functions of those agents' states, which may be parameterized by parameters like coupling strength, etc. Hence, for the purpose of controlling the agents, understanding the functions become one key point and the functions or parameters need to be estimated when uncertainties exist in such local interactions. And this problem intrigues our desire to do this work along the so-called new direction of decentralized filtering for multi-agent systems.

For the purpose of solving this problem and making the functions clear, estimation theories are considered. One popular theory is Kalman filter [21–23] which was put forward by R. E. Kalman and later applied first in modified form (now known as EKF) in the Apollo program and used for orbital prediction. Kalman filter, as one efficient recursive filter, can estimate the states of dynamic system well from a series of measurements which contain noises. Its computation loop composed of two main steps: first, estimate current states by using previous states through states equations; second, correct the states estimation obtained by the first step through observed outputs. Kalman filter is the optimal estimator when the noise is Gaussian, and it is the optimal linear estimator when the noise is not Gaussian. It is widely applied in vehicle navigation [24, 25], target tracking, stock price prediction, and many other fields. With Kalman filter, some scholars begin to study estimation problems of nonlinear systems. Then, EKF (extended Kalman filter) [26] which was originally proposed by Stanley Schmidt and unscented Kalman filter (UKF) [27] which can reduce the linearization errors of EKF are put forward in succession. And two-stage Kalman filter (TSKF) [28] is originally proposed to reduce the computational complexity of the augmented state Kalman filter. In recent years, it is also applied in systems in which there are random biases. In this chapter, five methods, respectively, based on Kalman filter, EKF, TSKF, and UKF are introduced in identifying relations among agents mentioned above, to resolve the filtering problem of linear multi-agent systems with unknown coupling parameters.

The main contribution of this chapter is describing one kind of multi-agent system which has influence between agents by a mathematical model and investigating five estimation methods to estimate the unknown coupling parameters and states of this special multi-agent system. The rest of this chapter is organized as follows. In Sect. 13.2, the special multi-agent system model is introduced in details. Based on this model, in Sect. 13.3, five filter methods are presented for estimating the coupling parameters and states. And then, in Sect. 13.4, simulation about five methods will be given, including comparison of five methods and simulation graphs about EKF which can demonstrate which method is better to a certain extent. Finally, we shall finish this chapter with some concluding remarks in Sect. 13.5.

13.2 Problem Formulation

This chapter studies one kind of multi-agent dynamic systems, every agent of which can be represented by a discrete-time stochastic linear time-varying dynamics model. And each agent's states are influenced by its neighbor agents' states. As a preliminary study, we suppose that the influence can be represented by linear functions. That is, functions can be represented by using scale factors. One agent of a multi-agent system with N agents can be written as the equations below:

$$
\begin{aligned}
x_{i,k} &= A_{i,k-1}x_{i,k-1} + B_{i,k-1}u_{i,k-1} \\
&\quad + \sum_{j_p \in \mathcal{N}_i} q_{i,j_p}x_{j_p,k-1} + w_{i,k-1} \\
y_{i,k} &= C_{i,k}x_{i,k} + D_{i,k}u_{i,k} + v_{i,k} \\
\mathcal{N}_i &= \{j_1, j_2, \ldots, j_{n_i}\}
\end{aligned} \tag{13.1}
$$

In the equations, x_i is the state of agent i, and x_{j_p} is the state of agent j_p, $\sum_{j_p \in \mathcal{N}_i} q_{i,j_p}x_{j_p,k-1}$ denotes coupling terms which show influence between agents and their neighbor agents, where q_{i,j_p} denotes the coupling parameter of agent i and agent j_p. Since agents often have similar structures and same state dimension, q_{i,j_p} is a scalar constant. That is, there is one proportional factor between each pair of agents' states. In the first equation, w_i and v_i denote process noise and measure noise respectively. Suppose that the noises are white, zero mean, and uncorrelated, and the noises of different agents are uncorrelated. If the noises are correlated, there are also methods proposed in optimal state estimation written by Simon [22] to remove the correlations. In the second equation, y_i is output of agent i. There are also some conditions needed to be shown:

$$
\begin{aligned}
w_{i,k} &\sim \mathcal{N}(0, Q_{i,k}) \\
v_{i,k} &\sim \mathcal{N}(0, R_{i,k})
\end{aligned} \tag{13.2}
$$

That is, $w_{i,k}$ and $v_{i,k}$ have known covariance matrices $Q_{i,k}$ and $R_{i,k}$. And $x_{i,k} \in \boldsymbol{R}^n$, $x_{j_p,k} \in \boldsymbol{R}^n$, $u_{i,k-1} \in \boldsymbol{R}^r$, $w_{i,k-1} \in \boldsymbol{R}^n$, $v_{i,k-1} \in \boldsymbol{R}^m$, $y_{i,k} \in \boldsymbol{R}^m$. The whole prescription of the multi-agent dynamic system is shown above. Since the system has unknown parameters, we cannot estimate the states only by standard Kalman filter; hence, new filter algorithm needs to be studied to estimate the states and coupling parameters in the linear state-space model.

13.3 Algorithms

In this section, we are going to discuss five different methods to resolve the mentioned filtering problem for the linear multi-agent systems with unknown coupling parameters. These methods can be classified into decentralized filtering methods and centralized filtering methods, among which the former methods only use local information of neighbor agents and the latter methods need to use all information of all agents. Decentralized filtering methods are more efficient in terms of computation and communication cost, while centralized filtering methods may yield better filtering results due to the full use of information of all agents.

13.3.1 Method 1—DF-AKF (Decentralized Filtering with Augmented Kalman Filter)

Since the coupling parameters q_{i,j_p} are unknown, we consider the coupling parameter q_{i,j_p} as a part of states with this method.

First, consider the unknown parameters as additional states of our new agents, and a new linear state-space model can be obtained, that is to say, the augmented states of agent i are given as

$$x_{i,k}^* = \begin{bmatrix} x_{i,k} \\ q_{ij_1} \\ q_{ij_2} \\ \vdots \\ q_{ij_{n_i}} \end{bmatrix}_{(n+n_i)\times 1} \tag{13.3}$$

where q_{ij_k} denotes the unknown coupling parameter with neighbor agent k, and we put them in a column attaching to states of agent i. With the augmented states, we can rewrite the dynamics of original multi-agent system with the following new system model:

$$\begin{aligned} x_{i,k}^* &= A_{i,k-1}^* x_{i,k-1}^* + B_{i,k-1}^* \\ &\quad u_{i,k-1}^* + w_{i,k-1}^* \end{aligned} \tag{13.4}$$

$$\begin{aligned} y_{i,k}^* &= y_{i,k} \\ &= C_{i,k}^* x_{i,k} + D_{i,k}^* u_{i,k} + v_{i,k}^* \end{aligned} \tag{13.5}$$

where

$$A_{i,k-1}^* = \begin{bmatrix} A_{i,k-1} & x_{j_1,k-1} & x_{j_2,k-1} & \dots & x_{j_{n_i},k-1} \\ 0_{1\times n} & 1 & 0 & \dots & 0 \\ 0_{1\times n} & 0 & 1 & \dots & 0 \\ \vdots & \vdots & \vdots & \ddots & \vdots \\ 0_{1\times n} & 0 & 0 & \dots & 1 \end{bmatrix} \tag{13.6}$$

$$B_{i,k-1}^* = \begin{bmatrix} B_{i,k-1} \\ 0_{1\times r} \\ 0_{1\times r} \\ \vdots \\ 0_{1\times r} \end{bmatrix}_{(n+n_i)\times r}$$

$$
\begin{aligned}
C^*_{i,k} &= \begin{bmatrix} C_{i,k} \\ 0_{m\times 1} \\ 0_{m\times 1} \\ \cdots \\ 0_{m\times 1} \end{bmatrix}^T_{m\times(n+n_i)} \\
D^*_{i,k} &= D_{i,k} \\
u^*_{i,k-1} &= u_{i,k-1} \\
w^*_{i,k-1} &= \begin{bmatrix} I_{n\times n} \\ 0_{1\times n} \\ 0_{1\times n} \\ \vdots \\ 0_{1\times n} \end{bmatrix}_{n\times(n+n_i)} \times w_{i,k-1} \\
v^*_{i,k} &= v_{i,k}
\end{aligned} \tag{13.7}
$$

Here, $I_{n\times n}$ denotes the identity matrix. Then according to initial conditions on the $w_{i,k}$, we can conclude that

$$
\begin{aligned}
w^*_{i,k} &\sim \mathcal{N}(0, Q^*_{i,k}) \\
v^*_{i,k} &\sim \mathcal{N}(0, R^*_{i,k}) \\
Q^*_{i,k} &= \begin{bmatrix} Q_{i,k} & 0_{n\times n_i} \\ 0_{n_i\times n} & 0_{n_i\times n_i} \end{bmatrix}_{(n+n_i)\times(n+n_i)} \\
R^*_{i,k} &= R_{i,k}
\end{aligned} \tag{13.8}
$$

In the matrix $A^*_{i,k-1}$, a_{ij} denotes the elements of the i_{th} row and j_{th} column. Every diagonal element is equal to one except for the diagonal elements which are taken up by elements of $A_{i,k-1}$. With the new dynamic system model, the coupling relation of neighbor agents still exists. The coefficient matrix includes $x_{j_p,k-1}$ which is still unknown. As agents of multi-agent systems can correspond with their neighbor agents, they can receive current states estimation of its neighbor agents. So another new idea is to use the posterior estimation of $x_{j_p,k-1}$ to replace the true value of $A^*_{i,k-1}$. Then, the filter process can be described as follows ($P^{*-}_{i,k}$ and $P^{*+}_{i,k}$ represent the new prior and posterior state estimation error covariance, in the same way, $x^{*-}_{i,k}$ and $x^{*+}_{i,k}$ denote prior and posterior state estimation):

$$
\begin{aligned}
P^{*-}_{i,k} &= A^{*+}_{i,k-1} P^{*+}_{i,k-1} A^{*+}_{i,k-1}{}^T \\
&\quad + Q^*_{i,k-1} \\
K^*_{i,k} &= P^{*-}_{i,k} C^*_{i,k} (C^*_{i,k} P^{*-}_{i,k} C^*_{i,k}{}^T \\
&\quad + R^*_{i,k})^{-1} \\
x^{*-}_{i,k} &= A^{*+}_{i,k-1} x^{*+}_{i,k-1} + B^*_{i,k-1}
\end{aligned} \tag{13.9}
$$

$$
\begin{aligned}
&\times u_{k-1}\\
x_{i,k}^{*+} &= x_{i,k}^{*-} + K_{i,k}^{*}(y_{i,k}^{*} - C_{i,k}^{*}x_{i,k}^{*-}\\
&\quad -D_{i,k}^{*}u_{i,k}^{*})\\
P_{i,k}^{*+} &= (I - K_{i,k}^{*}C_{i,k}^{*})P_{i,k}^{*-}(I + K_{i,k}^{*}\\
&\quad \times C_{i,k}^{*})^{T} + K_{i,k}^{*}R_{i,k}^{*}K_{i,k}^{*}
\end{aligned}
$$

where

$$
A_{i,k-1}^{*+} = \begin{bmatrix} A_{i,k-1} & x_{j_1,k-1}^{+} & x_{j_2,k-1}^{+} & \cdots & x_{j_{n_i},k-1}^{+} \\ 0_{1\times n} & 1 & 0 & \ldots & 0 \\ 0_{1\times n} & 0 & 1 & \ldots & 0 \\ \vdots & \vdots & \vdots & \ddots & \vdots \\ 0_{1\times n} & 0 & 0 & \ldots & 1 \end{bmatrix} \tag{13.10}
$$

Here $x_{j_1,k-1}^{+}, x_{j_2,k-1}^{+}, \ldots, x_{j_{n_i},k-1}^{+}$ are the posterior state estimations of agents $j_1, j_2, \ldots, j_{n_i}$. Then continue the new Kalman filter for every agent step by step in the corresponding order. The problem proposed above can be solved. That is, the estimation of $x_{i,k}^{*}$ and q_{i,j_p} can be obtained.

13.3.2 *Method 2—DF-TSKF (Decentralized Filtering with Two-Stage Kalman Filter)*

Since TSKF can reduce calculation and time of filtering process, in Method 2, we try to apply TSKF [29]. First, the system can be rewritten as follows:

$$
\begin{aligned}
x_{i,k} &= A_{i,k-1}x_{i,k-1} + B_{i,k-1}u_{i,k-1}\\
&\quad + \begin{bmatrix} x_{j_1,k-1} \\ x_{j_2,k-1} \\ \cdots \\ x_{j_{n_i},k-1} \end{bmatrix}^{T} \gamma_{i,k-1} + w_{i,k-1}\\
\gamma_{i,k} &= \gamma_{i,k-1} = \begin{bmatrix} q_{ij_1} \\ q_{ij_2} \\ \vdots \\ q_{ij_{n_i}} \end{bmatrix} \qquad (13.11)\\
y_{i,k} &= C_{i,k}x_{i,k} + D_{i,k}u_{i,k} + v_{i,k}\\
\mathcal{N}_i &= \{j_1, j_2, \ldots, j_{n_i}\} \qquad (13.12)
\end{aligned}
$$

In this system, Eq. (13.11) is added which includes no noise and u_i is zero. Then with the new model above, a two-stage Kalman filter can be implemented. It has been derived in [30] that TSKF is minimum mean square error (MMSE) when $Q = 0$. The estimation process is divided into two parts: the modified bias-free estimation and bias estimation.

The modified bias-free filter:

$$\begin{aligned}
x^{-}_{i,k/k-1} &= A_{i,k-1}x^{-}_{i,k-1/k-1} \\
&\quad +tu_{i,k-1} \\
x^{-}_{i,k/k} &= x^{-}_{i,k/k-1} + K^{x}_{i,k}(y_{i,k} \\
&\quad -C_{i,k}x^{-}_{i,k/k-1}) \\
P^{x}_{i,k/k-1} &= A_{i,k-1}P^{x}_{i,k-1/k-1}A^{T}_{i,k-1} \\
&\quad +Q^{x}_{i,k-1} \\
K^{x}_{i,k} &= P^{x}_{i,k/k-1}C^{T}_{i,k}(C_{i,k} \times \\
&\quad P^{x}_{i,k/k-1}C^{T}_{i} + R_{i,k})^{-1} \\
P^{x}_{i,k/k} &= (I - K^{x}_{i,k}C_{i,k})P^{x}_{i,k/k-1}
\end{aligned} \tag{13.13}$$

where

$$\begin{aligned}
U^{-}_{i,k} &= A_{i,k-1}V_{i,k-1} + \begin{bmatrix} \hat{x}_{j_1,k-1} \\ \hat{x}_{j_2,k-1} \\ \cdots \\ \hat{x}_{j_{n_i},k-1} \end{bmatrix}^{T} \\
U_{i,k} &= U^{-}_{i,k}P^{\gamma}_{i,k-1/k-1}(P^{\gamma}_{i,k/k-1})^{-1} \\
Q^{x}_{i,k-1} &= Q_{i,k-1} \\
tu_{i,k-1} &= (U^{-}_{i,k} - U_{i,k})\gamma^{-}_{i,k-1/k-1} \\
S_{i,k} &= C_{i,k}U_{i,k} \\
V_{i,k} &= U_{i,k} - K^{x}_{i,k}S_{i,k}
\end{aligned} \tag{13.14}$$

The bias filter:

$$\begin{aligned}
\gamma^{-}_{i,k/k-1} &= \gamma^{-}_{i,k-1/k-1} \\
\gamma^{-}_{i,k/k} &= \gamma^{-}_{i,k/k-1} + K^{\gamma}_{i,k}(y_{i,k} - C_{i,k} \\
&\quad x^{-}_{i,k/k-1} - S_{i,k}\gamma^{-}_{i,k/k-1}) \\
P^{\gamma}_{i,k/k-1} &= P^{\gamma}_{i,k-1/k-1} \\
P^{\gamma}_{i,k/k} &= (I - K^{\gamma}_{i,k}S_{k})P^{\gamma}_{i,k/k-1}
\end{aligned} \tag{13.15}$$

$$\begin{aligned}
K^{\gamma}_{i,k} &= P^{\gamma}_{i,k/k-1}S^{T}_{i,k}(C_{i,k}P^{x}_{i,k/k-1}C^{T}_{i,k} \\
&\quad +R_{i,k} + S_{i,k}P^{\gamma}_{i,k/k-1}S^{T}_{i,k})^{-1}
\end{aligned} \tag{13.16}$$

Combination:

$$
\begin{aligned}
\hat{x}_{i,k} &= x^{-}_{i,k/k} + V_{i,k}\gamma^{-}_{i,k/k} \\
\hat{P}^{x}_{i,k/k} &= P^{x}_{i,k/k} + V_{i,k}P^{\gamma}_{i,k/k}V^{T}_{i,k} \\
\hat{P}^{\gamma}_{i,k/k} &= P^{\gamma}_{i,k/k}
\end{aligned} \tag{13.17}
$$

Only if the coefficient matrix of γ in the states equation, that is, $x_{j_1,k-1}$, $x_{j_2,k-1}$, ..., $x_{j_{n_i},k-1}$ are known, two-stage Kalman filter can be implemented. So in this method, we use the estimated values, $\hat{x}_{j_p,k-1}$, to replace the true values. For every agent, TSKF is implemented and state estimations of neighbors are transmitted.

13.3.3 Method 3—CF-AKF (Centralized Filtering with Augmented Kalman Filter)

First, all agents are grouped to form a huge system:

$$
\begin{aligned}
\begin{bmatrix} x_{1,k} \\ x_{2,k} \\ \vdots \\ x_{N,k} \end{bmatrix} &= M_{k-1}\begin{bmatrix} x_{1,k-1} \\ x_{2,k-1} \\ \vdots \\ x_{N,k-1} \end{bmatrix} + B_{k-1}\begin{bmatrix} u_{1,k-1} \\ u_{2,k-1} \\ \vdots \\ u_{N,k-1} \end{bmatrix} \\
&\quad + \begin{bmatrix} w_{1,k-1} \\ w_{2,k-1} \\ \vdots \\ w_{N,k-1} \end{bmatrix}
\end{aligned} \tag{13.18}
$$

$$
\begin{bmatrix} y_{1,k} \\ y_{2,k} \\ \vdots \\ y_{N,k} \end{bmatrix} = O_k \begin{bmatrix} x_{1,k} \\ x_{2,k} \\ \vdots \\ x_{N,k} \end{bmatrix} + \begin{bmatrix} v_{1,k} \\ v_{2,k} \\ \vdots \\ v_{N,k} \end{bmatrix}
$$

where

$$
M_{k-1} = \begin{bmatrix} M_{1,1,k-1} & M_{1,2,k-1} & \cdots & M_{1,N,k-1} \\ M_{2,1,k-1} & M_{2,2,k-1} & \cdots & M_{2,N,k-1} \\ \vdots & \vdots & \vdots & \vdots \\ M_{N,1,k-1} & M_{N,2,k-1} & \cdots & M_{N,N,k-1} \end{bmatrix} \tag{13.19}
$$

$$
B_{k-1} = \begin{bmatrix} B_{1,k-1} & 0_{n\times r} & \cdots & 0_{n\times r} \\ 0_{n\times r} & B_{2,k-1} & \cdots & 0_{n\times r} \\ \vdots & \vdots & \vdots & \vdots \\ 0_{n\times r} & 0_{n\times r} & \cdots & B_{N,k-1} \end{bmatrix} \tag{13.20}
$$

$$O_{k-1} = \begin{bmatrix} O_{1,1,k} & O_{1,2,k} & \dots & O_{1,N,k} \\ O_{2,1,k} & O_{2,2,k} & \dots & O_{2,N,k} \\ \vdots & \vdots & \vdots & \vdots \\ O_{N,1,k} & O_{N,2,k} & \dots & O_{N,N,k} \end{bmatrix} \tag{13.21}$$

In the above equations, matrix $M_{i,j,k-1}$ and $O_{i,j,k}$, $i, j \in \{1, 2, 3, \cdots, N\}$ are given as follows:

$$M_{i,j,k-1} = \begin{cases} A_{i,k-1} & \text{if } i = j \\ q_{i,j_p} I_{n\times n} & \text{if } j = j_p \text{ and } j_p \in \mathcal{N}_i \\ 0_{n\times n} & \text{else} \end{cases}$$
$$O_{i,j,k-1} = \begin{cases} C_{i,k-1} & \text{if } i = j \\ 0_{m\times n} & \text{else} \end{cases} \tag{13.22}$$

where $I_{n\times n}$ denotes an identity matrix.

Second, use Kalman filter to estimate the huge system. But there are also unknown parts in the coefficient matrix M, such as $M_{i,j,k-1}$ when $j = j_p$. So just as Method 1, consider q_{i,j_p} as an element of states and transform the huge system to a new huge system in which coefficient matrix only contains the origin states and matrices, such as $x_{j_p,k-1}$, $A_{i,k-1}$. Then use the estimated values of $x_{j_p,k-1}$ to replace the true values in the new coefficient matrix. The filtering process is similar to Method 1. The difference is that this system is centralized and filtering is centralized, but in Method 1 the filtering is decentralized, that is, every agent has its own filtering process.

13.3.4 Method 4—CF-AEKF (Centralized Filtering with Augmented Extended Kalman Filter)

EKF is applied to solve the estimation problem. First, consider q_{i,j_p} as a part of states and the system as a nonlinear system.

$$X_k = \begin{bmatrix} x_{1,k} \\ x_{2,k} \\ \cdots \\ x_{N,k} \\ q_{1,j_1} \\ q_{1,j_2} \\ \cdots \\ q_{1,j_{n_1}} \\ q_{2,j_1} \\ \cdots \\ q_{2,j_{n_2}} \\ \cdots \\ q_{N,j_{n_N}} \end{bmatrix} \tag{13.23}$$

In brief:

$$
\begin{aligned}
X_k &= \begin{bmatrix} x_{1,k} \\ x_{2,k} \\ \dots \\ x_{N,k} \\ q_{1,j_1} \\ \vdots \\ q_{N,jn_N} \end{bmatrix} \\
&= f_{k-1}(X_{k-1},U_{k-1},W_{k-1}) \\
&= \begin{bmatrix} A_{1,k-1}x_{1,k-1}+\sum_{jp_1\in\mathcal{N}_1} q_{1,jp_1}x_{jp_1,k-1} \\ A_{2,k-1}x_{2,k-1}+\sum_{jp_2\in\mathcal{N}_2} q_{2,jp_2}x_{jp_2,k-1} \\ \vdots \\ A_{N,k-1}x_{N,k-1}+\sum_{jp_N\in\mathcal{N}_N} q_{N,jp_N}x_{jp_N,k-1} \\ q_{1,j_1} \\ \vdots \\ q_{N,jn_N} \end{bmatrix} \\
&+ \begin{bmatrix} B_{1,k}(k) & 0_{n\times r} & \dots & 0_{n\times r} \\ 0_{n\times r} & B_2(k) & \dots & 0_{n\times r} \\ \vdots & \vdots & \vdots & \vdots \\ 0_{n\times r} & 0_{n\times r} & \dots & B_N(k) \\ 0_{1\times r} & 0_{1\times r} & \dots & 0_{1\times r} \\ \vdots & \vdots & \vdots & \vdots \\ 0_{1\times r} & 0_{1\times r} & \dots & 0_{1\times r} \end{bmatrix} U_{k-1} \\
&+W(k-1)
\end{aligned}
\tag{13.24}
$$

$$
\begin{aligned}
Y_k &= C_kX_k+D_kU_k+V_k \\
&= \begin{bmatrix} C_{1,k} & C_{2,k} & \dots & C_{N,k} & 0 & \dots & 0 \end{bmatrix} X_k+ \\
&\quad \begin{bmatrix} D_{1,k} & D_{2,k} & \dots & D_{N,k} \end{bmatrix} U_k+V_k
\end{aligned}
\tag{13.25}
$$

where

$$
\begin{aligned}
U_{k-1} &= \begin{bmatrix} u_{1,k-1} & u_{2,k-1} & \dots & u_{N,k-1} \end{bmatrix}^T \\
W_{k-1} &= \begin{bmatrix} w_{1,k-1} & \dots & w_{N,k-1} & 0 & \dots & 0 \end{bmatrix}^T \\
V_k &= \begin{bmatrix} v_{1,k} & \dots & v_{N,k} \end{bmatrix}
\end{aligned}
\tag{13.26}
$$

and the noise covariance matrices become

$$Q_{k-1} = \begin{bmatrix} Q_{1,k-1} & 0_{n\times n} & \cdots & 0_{n\times n} & 0_{n\times 1} & \cdots & 0_{n\times 1} \\ 0_{n\times n} & Q_{2,k-1} & \cdots & 0_{n\times n} & 0_{n\times 1} & \cdots & 0_{n\times 1} \\ \vdots & \vdots & \vdots & \vdots & \vdots & \vdots & \vdots \\ 0_{n\times n} & 0_{n\times n} & \cdots & Q_{N,k-1} & 0_{n\times 1} & \cdots & 0_{n\times 1} \\ 0_{1\times n} & 0_{1\times n} & \cdots & 0_{1\times n} & 0 & \cdots & 0 \\ \vdots & \vdots & \vdots & \vdots & \vdots & \vdots & \vdots \\ 0_{1\times n} & 0_{1\times n} & \cdots & 0_{1\times n} & 0 & \cdots & 0 \end{bmatrix} \tag{13.27}$$

$$R_k = \begin{bmatrix} R_{1,k} & 0_{n\times n} & \cdots & 0_{n\times n} \\ 0_{n\times n} & R_{2,k} & \cdots & 0_{n\times n} \\ \vdots & \vdots & \vdots & \vdots \\ 0_{n\times n} & 0_{n\times n} & \cdots & R_{N,k} \end{bmatrix} \tag{13.28}$$

After the system is formed, the extended Kalman filtering can be applied and the process is as follows:

1. Compute partial derivative matrix:

$$A_{k-1} = \frac{\partial f_{k-1}}{\partial X}|_{\hat{X}^+_{k-1}} \tag{13.29}$$

   where $\hat{X}^+_{k-1}$ is the posterior estimation.
2. Perform the time update:

$$\begin{aligned} P_k^- &= A_{k-1} P^+_{k-1} A^T_{k-1} + Q_{k-1} \\ X_k^- &= f_{k-1}(X^+_{k-1}, U_{k-1}, \mathbf{0}) \end{aligned} \tag{13.30}$$

3. Perform measurement update:

$$\begin{aligned} K_k &= P_k^- C_k (C_k P_k^- C_k^T + R_k)^{-1} \\ X_k^+ &= X_k^- + K_k (Y_k - C_k X_k^-) \\ P_k^+ &= (I - K_k C_k) P_k^- \end{aligned} \tag{13.31}$$

13.3.5 Method 5—CF-AUKF (Centralized Filtering with Augmented Unscented Kalman Filter)

UKF is applied, and the transformed system is the same as the system considered in Method 4. For the time updating process, use sigma points to propagate from step $k-1$ to k and choose sigma points X^s_{k-1} with appropriate changes; dim denotes the dimension of X_k:

$$
\begin{aligned}
X_{k-1}^{s} &= X_{k-1}^{+} + \widetilde{X}^{s} \\
&\quad s = 1, 2, \ldots, 2dim \\
\widetilde{X}^{s} &= (\sqrt{dim \times P_{k-1}^{+}})_{s}^{T} \\
&\quad s = 1, 2, \ldots, dim \\
\widetilde{X}^{dim+s} &= -(\sqrt{dim \times P_{k-1}^{+}})_{s}^{T} \\
&\quad s = 1, 2, \ldots, dim
\end{aligned} \tag{13.32}
$$

where $(\sqrt{dim \times P_{k-1}^{+}})_{s}^{T}$ denotes the s_{th} row of $(\sqrt{dim \times P_{k-1}^{+}})^{T}$. Then use the nonlinear system equation f_{k-1}:

$$
\begin{aligned}
X_{k}^{s} &= f_{k-1}(X_{k-1}^{s}, U_{k-1}, \mathbf{0}) \\
X_{k}^{-} &= \frac{1}{2dim} \sum_{s=1}^{2dim} X_{k}^{s} \\
P_{k}^{-} &= \frac{1}{2dim} \sum_{s=1}^{2dim} (X_{k}^{s} - X_{k}^{-})(X_{k}^{s} \\
&\quad - X_{k}^{-})^{T} + Q_{k-1}
\end{aligned} \tag{13.33}
$$

The measurement update process:

$$
\begin{aligned}
X_{k}^{s} &= X_{k}^{-} + \widetilde{X}^{s} \quad s = 1, 2, \ldots, 2dim \\
\widetilde{X}^{s} &= (\sqrt{dim \times P_{k}^{-}})_{s}^{T} \quad s = 1, 2, \ldots, dim \\
\widetilde{X}^{dim+s} &= -(\sqrt{dim \times P_{k}^{-}})_{s}^{T} \quad s = 1, 2, \ldots, dim \\
\widehat{Y}_{k}^{s} &= C_{k} X_{k}^{s} + D_{k} U_{k} \\
\widehat{Y}_{k} &= \frac{1}{2dim} \sum_{s=1}^{2dim} \widehat{Y}_{k}^{s}
\end{aligned} \tag{13.34}
$$

Corresponding covariance matrices:

$$
\begin{aligned}
P_{y,k} &= \frac{1}{2dim} \sum_{s=1}^{2dim} (\widehat{Y}_{k}^{s} - \widehat{Y}_{k})(\widehat{Y}_{k}^{s} - \widehat{Y}_{k})^{T} \\
&\quad + R_{k} \\
P_{xy,k} &= \frac{1}{2dim} \sum_{s=1}^{2dim} (X_{k}^{s} - X_{k}^{-})(\widehat{Y}_{k}^{s} - \widehat{Y}_{k})^{T}
\end{aligned} \tag{13.35}
$$

$$
\begin{aligned}
K_k &= P_{xy,k} P_{y,k}^{-1} \\
X_k^+ &= X_k^- + K_k (Y_k - \widehat{Y}_k) \\
P_k^+ &= P_k^- - K_k P_{y,k} K_k^T
\end{aligned} \tag{13.36}
$$

Then continue the two processes above step by step, estimations of states and parameters can be accomplished.

13.4 Simulation Study

Through simulating the above filtering methods with a three-agent system, some results are displayed. Suppose that the system is linear and time-invariant where every agent is two-dimensional, and the output is one-dimensional. And every agent has influence on its next agent, that is, agent $i-1$ has influence on agent i. And the last agent of the multi-agent system has influence on the first agent. The unknown influence parameter about agent $i-1$ to agent i is represented by q_i. The model of agent i can be written as follows:

$$
\begin{aligned}
x_{i,k} &= A_i x_{i,k-1} + q_i x_{i-1,k-1} + w_{i,k-1} \\
y_{i,k} &= C_{i,k} x_{i,k} + v_{i,k}
\end{aligned} \tag{13.37}
$$

As we expect that this model is stable without inputs, A_i, q_i and C_i can be chose as follows:

$$
\begin{aligned}
A_i &= \begin{bmatrix} 0.6 & 0 \\ 0 & 0.6 \end{bmatrix} \\
C_i &= \begin{bmatrix} 0.8 \; 0.7 \end{bmatrix} \\
q_i &= 0.3
\end{aligned} \tag{13.38}
$$

For comparison of different methods, all methods are initialized as follows:

$$
x_{i,1}^+ = E[x_{i,1}] \tag{13.39}
$$

$$
q_{i,1}^+ = 2 \tag{13.40}
$$

That is, in the first step, estimation of state $x_i(1)$ of every agent is expectation of state $x_i(1)$. Then we can calculate the initial error covariance:

$$
\begin{aligned}
P_{x_{i,1}}^+ &= E[(x_{i,1} - x_{i,1}^+)(x_{i,1} - x_{i,1}^+)^T] \\
P_{q_{i,1}}^+ &= (q_i - q_{i,1}^+)^2
\end{aligned} \tag{13.41}
$$

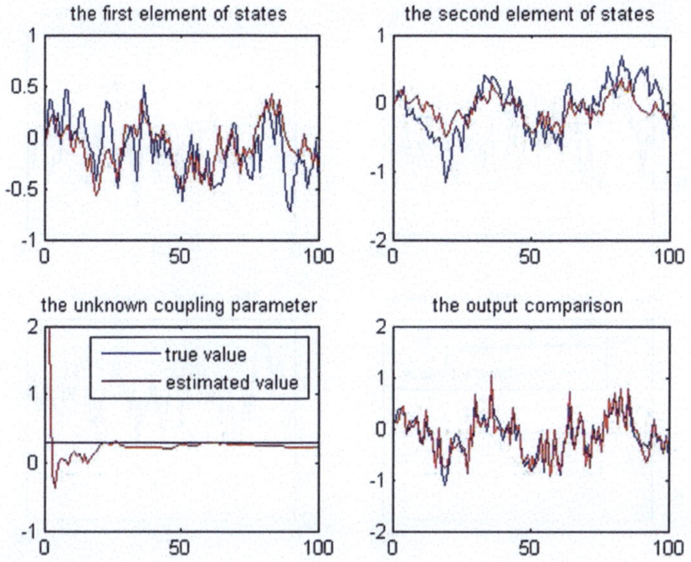

Fig. 13.1 The first agent

By programming, we choose steps as 100. Draw estimation graphs of Method 4. In the graphs, the red line denotes the estimated value and the blue line denotes the true value. Draw the state estimation and true value in each figure. Also, plot about the unknown parameter estimation and true value (Figs. 13.1, 13.2 and 13.3).

$$P(q) = \frac{1}{N}\sum_{i=1}^{N}(agent(i).qpre(T) - agent(i).q)^2$$

$$\begin{aligned} P(x) = \frac{1}{2N \times T}\sum_{i=1}^{N}\sum_{j=1}^{T}[(agent(i).xpre(1, j) - \\ agent(i).x(1, j))^2 + (agent(i).xpre(2, j) - \\ agent(i).x(2, j)]^2 \end{aligned} \tag{13.42}$$

That is, $P(q)$ represents mean value of error covariance estimation about q in the last step (at time T) and $P(x)$ denotes mean value of error covariance estimation about x of every step.

From the simulation results summarized in Table 13.1, we can draw the following conclusions:

- Centralized filtering methods 3–5 can obtain considerably smaller covariances than decentralized methods 1 and 2. This conclusion can be interpreted naturally

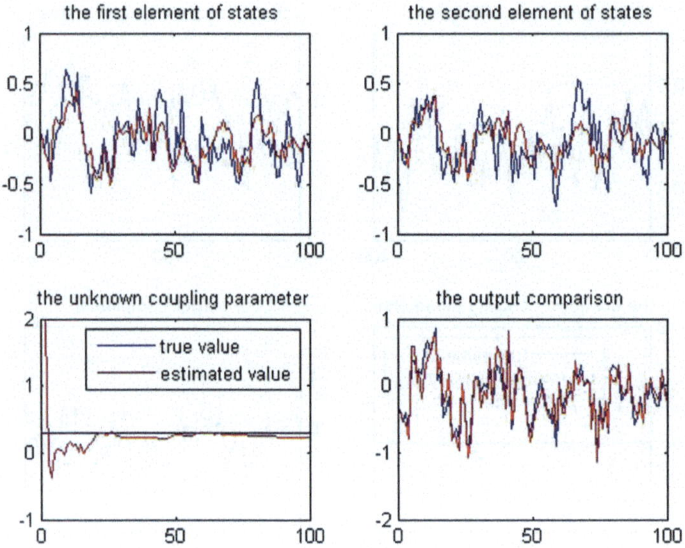

Fig. 13.2 The second agent

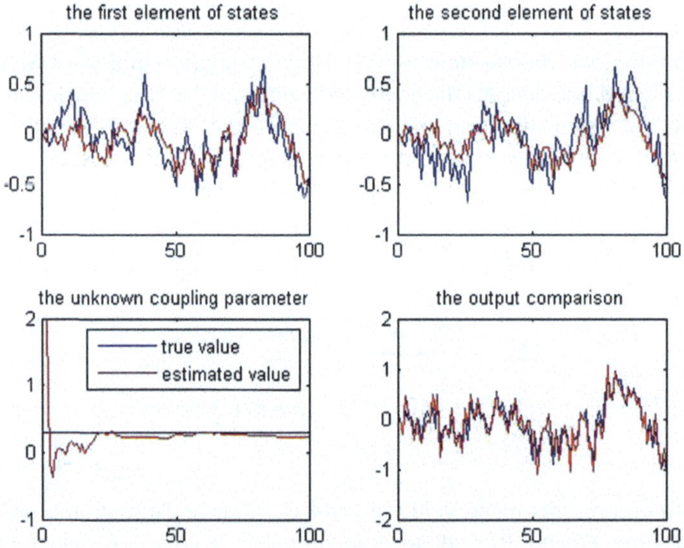

Fig. 13.3 The third agent

since centralized methods require to know information of all agents, which can improve the accuracy of estimation for each agent.

- Among decentralized filtering methods, Method 1 can obtain better results than Method 2 in terms of both accuracy and running time, which indicates that for linear

Table 13.1 The comparison of five methods

	Running Time (s)	$P(q)$	$P(x)$
Method 1 (DF-AKF)	0.120665	0.0088	0.0077
Method 2 (DF-TSKF)	0.195658	0.0227	0.0261
Method 3 (CF-AKF)	0.092705	0.0021	0.0071
Method 4 (CF-AEKF)	0.041510	0.0016	0.0055
Method 5 (CF-AUKF)	0.555756	0.0016	0.0065

multi-agent systems, DF-AKF is not only efficient but also easy to implement; hence, it is the desired method to use in practice, while DF-TSKF cannot generate state and parameter estimations as good as DF-AKF.

- Among centralized filtering methods, Method 4 can obtain better results than Method 3 and Method 5 in terms of both accuracy and running time, which suggests CF-AEKF is one desirable method due to its relatively low computation cost and high accuracy.
- Considering both centralized filtering methods and decentralized filtering methods, it looks like that Method 4 is the best one in terms of accuracy and running time. However, if we notice the fact that computation of agents in decentralized filtering methods can be made parallelly in nature, it may be more fair to divide the total running time by a factor (roughly equal to the number of agents, $N = 3$ here) by ignoring the overhead of implementation of parallel filtering. Thus, Method 1 may need significantly less running time than Method 4 and other centralized filtering methods, and this fact is consistent with the decentralized nature of Method 1. Besides, although Method 4 can yield the best estimates in terms of accuracy due to its full use of all agents' information, we can see that Method 1 in fact can obtain comparable estimation results as Method 4.

Summarizing the above discussions, we may conclude that CF-AEKF can yield the best estimation results in terms of accuracy, while DF-AKF may be the best method in trade-off of accuracy and efficiency, due to its nature of decentralized filtering and the elegant design of estimating the unknown parameters and states simultaneously.

13.5 Conclusion

In this chapter, we introduce a fundamental estimation problem for a class of multi-agent systems which contain uncertain coupling parameters. This problem is crucial for further studies on controlling agents, where only noisy measurement information is available, which are coupled together by uncertain interactions. That is, when agents of a multi-agent system have uncertain influence on their neighbor agents, how to achieve estimations to uncertain coupling parameters and states. For simplicity without loss of generality, the coupling uncertainty considered in this chapter is simplified as unknown proportional parameters between states of agents which have influence on its neighbor agents.

Then, five different methods of decentralized filtering and centralized filtering are introduced to solve this estimation problem and discussed in details. Finally, extensive simulations have been conducted for testing which method is better than others, judging from the estimation accuracy and running time (efficiency). The extensive simulations have shown that centralized filtering methods may provide better estimation accuracy, while decentralized filtering methods may be more efficient due to the nature of decentralized filtering. Among these methods, DF-AKF and CF-AEKF may be two good choices for resolving the problem studied in this chapter, while DF-AKF may be the best usable method which can take balanced trade-off among the accuracy and efficiency.

In the future, we still need to investigate the methods dealing with more complex situations in which influence relation needs to be represented by nonlinear functions and agents are not linear systems. And theoretic analysis of these methods should be conducted, such as stability analysis, which may be very challenging due to the complex nonlinear closed-loop system resulted by the coupling uncertainty and the filtering algorithm. There is still a long way ahead along this new direction of filtering for multi-agent systems with coupling uncertainties, especially decentralized filtering of multi-agent systems with parametric coupling uncertainties.

References

1. J.H. Holland, *Hidden Order: How Adaptation Builds Complexity* (Addison-Wesley, New York, 1996)
2. M.M. Waldrop, *Complexity: The Emerging Science at the Edge of Order and Chaos* (Touchstone Books, New York, 1993)
3. M. Gell-Mann, *The Quark and the Jaguar, Adventures in the Simple and the Complex* (W. H. Freeman and Company, New York, 1994)
4. X.P. Liu, Y.M. Tang, L.P. Zheng, Survey of complex system and complex system simulation. J. Syst. Simul. **20**(23), 6303–6315 (2008)
5. G. Schweitzer, What do we expect from intelligent robots?, in *Proceedings 1999 IEEE/RSJ International Conference on Intelligent Robots and Systems. Human and Environment Friendly Robots with High Intelligence and Emotional Quotients*, vol. 3 (IEEE, 1999), pp. 1271–1278
6. E.L. Hall, S.M.A. Ali, M. Ghaffari, *Engineering Robust Intelligent Robots*, vol. 7539 (SPIE - The International Society for Optical Engineering, 2010), pp. 753904–753918
7. L.T. Lin, H.J. Huang, J.M. Lin, A new intelligent traffic control system for Taiwan, in *2009 9th International Conference on Intelligent Transport Systems Telecommunications* (2009), pp. 138–142
8. A.K. Noor, S.L. Venneri, Ise: Intelligent synthesis environment for future aerospace systems. IEEE Aerosp. Electron. Syst. Mag. **23**(4), 31–44 (2008)
9. T. Al-Kanhal, M. Abbod, Multi-agent system for dynamic manufacturing system optimization. Lect. Notes Comput. Sci. **5103**(3), 634–643 (2008)
10. R. Lukomski, K. Wilkosz, Modeling of multi-agent system for power system topology verification with use of petri nets, in *2010 Modern Electric Power Systems* (2010), pp. 1–6

11. H.B. Ma, S.S. Ge, K.Y. Lum, WLS-based partially decentralized adaptive control for coupled ARMAX multi-agent dynamical system, in *Proceedings of the 2009 American Control Conference (ACC2009)* (St. Louis, Missouri, USA, 2009), pp. 10–12
12. H.B. Ma, Decentralized adaptive synchronization of a stochastic discrete-time multi-agent dynamic model. SIAM J. Control. Optim. **48**(2), 859–880 (2009). Published February 25, 2009
13. H.B. Ma, C.G. Yang, M.Y. Fu, Decentralized adaptive control of discrete-time multi-agent systems, in *Discrete Time Systems*, ed. by M.A. Jordan, chap. 14 (Vienna, Austria: I-Tech Education and Publishing, 2011), pp. 229–254
14. J.K. Liu, L.J. Er, Overview of application of multiagent technology. Control Decis. **16**(2), 133–140+180 (2001)
15. Y.S. Gao, J. Zhao, H.G. Cai, Multi-agent based multi-robot network control architecture. Comput. Eng. **32**(22), 14–16 (2006)
16. L.H. Chen, G.L. Zhang, J. Zeng, H.T. Song, Research on multi-robot control based on multi-agent. Microcomput. Inf. **23**, 180–181 (2009)
17. Z.S. Yang, J.P. Sun, C.X. Yang, Multi-agent urban expressway control system based on generalized knowledge-based model, in *Proceedings of the 2003 IEEE International Conference on Intelligent Transportation Systems*, vol. 2 (Shanghai, China) (IEEE, 2003), pp. 1759–1763
18. X.J. Kong, G.J. Shen, Y.X. Sun, Dynamic intelligent coordinated control with arterial traffic based on multi-agent. J. PLA Univ. Sci. Technol. (Nat. Sci. Ed.) **11**(5), 544–550 (2010)
19. J.C. Guo, H. Shi, A.D. Cao, Design and implementation of distributed multi-agent expert system. Comput. Meas. Control **12**(8), 717–719 (2004)
20. J.K. Liu, S.Q. Wang, The integrated research of intelligent cad design based on multiagent system. Comput. Eng. Des. **3**, 1–5 (2000)
21. R. E. Kalman, "A new approach to linear filtering and prediction problems," *Transactions of the ASME - Journal of Basic Engineering*, vol. 82 (Series D), pp. 35–45, 1960
22. D. Simon, *Optimal State Estimation* (Wiley and Sons Inc, New Jersey, 2006)
23. J. Wang, Q.H. Liang, K. Liang, A new extended Kalman filter based carrier tracking loop, in *Proceedings - 2009 3rd IEEE International Symposium on Microwave, Antenna, Propagation and EMC Technologies for Wireless Communications* (IEEE, 2009), pp. 1181–1184
24. T.C. Li, Analysis of inertial navigation system errors from van testing using an optimal Kalman filter/smoother, in *PLANS 2004. Position Location and Navigation Symposium* (2004), pp. 120–128
25. K.K. Choi, B.F. Thompson, Adaptive measurement covariance for multi-input Kalman filter-based spacecraft navigation. Adv. Astronaut. Sci. 163–171 (2008)
26. B.M. Quine, A derivative-free implementation of the extended Kalman filter. Automatica **42**(11), 1927–1934 (2006)
27. S.J. Julier, J.K. Uhlmann, Unscented filtering and nonlinear estimation. Proc. IEEE **92**(3), 401–422 (2004)
28. J.Y. Keller, Optimal two-stage Kalman filter in the presence of random bias. Automatica **33**(9), 1745–1748 (1997)
29. H.Z. Qiu, Solution of two-stage Kalman filter. Control Theroy Appl. **152**(2), 152–156 (2005)
30. M. Ignagni, Optimal and suboptimal separate-bias Kalman estimators for a stochastic bias. Autom. Control **45**(3), 547–551 (2000)

The manufacturer's authorised representative in the EU is Springer Nature Customer Service Centre GmbH, Europaplatz 3, 69115 Heidelberg, Germany. If you have any concerns regarding our products, please contact ProductSafety@springernature.com

Printed and bound by CPI Group (UK) Ltd, Croydon, CR0 4YY
30/09/2026
02225452-0010